T013358

Keep.

26. APR 04

27. MAR 08

1 4 MAR 2011

2 6 APR 2011

WHJ 1207

Behavioral Genetics

Behavioral Genetics

THIRD EDITION

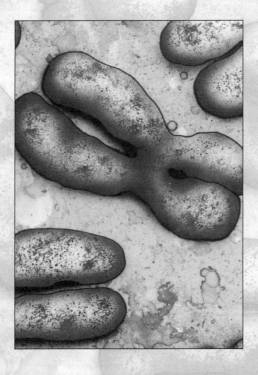

Robert Plomin
*Institute of Psychiatry,
London*

John C. DeFries
*University of Colorado,
Boulder*

Gerald E. McClearn
*Pennsylvania State
University*

Michael Rutter
*Institute of Psychiatry,
London*

W. H. Freeman and Company
New York

ACQUISITIONS EDITOR: Susan Finnemore Brennan

PROJECT EDITOR: Mary Louise Byrd

TEXT AND COVER DESIGNER: Victoria Tomaselli

ILLUSTRATION COORDINATOR: Bill Page

ILLUSTRATION: Network Graphics

PRODUCTION COORDINATOR: Sheila E. Anderson

COMPOSITION: W. H. Freeman Electronic Publishing Center/Susan Cory

MANUFACTURING: Vail-Ballou Press

Library of Congress Cataloging-in-Publication Data

Behavioral genetics / Robert Plomin . . . [et al.]. — 3rd ed.
      p.   cm.
    Previous ed. cataloged under: Plomin, Robert, 1948– .
    Includes bibliographical references and index.
    ISBN 0-7167-2824-9
    1. Behavioral genetics.   I. Plomin, Robert, 1948– .
QH457.P56   1997
591.5—dc21   96-45063
                                                        CIP

First printing 1997

C O N T E N T S

CHAPTER NINE

Specific Cognitive Abilities   154

CHAPTER TEN

Psychopathology   169

CHAPTER ELEVEN

Personality and Personality Disorders   195

ABOUT THE AUTHORS

Robert Plomin is professor of behavioral genetics at the Institute of Psychiatry in London, where he is deputy director of the Social, Genetic and Developmental Psychiatry Research Centre at the Institute. The goal of the Research Centre is to bring together genetic and environmental research strategies to investigate behavioral development, a theme that characterizes his research. With Michael Rutter, who directs the Research Centre, Plomin has begun a study of all twins born in England during the period 1994-96, focusing on developmental delays in early childhood and their association with behavioral problems. After receiving his doctorate in psychology from the University of Texas, Austin, in 1974, he worked with John DeFries and Gerald McClearn at the Institute for Behavioral Genetics at the

University of Colorado, Boulder. Together, they initiated several large longitudinal twin and adoption studies of behavioral development throughout the life span. From 1986 until 1994, he worked with McClearn at Pennsylvania State University. They launched a study of elderly twins reared apart and twins reared together to study aging, and they developed mouse models to identify genes in complex behavioral systems. Plomin's current interest is in harnessing the power of molecular genetics to identify genes for psychological traits. He has been president of the Behavior Genetics Association.

John C. DeFries is professor of psychology and director of the Institute for Behavioral Genetics, University of Colorado, Boulder. After receiving his doctorate in agriculture (with specialty training in quantitative genetics) from the University of Illinois in 1961, he remained on the faculty of that institution for six years. In 1962, he began research on mouse behavioral genetics and, the following year, was a research fellow in genetics at the University of California, Berkeley, where he conducted research in the laboratory of G. E. McClearn. After returning to Illinois in 1964, DeFries initiated an extensive genetic analysis of open-field behavior in laboratory mice that included a classic bidirectional selection experiment with replicate selected and control lines. Three years later, he joined the Institute for

Behavioral Genetics, which McClearn had founded in 1967. DeFries and Steven G. Vandenberg founded the journal *Behavior Genetics* in 1970; and DeFries and Robert Plomin founded the Colorado Adoption Project in 1975. For over two decades, DeFries's major research interest has concerned the genetics of reading disabilities, and he is currently director of the Colorado Learning Disabilities Research Center. He served as president of the Behavior Genetics Association in 1982 to 1983, receiving the association's Th. Dobzhansky Award for Outstanding Research in 1992, and he became a Fellow of the American Association for the Advancement of Science (Section J, Psychology) in 1994.

Gerald E. McClearn is Evan Pugh Professor and director of the Center for Developmental and Health Genetics in the College of Health and Human Development at Pennsylvania State University. After receiving his doctorate from the University of Wisconsin in 1954, he taught at Yale University, Allegheny College, and the University of California, Berkeley, before moving to the University of Colorado in 1965. There he founded the Institute for Behavioral Genetics in 1967. In 1981 McClearn moved to Penn State, where he has served as associate dean for research and dean of the College of Health and Human Development. He was also founding head of the Program in Biobehavioral Health and founding director of the Center for Developmental and Health Genetics. His research with colleagues at Penn State on mice has two main emphases: drug-related processes and behavioral and physiological aging. With Robert Plomin and other colleagues at Penn State and in Sweden, he has been involved for the past 12 years in large-scale studies of genetic and environmental influences on pattern and rate of aging in Swedish twins. McClearn has been president of the Behavior Genetics Association, and he received a MERIT Award from the National Institute on Aging in 1994.

Michael Rutter is professor of child psychiatry at the Institute of Psychiatry in London, where he is director of both the Social, Genetic and Developmental Psychiatry Research Centre and the Medical Research Council Child Psychiatry Research Unit. He qualified in medicine at the University of Birmingham in 1955, and after experience in internal medicine, neurology, and pediatrics, trained in psychiatry at the Maudsley Hospital. Most of his initial research used largely epidemiological and longitudinal research strategies to investigate psychosocial risk and protective processes. Yet his interest in genetics goes back to the early 1960s, when he studied a small number of twin pairs as part of the New York longitudinal study of temperament. In the mid-1970s, together with Susan Folstein, Rutter undertook the first systematic twin study of autism. Since then, he has undertaken further twin, family-genetic, and adoptee studies of autism, as well as a family-genetic study of childhood depression. He has been a coinvestigator on the Virginia Twin Study of Adolescent Behavioral Development since the outset, and more recently he has collaborated with Robert Plomin in the longitudinal study of all twins born in England during the period 1994 to 1996. He was elected Fellow of the Royal Society in 1987, was knighted in 1992, and has honorary doctorates from universities in the United Kingdom, Europe, and North America. He is a foreign associate member of the U.S. National Academy of Education and of the U.S. Institute of Medicine, as well as foreign honorary member of the American Academy of Arts and Sciences. In 1995 Rutter received the Distinguished Scientific Contribution Award from the American Psychological Association.

PREFACE

One of the most dramatic developments in psychology during the past few decades is the increasing recognition and appreciation of the important contribution of genetic factors. Genetics is not a neighbor chatting over the fence with some helpful hints—it is central to psychology and other behavioral sciences. In fact, genetics is central to all the life sciences. Genetics bridges the biological and behavioral sciences and helps to give psychology, the science of behavior, a place in the biological sciences. Genetics includes diverse research strategies such as twin and adoption studies (called quantitative genetics) that investigate the relative influence of genetic and environmental factors as well as strategies to identify specific genes (called molecular genetics). Behavioral genetics, like medical genetics or anthropological genetics, is a specialty that applies these genetic research strategies to the study of behavior. It includes subspecialties that focus on specific domains of behavior, such as psychiatric genetics (the genetics of mental illness) and psychopharmacogenetics (the genetics of behavioral responses to drugs).

The goal of this book is to share with you our excitement about behavioral genetics, a field in which we believe some of the most important discoveries in the behavioral sciences have been made in recent years.

This is the third edition of a textbook (see Plomin, DeFries, & McClearn, 1980, 1990d) that followed an earlier version (McClearn & DeFries, 1973). This edition is more of a sequel than a revision. The previous editions focused on the methods of behavioral genetics. This edition is entirely rewritten, with a topical rather than methodological focus. It tells what we know about genetics in psychology and psychiatry rather than how we know it. Its goal is not to train students to become behavioral geneticists but rather to introduce students in the behavioral, biological, and social sciences to the field of behavioral genetics.

We begin with an introductory chapter that will, we hope, whet your appetite for learning about genetics in the behavioral sciences. The next few chapters present the basic rules of heredity, its DNA basis, and the methods used to find genetic influence and to identify specific genes. The rest of the text highlights what is known about genetics in psychology and psychiatry. The areas about which most is known are cognitive disabilities and abilities, psychopathology, and personality. We also consider areas of psychology more recently introduced to genetics: health psychology, aging, and evolutionary psychology. These topics are followed by a chapter on environment, as viewed from the perspective of genetics. At first, a chapter on the environment might seem odd in a textbook on genetics, but in fact genetic research has made important discoveries about how the environment affects psychological development. The last chapter looks to the future: behavioral genetics in the twenty-first century. Throughout these chapters, quantitative genetics and molecular genetics are interwoven. One of the most exciting developments in behavioral genetics is the ability to begin to identify specific genes that influence behavior.

Because behavioral genetics is an interdisciplinary field that combines genetics and the behavioral sciences, it is complex. We have tried to write about it as simply as possible without sacrificing honesty of presentation. Although our coverage is representative, it is by no means exhaustive or encyclopedic. History and methodology are relegated to boxes and appendixes to keep the focus on what we now know about genetics and behavior. Appendixes present an overview of statistics, quantitative genetic theory, and a new type of quantitative genetic analysis called model fitting. The text is sprinkled with brief auto-biographical "close-ups" of representative researchers in the field to personalize the research. We are grateful to our colleagues for contributing autobiographical statements and photographs.

This edition benefited greatly from the advice of many colleagues, too many to name. We are grateful to Adele Summers, who organized us as well as the book and prepared many of the line illustrations. We also appreciate the effort and support of our editor, Susan Finnemore Brennan, who encouraged us to undertake this revision and accelerated its publication.

Behavioral Genetics

CHAPTER ONE

Overview

Some of the most important recent discoveries about behavior involve genetics. Consider the terrible memory loss and confusion of Alzheimer's disease, which strikes as many as one in five individuals in their eighties (Chapter 7). Although Alzheimer's disease rarely occurs before the age of 60, some early-onset cases run in families in a simple manner that suggests the influence of only one gene; and, in 1992, a single gene on chromosome 14 was found to be responsible for many of these early-onset cases.

The gene on chromosome 14 is not responsible for the much more common form of Alzheimer's disease that occurs after 60 years of age. Like most behavioral disorders, late-onset Alzheimer's disease is not caused by a single gene. Still, genetic studies comparing the risk for identical twins and fraternal twins indicate genetic influence. Identical twins are identical genetically, but fraternal twins, like other nonidentical siblings, are only 50 percent similar genetically. Twin studies have shown that if you had an identical twin who had Alzheimer's disease, your risk would be 60 percent, but if you had a fraternal twin who was affected, your risk would be 30 percent. These findings suggest genetic influence.

Even for complex disorders like late-onset Alzheimer's, it is now possible to identify genes that contribute to the risk for the disorder. In 1993, a gene that predicts risk for late-onset Alzheimer's disease far better than any other known risk factor was identified. If you inherit one copy of a particular form (*allele*) of the gene, your risk for Alzheimer's disease is about four times greater than if you have another allele. If you inherit two copies of the allele (one from each of your parents), your risk is much greater.

Another example of recent genetic discoveries involves mental retardation (Chapter 7). It has been known for decades that the single most important cause of mental retardation, which accounts for more than a quarter of institutionalized mentally retarded individuals, is the inheritance of an entire extra chromosome 21. Instead of inheriting only one pair of chromosomes 21, one from the mother and one from the father, an entire extra chromosome is inherited, usually from the mother. Often called Down syndrome, trisomy-21 is one of the major reasons why women worry about pregnancy later in life, because it occurs much more frequently when mothers are over 40 years old. The extra chromosome can be detected in the first few months of pregnancy by a procedure called amniocentesis.

In 1991, researchers identified a single gene that is the second most common cause of mental retardation. This form of mental retardation is called fragile X retardation. The gene that causes the disorder is on the X chromosome. The name *fragile X* is based on the finding that a chromosome carrying the fragile X allele tends to break when cells that carry it are grown on a special medium. Fragile X mental retardation occurs nearly twice as often in males as in females because males have only one X chromosome. If a boy has the fragile X allele on his X chromosome, he will develop the disorder. Females have two X chromosomes, so they may inherit the fragile X allele on both X chromosomes and develop the disorder. However, females with one fragile X allele can also be affected to some extent. The fragile X gene is especially interesting because it involves a newly discovered type of genetic defect in which a short sequence of DNA mistakenly repeats hundreds of times. This type of genetic defect is now also known to be responsible for several other previously puzzling diseases (Chapter 3).

Until the 1970s, the origins of mental illnesses such as schizophrenia and autism (Chapter 10) were thought by many people to be environmental, with theories putting the blame on poor parenting. However, genetic research has convincingly demonstrated that genetic factors contribute importantly to most mental illness. For schizophrenia, if one member of an identical twin pair is schizophrenic, the chances are 45 percent that the other twin is also affected. For fraternal twins, the chances are 17 percent. It has recently been discovered that autism is one of the most heritable disorders, with 60 percent risk for identical twins and 10 percent for fraternal twins. The race is now on to find specific genes that influence mental illness.

Not only does genetics contribute to disorders such as dementia, mental retardation, and mental illness, it also plays an important role in normal variation. For example, the normal variation in height is largely due to genetic differences among individuals. You might be more surprised to learn that differences in weight are almost as heritable as height (Chapter 12). Even though we can control how much we eat and are free to go on crash diets, differences among us in weight are much more a matter of nature (genetics) than nurture (environment). What about behavior? Genetics contributes to the normal variations in cognitive abilities (Chapters 8 and 9), personality (Chapter 11), school achievement (Chapter 9), self-esteem (Chapter 11), and drug use (Chapter 12). Genetic factors are often as important as all other factors put together.

Genetic research on behavior goes beyond just demonstrating the importance of genetics to the behavioral sciences and allows us to ask questions about *how* genes influence behavior. For example, does genetic influence change during development? Consider cognitive ability, for example; you might think that as time goes by we increasingly accumulate Shakespeare's "slings and arrows of outrageous fortune." That is, environmental differences might become increasingly important during one's life span, whereas genetic differences might become less important. However, genetic research shows just the opposite: Genetic influence on cognitive ability increases throughout the individual's life span, reaching levels later in life that are nearly as great as the genetic influence on height (Chapter 8). This finding is an example of developmental genetic analysis.

School achievement and the results of tests you took to apply to college are influenced almost as much by genetics as are the results of tests of cognitive abilities such as intelligence (IQ) tests. Even more interesting, the substantial overlap between such achievement and ability on tests is nearly all genetic in origin (Chapter 9). This finding is an example of what is called multivariate genetic analysis.

Genetic research is also changing the way we think about the environment (Chapter 14). For example, we used to think that growing up in the same family makes brothers and sisters similar psychologically. However, for most behavioral dimensions and disorders, it is genetics that accounts for similarity among siblings. Although the environment is important, environmental influences make siblings growing up in the same family different, not similar. This genetic research has sparked an explosion of environmental research looking for the environmental reasons why siblings in the same family are so different.

Recent genetic research has also shown a surprising result that emphasizes the need to take genetics into account when studying the environment: Many environmental measures used in the behavioral sciences show genetic influence! For example, research in developmental psychology often involves measures of parenting that are assumed to be measures of the family environment. However, genetic research during the past decade has convincingly shown genetic influence on parenting measures. How can this be? One way is that genetic differences influence parents' behavior toward their children. Genetic differences among children can also make a contribution. For example, parents who have more books in their home have children who do better in school, but this correlation does not necessarily mean that having more books in the home is an environmental *cause* for children doing better in school. Genetic factors could affect parental traits that relate both to the number of books parents have in their home and to their children's achievement at school. Genetic involvement has also been found for many other ostensible measures of the environment, including childhood accidents, life events, and social support. To some extent, people create their own experiences for genetic reasons.

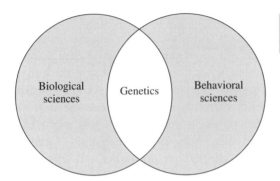

Figure 1.1 Genetics bridges the biological and behavioral sciences.

These are examples of what you will learn about in this book. The simple message is that genetics plays a major role in behavior. Genetics brings together the biological and behavioral sciences (Figure 1.1). Although research in behavioral genetics has been conducted for many years, the field-defining text was published only in 1960 (Fuller & Thompson, 1960). Since that date, discoveries in behavioral genetics have grown at a rate that few other fields in the behavioral sciences can match.

Recognition of the importance of genetics is one of the most dramatic changes in the behavioral sciences during the past two decades. Seventy years ago, John Watson's (1930) behaviorism detached psychology from its budding interest in heredity. A preoccupation with the environmental determinants of behavior continued until the 1970s, when a shift began toward the more balanced contemporary view that recognizes genetic as well as environmental influences. This shift toward genetics in the behavioral sciences can be seen in the increasing numbers of research projects and publications that involve genetics. One concrete sign of this shift is that, at its 1992 centennial conference, the American Psychological Association identified genetics as one of the themes that best represents the future of psychology (Plomin & McClearn, 1993a).

CHAPTER TWO

Mendel's Laws of Heredity

Huntington's disease (HD) begins with personality changes, forgetfulness, and involuntary movements. It typically strikes in middle adulthood; and during the next 15 to 20 years, it leads to complete loss of motor control and intellectual function. No treatment has been found to halt or delay the inexorable decline. This is the disease that killed the famous depression-era folksinger Woody Guthrie. Although it affects only about 1 in 20,000 individuals, a quarter of a million people in the world today will eventually develop Huntington's disease.

When the disease was traced through many generations, it showed a consistent pattern of heredity. Afflicted individuals had one parent who also had the disease, and approximately half of the children of an affected parent developed the disease. (See Figure 2.1 for an explanation of symbols traditionally used to describe family trees, called *pedigrees*. Figure 2.2 shows an example of a Huntington's disease pedigree.) What rules of heredity are at work? Why does this lethal condition persist in the population? We will answer these questions in the next section, but first, consider another inherited disorder.

In the 1930s, a Norwegian biochemist discovered an excess of phenylalanine in the urine of a pair of mentally retarded siblings. Phenylalanine is one of the essential amino acids, which are the building blocks of proteins. It is in many foods in the normal human diet. Other retarded individuals were soon found with this same excess. This type of mental retardation came to be known as *phenylketonuria (PKU)*.

Although the frequency of PKU is only about 1 in 10,000, PKU used to account for about 1 percent of the institutionalized mentally retarded population. PKU has a pattern of inheritance very different from that of Huntington's disease. PKU individuals do not usually have affected parents. Although this might make it seem at first glance as if PKU is not inherited, PKU does in fact

Male

Female

Mating

Parents

Children

Affected

Carriers

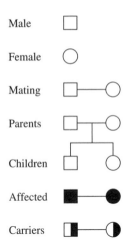

Figure 2.1 Symbols used to describe family pedigrees.

"run in families." If one child in a family has PKU, the risk for other children is about 25 percent, even though the parents themselves may not be affected (see Figure 2.3). One more piece of the puzzle is the observation that when parents are genetically related ("blood" relatives), typically in marriages between cousins, they are more likely to have children with PKU. How does heredity work in this case?

Mendel's First Law of Heredity

Although Huntington's disease and phenylketonuria, two examples of hereditary transmission of mental disorders, may seem complicated, they can be explained by a simple set of rules about heredity. The essence of these rules was worked out more than a century ago by Gregor Mendel (1866).

Mendel was a monk and studied inheritance in pea plants in the garden of his monastery in what is now the Czech Republic (see Box 2.1). On the basis of his many experiments, Mendel concluded that there are two "elements" of heredity for each trait in each individual and that these two elements separate, or segregate, during reproduction. Offspring receive one of the two elements from each parent. In addition, Mendel concluded that one of these elements can "dominate" the other, so that an individual with just one dominant element will display the trait. A nondominant, or *recessive*, element is expressed only if both elements are recessive. These conclusions are the essence of Mendel's first law, the *law of segregation*.

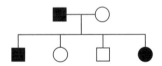

Figure 2.2 Huntington's disease. HD individuals have one HD parent. About 50 percent of the offspring of HD parents will have HD.

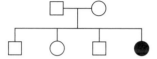

Figure 2.3 Phenylketonuria. PKU individuals do not typically have parents with PKU. If one child has PKU, the risk for other siblings is 25 percent. As explained later, parents in such cases are carriers for one allele of the PKU gene but a child must have two alleles in order to be afflicted with recessive disorders such as PKU.

No one paid any attention to Mendel's law of heredity for 40 years. Finally, in the early 1900s, several scientists recognized that Mendel's law is a general law of inheritance, not one peculiar to the pea plant. Mendel's "elements" are now known as *genes*, the basic units of heredity. Many genes have only one form throughout a species, for example, in all pea plants or in all people. Heredity focuses on genes that have different forms, differences that cause some pea seeds to be wrinkled or smooth, or that cause some people to have Huntington's disease or PKU. The alternate forms of a gene are called *alleles*. An individual's combination of alleles is its *genotype*, whereas the observed traits are its *phenotype*. The fundamental issue of heredity in the behavioral sciences is the extent to which differences in genotype account for differences in phenotype, observed differences among individuals.

This chapter began with two very different examples of inherited disorders. How can Mendel's law of segregation explain both examples?

Huntington's Disease

Figure 2.4 shows how Mendel's law explains the inheritance of Huntington's disease. HD is caused by a dominant allele. Affected individuals have one dominant allele (*H*) and one recessive, normal allele (*h*). (It is rare that an HD individual has two *H* alleles, an event that would require that both parents have HD.) Unaffected individuals have two normal alleles.

As shown in Figure 2.4, a parent with HD whose genotype is *Hh* produces gametes (egg or sperm) with either the *H* or the *h* allele. The unaffected (*hh*) parent's gametes all have an *h* allele. The four possible combinations of these gametes from the mother and father result in the offspring genotypes shown at the bottom of Figure 2.4. Offspring will always inherit the normal *h* allele from the unaffected parent, but they have a 50 percent chance of inheriting the *H* allele from the HD parent. This pattern of inheritance explains why HD individuals always have a parent with HD and why 50 percent of the offspring of an HD parent develop the disease.

Why does this lethal condition persist in the population? If HD had its effect early in life, HD individuals would not live to reproduce. In one generation,

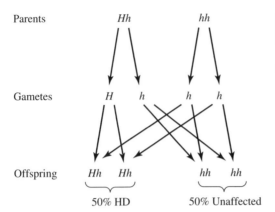

Parents

Gametes

Offspring

50% HD   50% Unaffected

Figure 2.4. Huntington's disease is due to a single gene, with the allele for HD dominant. *H* represents the dominant HD allele and *h* is the normal recessive allele. Gametes are sex cells (eggs and sperm), and each carries just one allele. The risk of HD in the offspring is 50 percent.

HD would no longer exist because any individual with the HD allele would not live long enough to reproduce. The dominant allele for HD is maintained from one generation to the next because its lethal effect is not expressed until after the reproductive years.

A particularly traumatic feature of HD is that offspring of HD parents know they have a 50 percent chance of developing the disease and of passing on the HD gene. In 1983, DNA markers were used to show that the gene for HD is on a particular chromosome. As will be discussed in greater detail in Chapter 4, the genetic material is chemical in nature, being composed of a linear array of nucleotides. The basic molecule is deoxyribonucleic acid (DNA). The function of a gene is determined by its *sequence* (linear order) of nucleotides of DNA. Particular DNA sequences can be detected by special analytical methods, and these sequences can serve as "markers" in genetic analysis. Using these methods, in 1993 investigators found the HD gene itself on chromosome 4. So now it is possible to determine for certain whether a person has the HD gene.

This genetic advance raises its own problems. If one of your parents had HD, you would be able to find out whether you did or did not have the HD allele. You would have a 50 percent chance of finding that you did not have the HD allele, but you would also have a 50 percent chance of finding that you did have the HD allele and would eventually die from it. In fact, most people at risk for HD decide *not* to take the test. Finding the gene does, however, make it possible to determine whether a fetus has the HD allele and holds out the promise of future interventions that can correct the HD defect (Chapter 6).

Phenylketonuria

Mendel's law also explains the inheritance of PKU. Unlike HD, PKU is due to a recessive allele. For offspring to be affected, they must have two copies of the allele. Those offspring with only one copy of the allele are unafflicted by the disorder, but they are called *carriers* because they carry the allele and can pass it on to their offspring. Figure 2.5 illustrates the inheritance of PKU from two unaffected carrier parents. Each parent has one PKU allele and one normal allele. Offspring have a 50 percent chance of inheriting the PKU allele from one parent and a 50 percent chance of inheriting the PKU allele from the other parent. The chance of both of these things happening is 25 percent. If you flip a coin, the chance of heads is 50 percent. The chance of getting two heads in a row is 25 percent (i.e., 50 percent times 50 percent).

This pattern of inheritance explains why unaffected parents have children with PKU and why the risk of PKU in offspring is 25 percent when both parents are carriers. For PKU and other recessive disorders, identification of the genes makes it possible to determine whether potential parents are carriers. Identification of the PKU gene also makes it possible to determine whether a particular pregnancy involves an affected fetus. In fact, all newborns in most countries are screened for elevated phenylalanine levels, because early diagnosis can prevent retardation through low phenylalanine diets.

Figure 2.5 also shows that 50 percent of children born of two carrier parents are likely to be carriers and 25 percent will inherit the normal allele from both parents. If you understand how a recessive trait such as PKU is inherited, you should be able to work out the risk for PKU in offspring if one parent has PKU and the other parent is a carrier. The risk is 50 percent.

We have yet to explain why recessive traits like PKU are seen more often in offspring whose parents are genetically related. Although PKU is rare (1 in 10,000), about 1 in 50 individuals are carriers of one PKU allele (see Box 2.2). If you are a PKU carrier, your chance of marrying someone who is also a carrier is 2 percent. However, if you marry someone genetically related to you, the PKU allele must be in your family, so the chances are much greater than 2 percent that your spouse will also carry the PKU allele.

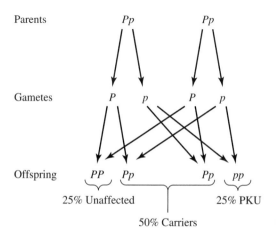

Figure 2.5 PKU is inherited as a single gene. The allele that causes PKU is recessive. *P* represents the normal dominant allele, and *p* is the recessive allele for PKU. Parents are carriers and the risk of PKU for their offspring is 25 percent.

BOX 2.1

Gregor Mendel's Luck

Gregor Mendel (1822–1884) was a lucky monk.

Before Mendel, much of the research on heredity involved crossing plants of different species. But the offspring of these matings were usually sterile, which meant that succeeding generations could not be studied. Another problem with research before Mendel was that features of the plants that were investigated were complexly determined. Mendel's success can be attributed in large part to the absence of these problems.

Mendel crossed different varieties of pea plants of the same species; thus the offspring were fertile. In addition, he picked simple either-or traits, qualitative traits, that happened to be due to single genes. He was also lucky in that in his chosen traits one allele completely dominated expression of

Gregor Johann Mendel. A photograph taken at the time of his research. (Courtesy of V. Orel, Mendel Museum, Brno, Czechoslovakia.)

the other allele, which is not always the case. However, one feature of Mendel's research was not due to luck. He counted all offspring rather than being content, as researchers before him had been, with a verbal summary of the typical result.

Mendel studied seven qualitative traits of the pea plant such as whether the seed was smooth or wrinkled. He obtained 22 varieties of the pea plant that differed in these seven characteristics. All the varieties were true-breeding plants: those that always yield the same result when crossed with the same kind of plant. Mendel presented the results of eight years of research on the pea plant in his 1866 paper. This paper now forms one cornerstone of genetics.

In one experiment, Mendel crossed true-breeding plants with smooth seeds to true-breeding plants with wrinkled seeds. Later in the summer, when he opened the pods containing their offspring (called the F_1, or first filial generation), he found that all of them had smooth seeds. This result indicated that the then-traditional view of blending inheritance was not correct. That is, the F_1 did not have seeds that were even moderately wrinkled. These F_1 plants were fertile, which allowed Mendel to take the next step of crossing plants of the F_1 generation with one another and studying their offspring, F_2. The results were striking: Of the 7234 seeds from the F_2, 5474 were smooth and 1850 were wrinkled. That is, $\frac{3}{4}$ of the offspring

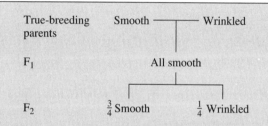

had smooth seeds and $\frac{1}{4}$ had wrinkled seeds. This result indicates that the factor responsible for wrinkled seeds had not been lost in the F_1 generation but had merely been dominated by the factor causing smooth seeds. The figure above summarizes Mendel's results.

Given these observations, Mendel deduced a simple explanation involving two hypotheses. First, each individual has two hereditary "elements," now called alleles (alternate forms of a gene). For Mendel's pea plants, these alleles determined whether the seed was wrinkled or smooth. Thus, each parent has two alleles but transmits only one of the alleles to each offspring. The second hypothesis was that one allele could dominate the other. These two hypotheses neatly explain the data (see the figure below).

The true-breeding parent plant with smooth seeds has two alleles for smooth seeds (*SS*). The true-breeding parent plant with wrinkled seeds has two alleles for wrinkled seeds (*ss*). First generation (F_1) offspring receive one allele from each parent and are therefore *Ss*. Because *S* dominates *s*, F_1 plants will have smooth seeds. The real test is the F_2 population. Mendel's theory predicts that when F_1 individuals are crossed with other F_1 individuals, $\frac{1}{4}$ of the F_2 should be *SS*, $\frac{1}{2}$ *Ss*, and $\frac{1}{4}$ *ss*. Assuming *S* dominates *s*, then *Ss* should have smooth seeds like the *SS*. Thus, $\frac{3}{4}$ of the F_2 should have smooth seeds and $\frac{1}{4}$ wrinkled, which is exactly what Mendel's data indicated. Mendel also discovered that the inheritance of one trait is not affected by the inheritance of another trait. Each trait is inherited in the expected 3:1 ratio.

Mendel was not so lucky in terms of acknowledgment of his work during his lifetime. When Mendel published the paper about his theory of inheritance in 1866, reprints were sent to scientists and libraries in Europe and the United States, and one even landed in Darwin's office. However, Mendel's findings on the pea plant were ignored by most biologists, who were more interested in evolutionary processes that could account for change rather than continuity. Mendel died in 1884, without knowing the profound impact that his experiments would have during the twentieth century.

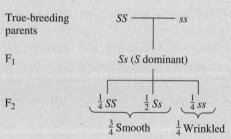

It is very likely that we all carry at least one harmful recessive gene of some sort. However, the risk that our spouses are also carriers for the same disorder is small unless we are genetically related to them. In contrast, about half of the children born to incestuous relationships between father and daughter show severe genetic abnormalities, often including childhood death or mental retardation. This pattern of inheritance explains why most severe genetic disorders are recessive: Recessive alleles are transmitted by carriers who do not show the disorder. In this way, recessive alleles escape detection.

It should be noted that exceptions can occur in the inheritance of even simple single-gene disorders such as PKU. For example, a new PKU mutation can emerge in individuals with no family history. In fact, some single-gene disorders are largely caused by new mutations. In addition, age of onset may vary for single-gene disorders, as it does in the case of HD. The degree to which the disorder is expressed may also vary.

SUMMING UP

Mendel's theory of heredity can explain dominant (Huntington's disease) and recessive (PKU) patterns of inheritance. A gene may exist in two or more different forms (alleles). The two alleles, one from each parent, separate (segregate) during gamete formation. This is Mendel's first law, the law of segregation.

Mendel's Second Law of Heredity

Not only do the alleles for Huntington's disease segregate independently during gamete formation, they also are inherited independently from the alleles for PKU. This finding makes sense because Huntington's disease and PKU are caused by different genes and each of the two genes is inherited independently. Mendel experimented systematically with crosses between varieties of pea plants that differed on two or more traits. He found that alleles for each gene assort independently of the other gene. In other words, the inheritance of one gene is not affected by the inheritance of another gene. This is Mendel's *law of independent assortment*.

What is most important to us about Mendel's second law are its exceptions. We now know that genes are not just floating around in eggs and sperm. They are carried on *chromosomes*. The term *chromosome* literally means "colored body," because in certain laboratory preparations these structures stain differently from the rest of the nucleus of the cell. Genes are located at places called *loci* (singular, *locus*, from the Latin meaning "place") on chromosomes. Eggs contain just one chromosome from each pair of the mother's set of chromosomes and sperm contain just one from each pair of the father's set. An egg fertilized

BOX 2.2

How Do We Know That 1 in 50 People Are Carriers for PKU?

If you randomly mate F_2 plants to obtain an F_3 generation, the frequencies of the S and s alleles will be the same as in the F_2 generation, as will the frequencies of the SS, Ss, and ss genotypes. Shortly after the rediscovery of Mendel's law in the early 1900s, this implication of Mendel's law was formalized and eventually called the *Hardy-Weinberg equilibrium*. The frequencies of alleles and genotypes do not change across generations unless forces such as natural selection or migration change them. This rule is the basis for a field called *population genetics*, which studies forces that change gene frequencies (see Chapter 13).

Hardy-Weinberg equilibrium also makes it possible to estimate frequencies of alleles and genotypes. The frequencies of the dominant and recessive alleles are usually referred to as p and q, respectively. Eggs and sperm have just one allele for each gene. The chance that any particular egg or sperm has the dominant allele is p. Because sperm and egg unite at random, the chance that a sperm with the dominant allele fertilizes an egg with the dominant allele is the product of the two frequencies, $p \times p = p^2$. Thus, p^2 is the frequency of offspring with two dominant alleles (called the *homozygous dominant* genotype). In the same way, the *homozygous recessive* genotype has a frequency of q^2. As shown in the following table, the frequency of offspring with one dominant allele and one recessive allele (called the *heterozygous* genotype) is $2pq$. In other words, if a population is in Hardy-Weinberg equilibrium, the frequency of the offspring genotypes is $p^2 + 2pq + q^2$. In populations with random mating, the

		Eggs	
Frequencies		p	q
Sperm	p	p^2	pq
	q	pq	q^2

expected genotypic frequencies are merely the product of $p + q$ for the mothers' alleles and $p + q$ for the fathers' alleles. That is, $(p + q)^2 = p^2 + 2pq + q^2$.

For PKU, q^2, the frequency of PKU individuals (homozygous recessive) is .0001. If you know q^2, you can estimate the frequency of the PKU allele and PKU carriers, assuming Hardy-Weinberg equilibrium. The frequency of the PKU allele is q, which is the square root of q^2. The square root of .0001 is .01, which means that 1 in 100 alleles in the population are the recessive PKU alleles. If there are only two alleles at the PKU locus, then the frequency of the dominant allele (p) is $1 - .01 = .99$. What is the frequency of carriers? Because carriers are heterozgyous genotypes with one dominant allele and one recessive allele, the frequency of carriers of the PKU allele is 1 in 50 (that is, $2pq = 2 \times .99 \times .01 = .02$).

by a sperm thus has the full chromosome complement, which, in humans, is 23 pairs of chromosomes. Chromosomes are discussed in more detail in Chapter 4.

When Mendel studied the inheritance of two traits (call them A and B) at the same time, he crossed true-breeding parents that showed the dominant trait for both A and B with parents that showed the recessive forms for A and B. He found second generation (F_2) offspring of all four possible types: dominant for A and B, dominant for A and recessive for B, recessive for A and dominant for B, and recessive for A and B. The frequencies of the four types of offspring were as expected if A and B were inherited independently. Mendel's law is violated, however, when genes for two traits are close together on the same chromosome. If Mendel had studied the joint inheritance of two such traits, the results would have surprised him. The two traits would not have been inherited independently.

Figure 2.6 illustrates what would happen if the genes for traits A and B were very close together on the same chromosome. Instead of finding all four

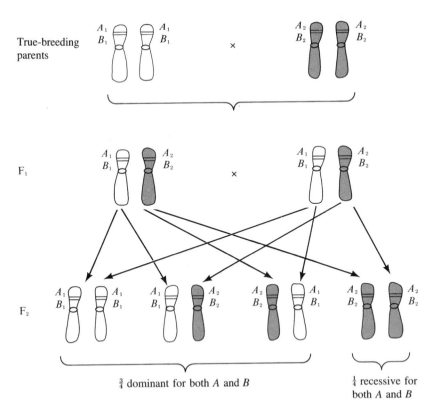

Figure 2.6 An exception to Mendel's second law occurs if two genes are closely linked on the same chromosome. The A_1 allele and the B_1 allele are dominant; the A_2 and B_2 alleles are recessive.

types of F_2 offspring, Mendel would have found only two types: dominant for both A and B and recessive for both A and B.

The reason why such violations of Mendel's second law are important is that they make it possible to map genes to chromosomes. If the inheritance of a particular pair of genes violates Mendel's second law, then it must mean that they tend to be inherited together and thus reside on the same chromosome. This phenomenon is called *linkage*. However, it is actually not sufficient for two linked genes to be on the same chromosome; they must also be very close together on the chromosome. Unless genes are near each other on the same chromosome, they will recombine by a process in which chromosomes exchange parts. Recombination occurs during meiosis in the ovaries and testes when gametes are produced.

Figure 2.7 illustrates recombination for three loci (A, C, B) on a single chromosome. The maternal chromosome, carrying the alleles A_1, C_1, and B_2, is represented in white; the paternal chromosome with alleles A_2, C_2, and B_1 is gray. During meiosis, each chromosome duplicates to form sister chromatids (Figure 2.7b). These sister chromatids may cross over one another (Figure 2.7c). This overlap happens an average of one time for each chromosome during meiosis. During this stage, the chromatids can break and rejoin (Figure 2.7d). Each of the chromatids will be transmitted to a different gamete (Figure 2.7e). Consider only the A and B loci for the moment. As shown in Figure 2.7e, one gamete will carry the genes A_1 and B_2, as in the mother, and one will carry A_2 and B_1, as in the father. The other two will carry A_1 with B_1 and A_2 with B_2. For the latter two pairs, recombination has taken place.

The probability of recombination between two loci on the same chromosome is a function of the distance between them. In Figure 2.7, for example, the A and C loci have not recombined. All gametes are either A_1C_1 or A_2C_2, as in the parents, because the crossover did not occur between these loci. Crossover could occur between the A and C loci, but it would happen less frequently than between A and B.

These facts have been used to "map" genes on chromosomes. The distance between two loci can be estimated by the number of recombinations per 100 gametes. This distance is called a map unit or *centimorgan*, named after T. H. Morgan, who first identified linkage groups in the fruit fly *Drosophila* (Morgan et al., 1915). If two loci are far apart, like the A and B loci, recombination will separate the two loci as often as if the loci were on different chromosomes, and they will not appear to be linked.

To identify the location of a gene on a particular chromosome, *linkage analysis* can be used. Linkage analysis refers to techniques that use information about violations of independent assortment to identify the chromosomal location of a gene. DNA markers serve as signposts on the chromosomes, as discussed in Chapter 6. Since 1980, the power of linkage analysis has greatly increased, with the discovery of thousands of these markers. Linkage analysis looks for a

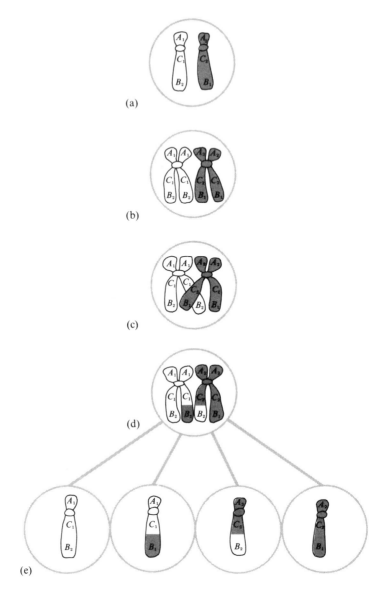

(a)

(b)

(c)

(d)

(e)

Figure 2.7 Illustration of recombination. The maternal chromosome, carrying the alleles A_1, C_1, and B_2, is represented in white; the paternal chromosome, with alleles A_2, C_2, and B_1, is gray. The right chromatid (the duplicated chromosome produced during meiosis) of the maternal chromosome crosses over (recombines) with the left chromatid of the paternal chromosome.

violation of independent assortment between a trait and a DNA marker. In other words, linkage analysis assesses whether the DNA marker and the trait co-assort in a family more often than expected by chance.

SUMMING UP

Mendel also showed that the inheritance of one gene is not affected by the inheritance of another gene. This is Mendel's second law, the law of independent assortment. Violation of Mendel's second law indicates that genes are inherited together on the same chromosome. This inheritance pattern is the basis for linkage analysis, which makes it possible to assign genes to specific chromosomes.

In 1983, the gene for Huntington's disease was shown to be linked to a DNA marker near the tip of one of the larger chromosomes (chromosome 4; see Chapter 6) (Gusella et al., 1983). This was the first time that the new DNA markers had been used to demonstrate a linkage for a disorder for which no chemical mechanism was known. DNA markers that are closer to the Huntington's gene have since been developed and have made it possible to pinpoint the gene. As noted earlier, the gene itself was finally located precisely in 1993.

Once a gene is found, two things are possible. First, the DNA variation responsible for the disorder can be identified. This identification provides a DNA test that is directly associated with the disorder in individuals and is more than just a risk estimate calculated on the basis of Mendel's laws. That is, the DNA test can be used to diagnose the disorder in individuals regardless of information about other family members. Second, the protein coded by the gene can be studied; this investigation is a major step toward understanding how the gene has its effect and thus can possibly lead to a therapy.

Although the disease process of the Huntington's gene is not yet understood, Huntington's disease, like fragile X mental retardation mentioned in Chapter 1, also involves a newly discovered type of genetic defect in which a short sequence of DNA is repeated hundreds of times (see Chapter 4).

Finding the PKU gene was easier because its enzyme product was known, as described in Chapter 1. In 1984, the gene for PKU was found and shown to be on chromosome 12 (Lidsky et al., 1984). For decades, PKU infants have been identified by screening for the physiological effect of PKU—high phenylalanine levels—but this test is not highly accurate. Developing a DNA test for PKU has been hampered by the finding that there are many different mutations at the PKU locus and that these mutations differ in the magnitude of their effects. This diversity contributes to the variation in phenylalanine levels among PKU individuals.

Of the several thousand single-gene disorders known (about half of which involve the nervous system), the precise chromosomal location has been identified for several hundred genes. The gene itself and the specific mutation have been found for more than a hundred disorders, and this number is rapidly increasing. One of the goals of the Human Genome Project is to identify all genes.

Fast progress toward this goal holds the promise of identifying genes even for complex behaviors influenced by multiple genes as well as environmental factors.

Summary

Huntington's disease (HD) and phenylketonuria (PKU) are examples of dominant and recessive disorders, respectively. They follow the basic rules of heredity described by Mendel more than a century ago. A gene may exist in two or more different forms (alleles). One allele can dominate the expression of the other. The two alleles, one from each parent, separate (segregate) during gamete formation. This rule is Mendel's first law, the *law of segregation*. The law explains many questions: why 50 percent of the offspring of an HD parent are eventually afflicted, why this lethal gene persists in the population, why PKU children usually do not have PKU parents, and why PKU is more likely when parents are genetically related.

Mendel's second law is the *law of independent assortment:* The inheritance of one gene is not affected by the inheritance of another gene. However, genes that are closely linked on the same chromosome can co-assort, thus violating Mendel's law of independent assortment. Such violations make it possible to map genes to chromosomes by using linkage analysis. For Huntington's disease and PKU, linkage has been established and the genes responsible for the disorders have been identified.

CHAPTER THREE

Beyond Mendel's Laws

Color blindness shows a pattern of inheritance that does not appear to conform to Mendel's laws. The most common color blindness involves difficulty in distinguishing red and green, a condition caused by a lack of certain color-absorbing pigments in the retina of the eye. It occurs more frequently in males than in females. More interesting, when the mother is color blind and the father is not, all of the sons but none of the daughters are color blind (see Figure 3.1a). When the father is color blind and the mother is not, offspring are seldom affected (Figure 3.1b). But something puzzling happens to these apparently normal daughters of a color-blind father: Half of their sons are likely to be color blind. This is the well-known skip-a-generation phenomenon—fathers have it, their daughters do not, but the grandsons do. What could be going on here in terms of Mendel's laws of heredity?

Genes on the X Chromosome

There are two chromosomes called the sex chromosomes because they differ for males and females. Females have two X chromosomes, and males have only one X chromosome and a smaller chromosome called Y. The Y chromosome has very few genes other than a gene responsible for maleness.

Color blindness is caused by a recessive allele on the X chromosome. But males have only one X chromosome; so, if they have one allele for color blindness (*c*) on their single X chromosome, they are color blind. For females to be color blind, they must inherit the *c* allele on both of their X chromosomes. For this reason, the hallmark of a sex-linked (meaning *X-linked*) recessive gene is a greater incidence in males. For example, if the frequency of an X-linked recessive allele for a disorder were 10 percent, then the expected frequency of the disorder in males would be 10 percent, but the frequency in females would be only 1 percent (i.e., $.10^2 = .01$).

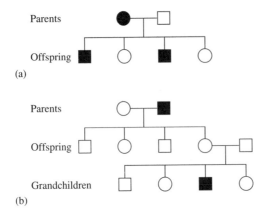

Parents

Offspring

(a)

Parents

Offspring

Grandchildren

(b)

Figure 3.1 Inheritance of color blindness. (a) A color-blind mother and unaffected father have color-blind sons but unaffected daughters. (b) An unaffected mother and color-blind father have unaffected offspring, but daughters have sons with 50 percent risk for color blindness. (See Figure 2.1 for symbols used to describe family pedigrees.)

Figure 3.2 illustrates the inheritance of the sex chromosomes. Both sons and daughters inherit one X chromosome from their mother. Daughters inherit their father's single X chromosome and sons inherit their father's Y chromosome. Sons cannot inherit an allele on the X chromosome from their father. For this reason, another sign of an X-linked recessive trait is that father-son resemblance is negligible. Daughters inherit an X-linked allele from their father, but they do not express a recessive trait unless they receive another such allele on the X chromosome from their mother.

Inheritance of color blindness is explained in Figure 3.3. In the case of a color-blind mother and unaffected father (Figure 3.3a), the mother has the c allele on both of her X chromosomes and the father has the normal allele (C) on his single X chromosome. Thus, sons must inherit an X chromosome from their mother

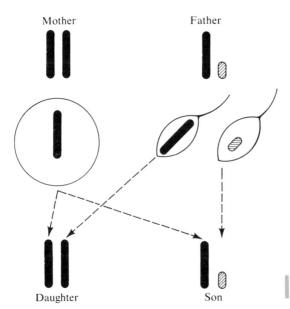

Mother

Father

Daughter

Son

Figure 3.2 Inheritance of X and Y chromosomes.

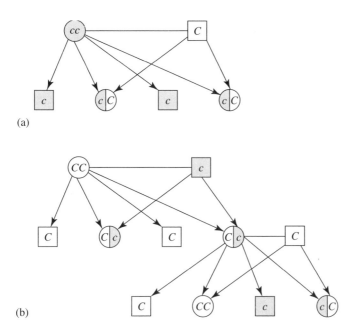

(a)

(b)

Figure 3.3 Color blindness is inherited as a recessive gene on the X chromosome. *c* refers to the recessive allele for color blindness, and *C* is the normal allele. (a) Color-blind mothers are homozygous recessive (*cc*). (b) Color-blind fathers have a *c* allele on their single X chromosome, which is transmitted to daughters but not to sons.

with the *c* allele and are color blind. Daughters carry one *c* allele from their mother but are not color blind because they have inherited a normal, and dominant, *C* allele from their father. They carry the *c* allele without showing the disorder, so they are called *carriers*, a status indicated by the half-shaded circles in Figure 3.3.

In the second example (Figure 3.3b), the father is color blind but the mother is neither color blind nor a carrier of the *c* allele. None of the children are color blind, but the daughters are all carriers because they must inherit their father's X chromosome with the recessive *c* allele. You should now be able to predict the risk of color blindness for offspring of these carrier daughters. As shown in the bottom row of Figure 3.3b, when a carrier daughter (*Cc*) has children by an unaffected male (*C*), half of her sons but none of her daughters are likely to be color blind. Half of the daughters are carriers. This pattern of inheritance explains the skip-a-generation phenomenon. Color-blind fathers have no color-blind sons or daughters, but their daughters are carriers of the *c* allele. The daughters' sons have a 50 percent chance of being color blind.

The sex chromosomes are inherited differently for males and females, so detecting X linkage is much easier than identifying a gene's location on other chromosomes. Color blindness was the first reported human X linkage. More than 100 genes have been identified as X-linked. The Y chromosome has few genes other than the gene for determining sex.

Recessive genes on the X chromosome, such as the gene for color blindness, affect more males than females and appear to skip a generation.

Other Exceptions to Mendel's Laws

Several other genetic phenomena do not appear to conform to Mendel's laws in the sense that they are not inherited in a simple way through the generations.

New Mutations

The most common case of exceptions to Mendel's laws involves new DNA mutations that do not affect the parent because they occur during the formation of the parent's eggs or sperm. But this situation is not really a violation of Mendel's laws, because the new mutations are passed on according to Mendel's laws, even though affected individuals have unaffected parents. Many genetic diseases involve such spontaneous mutations, which are not inherited from the preceding generation. In addition, DNA mutations frequently occur in cells other than those that produce eggs or sperm and are not passed on to the next generation. This is the cause of many cancers, for example. Although these mutations affect DNA, they are not heritable because they do not occur in the gametes.

Changes in Chromosomes

Changes in chromosomes are an important source of mental retardation, as discussed in Chapter 7. For example, Down syndrome occurs in about 1 in 1000 births and accounts for more than a quarter of individuals with mild to moderate retardation and about 10 percent of institutionalized mentally retarded individuals. It was first described by Langdon Down in 1866, the same year that Mendel published his classic paper. For many years, the origin of Down syndrome defied explanation because it does not "run in families." Another puzzling feature is that it occurs much more often in the offspring of women who have a child after 35 years of age. This relationship to maternal age suggested environmental explanations.

Instead, in the late 1950s, Down syndrome was shown to be caused by the presence of an entire extra chromosome with its thousands of genes. As explained in Chapter 4, during the formation of eggs and sperm, each of the 23 pairs of chromosomes separates and egg and sperm carry just one member of each pair. When the sperm fertilizes the egg, the pairs are reconstituted, with one chromosome of each pair coming from the father and the other coming from the mother. But sometimes the initial division is not even. When this accident happens, one egg or sperm might have both members of a particular chromosome pair and another egg or sperm might have neither. This failure to apportion the chromosomes equally is called *nondisjunction* (Figure 3.4). Non-disjunction is a major reason why more than half of all conceptions abort spontaneously in the first few weeks of prenatal life. A few fetuses with an abnormal chromosomal constitution survive. One example is an individual with Down syndrome, which is caused by the presence of three copies (called *trisomy*) of one of the smallest chromosomes. No individuals have been found with only one of these chromosomes (*monosomy*), which might occur when nondisjunction leaves an egg or sperm with no copy of the chromosome and another egg or sperm with two copies. It is assumed that this monosomy is lethal. Apparently, too little genetic material is more damaging than extra material. Because most cases of Down syndrome are created anew by nondisjunction each generation, Down syndrome generally is not familial.

Nondisjunction also explains why the incidence of Down syndrome is higher among the offspring of older mothers. All the immature eggs of a female

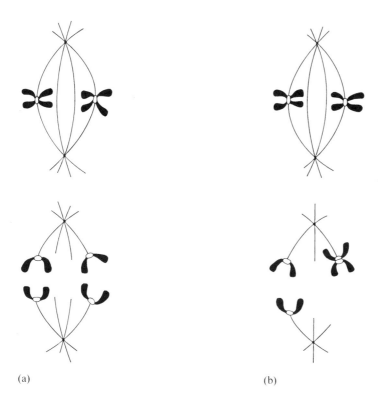

(a)   (b)

Figure 3.4 An exception to Mendel's laws of heredity: nondisjunction of chromosomes. (a) When eggs and sperm are formed, chromosomes for each pair (two pairs are shown here) line up and then split, and each new egg or sperm has just one member of each chromosome pair. (b) Sometimes this division does not occur properly, so one egg or sperm has both members of a chromosome pair and the other egg or sperm has neither.

mammal are present before birth. These eggs have both members of each pair of chromosomes. Each month, one of the immature eggs goes through the final stage of cell division. Nondisjunction is more likely to occur as the female grows older and activates immature eggs that have been dormant for decades. In contrast, fresh sperm are produced all the time. For this reason, the incidence of Down syndrome is not related to age of fathers.

Many women worry about reproducing later in life because of chromosomal abnormalities such as Down syndrome. Much of the worry of pregnancies later in life can be relieved by amniocentesis, a procedure that examines the chromosomes of the fetus.

Expanded Triplet Repeats

Mutations and chromosomal abnormalities have been known for a long time. Two other exceptions to Mendel's rules were discovered only recently. One involves repeat sequences of DNA. Although we do not know why, some very short segments of DNA—two, three, or four nucleotide bases of DNA (Chapter 4)—repeat a few times to a few dozen times. Different repeat sequences can be found in as many as 50,000 places in the human genome. Each repeat sequence has several, often a dozen or more, alleles that consist of various numbers of the same repeat sequence; these alleles are usually inherited from generation to generation according to Mendel's laws. For this reason, and because there are so many of them, repeat sequences are widely used as DNA markers of the genome in linkage studies.

Sometimes the number of repeats at a particular locus increases and causes problems. For example, most cases of Huntington's disease involve a repeat in the Huntington's gene on chromosome 4. It is called a triplet repeat because the repeated unit is a certain sequence of three nucleotide bases of DNA. Normal chromosomes contain between 11 and 34 copies of the triplet repeat, but Huntington's chromosomes have more than 40 copies. The expanded number of triplet repeats is unstable and can increase in subsequent generations. This phenomenon might explain a non-Mendelian process called *genetic anticipation*, in which symptoms appear at earlier ages and with greater severity in successive generations. For Huntington's disease, longer expansions lead to earlier onset of the disorder and greater severity. Despite this non-Mendelian twist, Huntington's disease generally follows Mendel's laws of heredity as a single-gene dominant disorder. Schizophrenia and manic-depressive psychosis also appear to show some genetic anticipation, an observation suggesting the possibility that a triplet repeat might also affect these disorders.

Fragile X mental retardation, the most common cause of mental retardation after Down syndrome, is caused by an expanded triplet repeat that violates Mendel's laws. Although this type of mental retardation was known to occur almost twice as often in males as in females, its pattern of inheritance did not conform to sex linkage because it is caused by an unstable expanded repeat. Parents who inherit X chromosomes with a normal number of repeats (6 to 54 repeats) at a particular locus sometimes produce eggs or sperm with an expanded number of repeats (up to 200 repeats), called a *premutation*. This premutation does not cause retardation in the offspring, but it is unstable and often leads to much greater expansions (200 or more repeats) in the next generation, which do cause retardation (see Figure 3.5).

Genomic Imprinting

Another example of exceptions to Mendel's laws is called *genomic imprinting*, or *gametic imprinting*. In genomic imprinting, the expression of a gene depends on

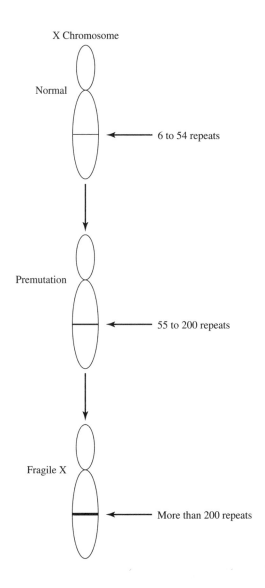

X Chromosome

Normal

6 to 54 repeats

Premutation

55 to 200 repeats

Fragile X

More than 200 repeats

Figure 3.5 Fragile X mental retardation involves a triplet repeat sequence of DNA on the X chromosome that can expand over generations.

whether it is inherited from the mother or from the father, even though, as usual, one allele is inherited from each parent. The mechanism by which one parent's allele is marked (imprinted) is not known. Several such genes have been described in mice and humans (Barlow, 1995). The most striking example of genomic imprinting involves deletions of a small part of chromosome 15 that lead to two very different disorders, depending on whether a deletion is inherited from the mother or the father. When it is inherited from the mother, it causes what is known as Angelman syndrome, which involves severe mental retardation and other manifestations such as an awkward gait and frequent inappropriate laughter. When a deletion is inherited from the father, it causes other behavioral problems such as overeating, temper outbursts, and depression, as well as physical problems such as obesity and short stature (Prader-Willi syndrome).

In addition to genes acting differently, depending on whether they are inherited from the mother or the father, the expansion of repeat sequences discussed in the previous section sometimes depends on the sex of the parent. For example, the fragile X repeat sequence expands in mothers, whereas the Huntington's disease repeat sequence expands in fathers.

SUMMING UP

Other exceptions to Mendel's laws include new mutations, changes in chromosomes, expanded triplet sequences, and genomic imprinting. Many genetic diseases involve spontaneous mutations that are not inherited from generation to generation. Changes in chromosomes include nondisjunction, which is the single most important cause of mental retardation, the trisomy of Down syndrome. Expanded triplet repeats are responsible for the next most important cause of mental retardation, fragile X, and for Huntington's disease. Genomic imprinting occurs when the expression of a gene depends on whether it is inherited from the mother or from the father, as in Angelman and Prader-Willi syndromes.

Complex Traits

Most psychological traits show patterns of inheritance that are much more complex than those of Huntington's disease or PKU. Consider schizophrenia and general cognitive ability.

Schizophrenia

Schizophrenia (Chapter 10) is a severe mental disorder characterized by thought disorders. Nearly 1 in 100 people are afflicted by this disorder throughout the world, 100 times more than HD or PKU. Schizophrenia shows no simple pattern of inheritance like HD, PKU, or color blindness, but it is familial (Figure 3.6). A special incidence figure used in genetic studies is called a *morbidity risk esti-*

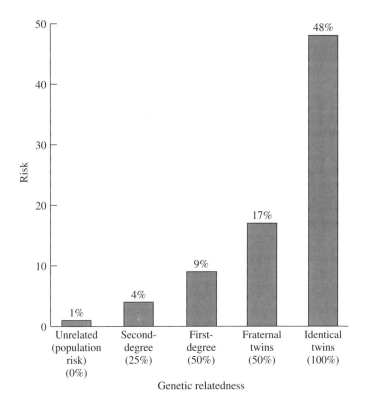

Figure 3.6 Risk for schizophrenia increases with genetic relatedness. (Data adapted from Gottesman, 1991.)

mate, which is an estimate of the risk of being affected. If you have a second-degree relative (grandparent or aunt or uncle) who is schizophrenic, your risk for schizophrenia is about 4 percent, four times greater than the risk in the general population. If a first-degree relative (parent or sibling) is schizophrenic, your risk is about 9 percent. If several family members are affected, the risk is greater. If your fraternal twin has schizophrenia, your risk is somewhat higher than for other siblings, at about 17 percent, even though fraternal twins are no more similar genetically than siblings. Most striking, the risk is about 48 percent for an identical twin whose co-twin is schizophrenic. Identical twins develop from one embryo, which in the first few days of life splits into two embryos, each with the same genetic material (Chapter 5).

Are Mendel's laws of heredity at all applicable to a complex disorder such as schizophrenia or to continuously distributed traits such as cognitive abilities?

General Cognitive Ability

Many psychological traits are quantitative dimensions, as are physical traits such as height and biomedical traits such as blood pressure. Quantitative dimensions are often continuously distributed in the familiar bell-shaped curve, with most people in the middle and fewer people toward the extremes. Appendix A provides an overview of the statistics of individual differences.

For example, as discussed in Chapter 8, an intelligence test score from a general test of intelligence is a composite of diverse tests of cognitive ability and is used to provide an index of general cognitive ability. Intelligence test scores are largely normally distributed.

Because general cognitive ability is a quantitative dimension, it is not possible to count "affected" individuals. Nonetheless, it is clear that general cognitive ability runs in families. For example, parents with high intelligence test scores tend to have children with higher than average scores. Like schizophrenia, transmission of general cognitive ability does not seem to follow simple Mendelian rules of heredity.

The statistics of quantitative traits are needed to describe family resemblance. Over a hundred years ago, Francis Galton, the father of behavioral genetics, tackled this problem of describing family resemblance for quantitative traits. He developed a statistic that he called co-relation and that has become the widely used correlation coefficient. More formally, it is called the Pearson product-moment correlation, named after Karl Pearson, Galton's colleague. The *correlation* is an index of resemblance that goes from .00, indicating no resemblance, to 1.0, indicating perfect resemblance. (See Appendix A for an overview of correlation.)

Correlations for intelligence test scores show that resemblance of family members depends on the closeness of the genetic relationship (Figure 3.7). The correlation of intelligence test scores for pairs of individuals taken at random from the population is .00. The correlation for cousins is about .15. For half siblings, who have just one parent in common, the correlation is about .30. For full siblings, who have both parents in common, the correlation is about .45; this correlation is similar to that between parents and offspring. Scores for fraternal twins correlate about .60, and identical twins correlate about .85. Husbands and wives correlate about .40, which has implications for interpreting sibling and twin correlations, as discussed in Chapter 8.

How do Mendel's laws of heredity apply to continuous dimensions such as general cognitive ability?

Pea Size

Let us return to pea plants. A large part of Mendel's success in working out the laws of heredity came from choosing simple traits that are either-or qualitative traits. If Mendel had studied, for instance, size of the pea seed as indexed by its diameter, he would have found very different results. First, pea seed size, like most traits, is continuously distributed. If he had taken plants with big seeds and crossed them with plants with small seeds, the seed size of their offspring would

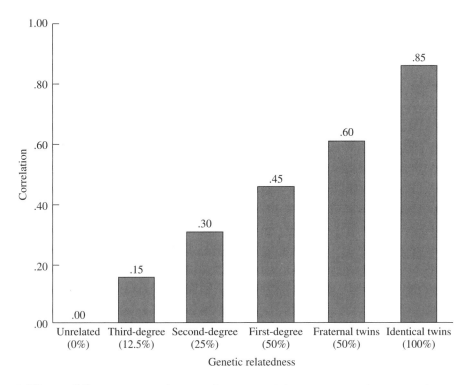

Figure 3.7 Resemblance for general cognitive ability increases with genetic relatedness. (Data adapted from Bouchard & McGue, 1981, as modified by Loehlin, 1989.)

not be either big or small. In fact, the seeds would have varied in size from small to large, with most offspring seeds of average size.

Only ten years after Mendel's report, Francis Galton studied pea seed size and concluded that it is inherited. For example, parents with large seeds were likely to have offspring with larger than average seeds. In fact, Galton developed the fundamental statistics of regression and correlation mentioned above in order to describe the quantitative relationship between pea seed size in parents and offspring (Appendix A). He plotted parent and offspring seed sizes and drew the regression line that best fits the observed data (Figure 3.8). The slope of the regression line is .33. This means that, for the entire population, as parental size increases by one unit, the average offspring size increases one-third of one unit.

Galton also demonstrated that human height shows the same pattern of inheritance. Children's height correlates with the average height of their parents. Tall parents have taller than average children. Children with one tall and one short parent are likely to be of average height. Inheritance of this trait is quantitative rather than qualitative. Quantitative inheritance is the way in which nearly all psychological as well as physical dimensions and disorders are inherited.

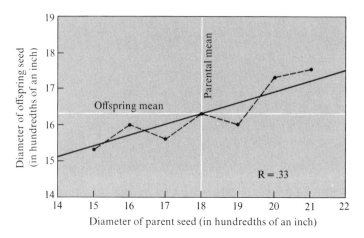

Figure 3.8 First regression line (solid line), drawn by Galton in 1877 to describe the quantitative relationship between pea seed size in parents and offspring. The dashed line connects actual data points. (Courtesy of the Galton Laboratory.)

Does quantitative inheritance violate Mendel's laws? When Mendel's laws were rediscovered in the early 1900s, many scientists thought this must be the case. They thought that heredity must involve some sort of blending, because offspring resemble the average of their parents. Mendel's laws were dismissed as a peculiarity of pea plants. However, recognizing that quantitative inheritance does *not* violate Mendel's laws is fundamental to an understanding of behavioral genetics.

SUMMING UP

Schizophrenia and general cognitive ability are examples of complex traits that are influenced by multiple genes and multiple environmental factors. Relatives resemble each other more for these traits the more genes they share: identical twins more than fraternal twins and first-degree relatives more than second-degree relatives. Complex, quantitative traits typical of behavioral disorders and dimensions do not violate Mendel's laws.

Multiple-Gene Inheritance

The traits that Mendel studied, as well as Huntington's disease and PKU, are examples in which a single gene is necessary and sufficient to cause the disorder. That is, you will only have Huntington's disease if you have the *H* allele (necessary); if you have the *H* allele, you will have Huntington's disease (sufficient). Other genes and environmental factors have little effect on its inheritance. In such cases, a dichotomous (either-or) disorder is found: You either have the specific allele, or not, and thus you have the disorder, or not. More than 2000 such single-gene disorders are known definitely and again as many are considered probable.

In contrast, more than just one gene is likely to affect complex disorders such as schizophrenia and continuous dimensions such as general cognitive ability. When Mendel's laws were rediscovered in the early 1900s, a bitter battle was fought between Mendelians and biometricians. Mendelians looked for single-gene effects, and biometricians argued that Mendel's laws could not apply to complex traits because they showed no simple pattern of inheritance. Mendel's laws seemed especially inapplicable to quantitative dimensions.

In fact, both sides were right and both were wrong. The Mendelians were correct in arguing that heredity works the way Mendel said it worked, but they were wrong in assuming that complex traits will show simple Mendelian patterns of inheritance. The biometricians were right in arguing that complex traits are distributed quantitatively, not qualitatively, but they were wrong in arguing that Mendel's laws of inheritance are particular to pea plants and do not apply to higher organisms.

The battle between the Mendelians and biometricians was resolved when biometricians realized that Mendel's laws of inheritance of single genes also apply to complex traits that are influenced by *several* genes. Each of the influential genes is inherited according to Mendel's laws.

Figure 3.9 illustrates this important point. The top distribution shows the three genotypes of a single gene with two alleles that are equally frequent in the population. As discussed in Box 2.1, 25 percent of the genotypes are homozygous for the A_1 allele (A_1A_1), 50 percent are heterozygous (A_1A_2), and 25 percent are homozygous for the A_2 allele (A_2A_2). If the A_1 allele were dominant, individuals with the A_1A_2 genotype would look just like individuals with the A_1A_1 genotype. In this case, 75 percent of individuals would have the observed trait (phenotype) of the dominant allele. For example, as discussed in Box 2.1, in Mendel's crosses of pea plants with smooth or wrinkled seeds, he found that in the F_2 generation, 75 percent of the plants had smooth seeds and 25 percent had wrinkled seeds.

However, not all alleles operate in a completely dominant or recessive manner. Many alleles are additive in that they each contribute something to the phenotype. In Figure 3.9a, each A_2 allele contributes equally to the phenotype, so if you have two A_2 alleles, you would have a higher score than if you had just one A_2 allele. Figure 3.9b adds a second gene that affects the trait. Again, each B_2 allele makes a contribution. Now there are nine genotypes and five phenotypes. Figure 3.9c adds a third gene, and there are 27 genotypes. Even if we assume that the alleles of the different genes equally affect the trait and that there is no environmental variation, there are still seven different phenotypes.

So, even with just three genes and two alleles for each gene, the phenotypes begin to approach a normal distribution in the population. When we consider

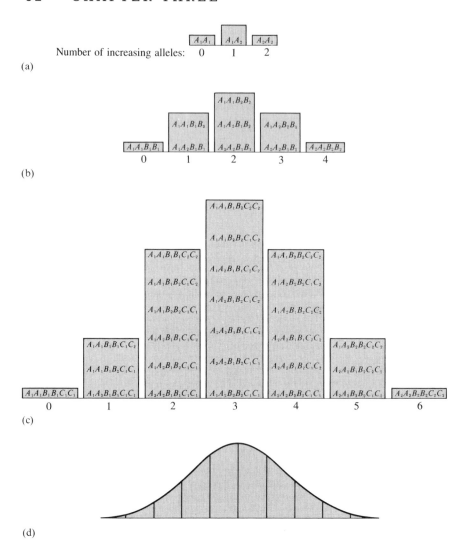

Figure 3.9 Single-gene and multiple-gene distributions for traits with additive gene effects. (a) A single gene with two alleles yields three genotypes and three phenotypes. (b) Two genes, each with two alleles, yield nine genotypes and five phenotypes. (c) Three genes, each with two alleles, yield twenty-seven genotypes and seven phenotypes. (d) Normal bell-shaped curve of continuous variation.

environmental sources of variability and the fact that the effects of alleles may not be equal, it is easy to see that the effects of even a few genes will lead to a quantitative distribution. Moreover, the complex behavioral traits that interest psychologists may be influenced by dozens or even hundreds of genes. Thus, it is not surprising to find continuous variation at the phenotypic level, even though each gene is inherited in accord with Mendel's laws.

Quantitative Genetics

The notion that multiple-gene effects lead to quantitative traits is the corner-stone of a branch of genetics called *quantitative genetics.*

Quantitative genetics was introduced in papers by R. A. Fisher (1918) and by Sewall Wright (1921). Their extension of Mendel's single-gene model to the multiple-gene model of quantitative genetics (Falconer & Mackay, 1996) is described in Appendix B. This multiple-gene model adequately accounts for the resemblance of relatives. If genetic factors affect a quantitative trait, phe-notypic resemblance of relatives should increase with increasing degrees of genetic relatedness. First-degree relatives, parents and offspring and full sib-lings, are 50 percent similar genetically. The simplest way to think about this is that offspring inherit half their genetic material from each parent. If one sib-ling inherits a particular allele from a parent, the other sibling has a 50 percent chance of inheriting that same allele. Other relatives differ in their degree of genetic relatedness.

Figure 3.10 illustrates degrees of genetic relatedness for the most common types of relatives, using male relatives as examples. Relatives are listed in relation

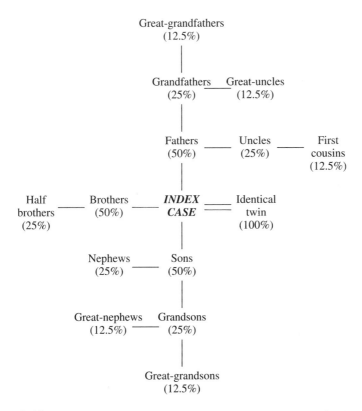

Figure 3.10 Genetic relatedness: Male relatives of male index case (*proband*), with degree of genetic relatedness in parentheses.

to an individual in the center, the index case. The illustration goes back three generations and forward three generations. First-degree relatives (e.g., fathers, sons), who are 50 percent similar genetically, are each just one step removed from the index case. Second-degree relatives (e.g., uncles) are two steps removed and are only half as similar genetically (i.e., 25 percent) as first-degree relatives are. Third-degree relatives (e.g., cousins) are three steps removed and only half as similar genetically (i.e., 12.5 percent) as second-degree relatives are. Identical twins are a special case, because they are the same person genetically.

For schizophrenia and general cognitive ability, phenotypic resemblance of relatives increases with genetic relatedness. For schizophrenia, as noted earlier (Figure 3.6), the chance that an individual picked at random from the population will develop schizophrenia is about 1 percent. If a second-degree relative (grandparent or aunt/uncle) is affected, the risk is 4 percent. If a first-degree relative (parent, sibling, or offspring) is affected, the risk doubles to about 9 percent. Finally, the risk shoots up to 48 percent for identical twins, which is much greater than the risk of 17 percent for fraternal twins. Comparison of traits in identical and fraternal twins to detect genetic influence is discussed in Chapter 5.

How can there be a dichotomous disorder if many genes cause schizophrenia? One possible explanation is that genetic risk is normally distributed but that schizophrenia is not seen until a certain threshold is reached. Another explanation is that disorders are actually dimensions artificially divided on the basis of a diagnosis. That is, there may be a continuum between what is normal and abnormal. These alternatives are described in Box 3.1.

For general cognitive ability, phenotypic resemblance also increases with genetic relatedness, as shown in Figure 3.7. Cousins (third-degree relatives) are only half as similar as half siblings (second-degree relatives), and half siblings are less similar than full siblings (first-degree relatives). Identical twins are more similar than fraternal twins.

These data are consistent with the hypothesis of genetic influence, but they do not *prove* that genetic factors are important. It is possible that familial resemblance increases with genetic relatedness for environmental reasons. First-degree relatives might be more similar because they live together. Second-degree and third-degree relatives might be less similar because of less similarity of rearing.

Two experiments of nature are the workhorses of behavioral genetics that disentangle genetic and environmental sources of family resemblance. One is the *twin study*, which compares the resemblance within pairs of identical twins, who are genetically identical, to the resemblance within pairs of fraternal twins, who, like other siblings, are 50 percent similar genetically. The second is the *adoption study*, which separates genetic and environmental influences. For example, when biological parents relinquish their children for adoption at birth, any resemblance between these parents and their adopted-away offspring can be attributed to shared heredity rather than to shared

BOX 3.1

Liability-Threshold Model of Disorders

If complex disorders such as schizophrenia are influenced by many genes, why are they diagnosed as qualitative disorders rather than assessed as quantitative dimensions? Theoretically, there should be a continuum of genetic risk, from people having none of the alleles that increase risk for schizophrenia to those having most of the alleles that increase risk. Most people should fall between these extremes, with only a moderate susceptibility to schizophrenia.

One model assumes that risk, or liability, is distributed normally but that the disorder occurs only when a certain threshold of liability is exceeded, as represented in the accompanying figure by the shaded area in (a). Relatives of an affected person have a greater liability, that is, their distribution of liability is shifted to the right, as in (b). For this reason, a greater proportion of the relatives of affected individuals exceed the threshold and manifest the disorder. If there is such a threshold, familial risk can be high only if genetic or shared environmental influence is substantial because many of an affected individual's relatives will fall just below the threshold and not be affected.

Liability and threshold are hypothetical constructs. However, it is possible to use the liability-threshold model to estimate correlations from family risk data (Falconer, 1965; Smith, 1974). For example, the correlation estimated for first-degree relatives for schizophrenia is .45, an estimate based on a population base rate of 1 percent and risk to first-degree relatives of 9 percent.

Although liability-threshold correlations are widely reported for psychological disorders, it should be emphasized that this statistic refers to hypothetical constructs of a threshold and an underlying liability derived from diagnoses, not to the risk for the actual diagnosed disorder. That is, in the previous example, the actual risk for schizophrenia for first-degree relatives for schizophrenia is 9 percent, even though the liability-threshold correlation is .45.

Alternatively, a second model assumes that disorders are actually continuous phenotypically. That is, the disorder might not just appear after a certain threshold is reached. Instead, symptoms might increase continuously from the normal to the abnormal. A continuum from normal to abnormal seems likely for common disorders such as depression and alcoholism. For example, people vary in the frequency and severity of their depression. Some people rarely get the blues; for others, depression completely disrupts their lives. Individuals diagnosed as depressed might be extreme cases that differ quantitatively, not qualitatively, from the rest of the population. In such cases, it may be possible to assess the continuum directly, rather than assuming a continuum from dichotomous diagnoses using the liability-threshold model. Even for less

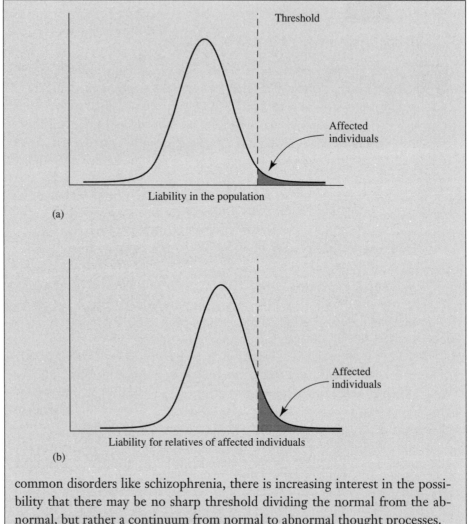

Threshold

Affected individuals

Liability in the population

(a)

Threshold

Affected individuals

Liability for relatives of affected individuals

(b)

common disorders like schizophrenia, there is increasing interest in the possibility that there may be no sharp threshold dividing the normal from the abnormal, but rather a continuum from normal to abnormal thought processes.

The relationship between dimensions and disorders is key and is discussed in later chapters. The best evidence for genetic links between dimensions and disorders will come as specific genes are found for behavior. For example, will a gene associated with diagnosed depression also relate to differences in mood within the normal range?

environment, if there is no selective placement. In addition, when these children are adopted, any resemblance between the adoptive parents and their adopted children can be attributed to shared environment rather than to shared heredity.

SUMMING UP

The battle between the Mendelians and biometricians was resolved when it was realized that Mendel's laws of the inheritance of single genes also apply to those complex traits that are influenced by several genes. Each of these genes is inherited according to Mendel's laws. This concept is the cornerstone of the field of quantitative genetics, which is a theory and set of methods (such as the twin and adoption methods) for investigating the inheritance of complex, quantitative traits.

As described in Chapter 5, results obtained from twin and adoption studies indicate that genetic factors play a major role in familial resemblance for schizophrenia and cognitive ability. The point of the present chapter is that the pattern of inheritance for complex disorders like schizophrenia and continuous dimensions like cognitive ability is different from that seen for single-gene traits, because multiple genes are involved. However, each gene is inherited according to Mendel's laws.

Chapter 4 briefly describes the DNA basis for Mendel's laws of heredity. Chapter 5 returns to research using the twin and adoption methods, which aim to assess the net effects of multiple genes and multiple experiences on complex behavioral traits of interest to psychologists.

Summary

Mendel's laws of heredity do not explain all genetic phenomena. For example, genes on the X chromosome, such as the gene for color blindness, require an extension of Mendel's laws. Other exceptions to Mendel's laws include new mutations, changes in chromosomes such as the chromosomal nondisjunction that causes Down syndrome, expanded DNA triplet repeat sequences responsible for Huntington's disease and fragile X mental retardation, and genomic imprinting.

Most psychological dimensions and disorders show more complex patterns of inheritance than do single-gene disorders such as Huntington's disease, PKU, or color blindness. Complex disorders such as schizophrenia and continuous dimensions such as cognitive ability are likely to be influenced by multiple genes as well as by multiple environmental factors. Quantitative genetic theory extends Mendel's single-gene rules to multiple-gene systems. The essence of the theory is that complex traits can be influenced by many genes but each gene is inherited according to Mendel's laws. Quantitative genetic methods, especially adoption and twin studies, can detect genetic influence for complex traits.

DNA: The Basis of Heredity

Mendel was able to deduce the laws of heredity even though he had no idea of how heredity works at the biological level. Quantitative genetics, such as twin and adoption studies, depends on Mendel's laws of heredity but does not require knowledge of the biological basis of heredity. However, it is important to understand the biological mechanisms underlying heredity for two reasons. First, understanding the biological basis of heredity makes it clear that the processes by which genes affect behavior is not mystical. Second, this understanding is crucial for appreciating the exciting advances in molecular genetic attempts to identify genes associated with behavior. Behavioral genetics includes molecular genetic as well as quantitative genetic research on behavior.

This chapter briefly describes the biological basis of heredity, how the process is regulated, how genetic variation arises, and how this genetic variation is detected using molecular genetic techniques. The biological basis of heredity includes the fact that genes are contained on structures called chromosomes. The linkage of genes that lie close together on chromosomes has made possible the mapping of the human genome. Moreover, abnormalities in chromosomes contribute importantly to behavioral disorders, especially mental retardation.

DNA

Nearly a century after Mendel did his experiments, it became apparent that DNA (deoxyribonucleic acid) is the molecule responsible for heredity. In 1953, James Watson and Francis Crick proposed a molecular structure for DNA that could explain how genes are replicated and how DNA codes for proteins. As shown in Figure 4.1, the DNA molecule consists of two strands that are held apart by pairs of four bases: adenine, thymine, guanine, and cytosine. As a result of the structural properties of these bases, adenine always pairs with thymine

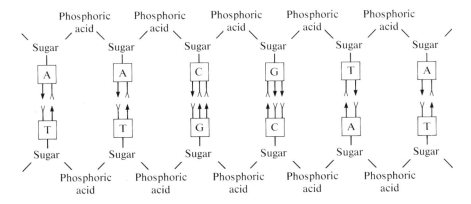

Figure 4.1 Flat representation of the four DNA bases in which adenine (A) always pairs with thymine (T) and guanine (G) always pairs with cytosine (C). (From *Heredity, Evolution, and Society* by I. M. Lerner. W. H. Freeman and Company. Copyright © 1968.)

and guanine always pairs with cytosine. The backbone of each strand consists of sugar and phosphate molecules. The strands coil around each other to form the famous double helix of DNA (Figure 4.2).

The specific pairing of bases in these two-stranded molecules allows DNA to carry out its two functions: to replicate itself and to synthesize proteins. Replication of DNA occurs during the process of cell division. The double helix of the DNA molecule unzips, separating the paired bases (Figure 4.3). The two strands unwind, and each strand attracts the appropriate bases to construct its complement. In this way, two complete double helices of DNA are created where there was previously only one. This process of replication is the essence of life that began billions of years ago when the first cells replicated themselves.

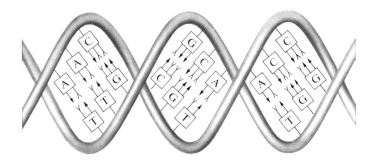

Figure 4.2 A three-dimensional view of a segment of DNA. (From *Heredity, Evolution, and Society* by I. M. Lerner. W. H. Freeman and Company. Copyright © 1968.)

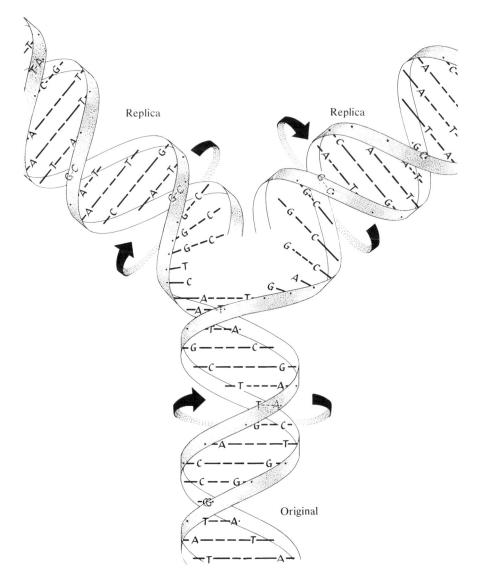

Figure 4.3 Replication of DNA. (After *Molecular Biology of Bacterial Viruses* by G. S. Stent. W. H. Freeman and Company. Copyright © 1963.)

Another major function of DNA is to synthesize proteins according to the genetic information that resides in the particular sequence of bases. DNA codes for sequences of the 20 amino acids making up the thousands of specific enzymes and proteins that are the stuff of living organisms. Box 4.1 describes this process, the so-called central dogma of molecular genetics.

What is the genetic code contained in the sequence of DNA bases, which is transcribed to messenger RNA (mRNA; see Box 4.1) and then translated into

TABLE 4.1

The Genetic Code

Amino Acid*	DNA Code
Alanine	CGA, CGG, CGT, CGC
Arginine	GCA, GCG, GCT, GCC, TCT, TCC
Asparagine	TTA, TTG
Aspartic acid	CTA, CTG
Cysteine	ACA, ACG
Glutamic acid	CTT, CTC
Glutamine	GTT, GTC
Glycine	CCA, CCG, CCT, CCC
Histidine	GTA, GTG
Isoleucine	TAA, TAG, TAT
Leucine	AAT, AAC, GAA, GAG, GAT, GAC
Lysine	TTT, TTC
Methionine	TAC
Phenylalanine	AAA, AAG
Proline	GGA, GGG, GGT, GGC
Serine	AGA, AGG, AGT, AGC, TCA, TCG
Threonine	TGA, TGG, TGT, TGC
Tryptophan	ACC
Tyrosine	ATA, ATG
Valine	CAA, CAG, CAT, CAC
(Stop signals)	ATT, ATC, ACT

The 20 amino acids are organic molecules that are linked together by peptide bonds to form polypeptides, which are the building blocks of enzymes and other proteins. The particular combination of amino acids determines the shape and function of the polypeptide.

amino acid sequences? The code consists of various sequences of three bases, which are called *codons* (see Table 4.1). For example, three adenines in a row (AAA) in the DNA molecule will be transcribed in mRNA as three uracils (UUU). This mRNA codon codes for the amino acid phenylalanine. Although there are 64 possible triplet codons ($4^3 = 64$), there are only 20 amino acids. Some amino acids are coded by as many as six codons. Three codons signal the end of a transcribed sequence.

This same genetic code applies to all living organisms. Breaking this code was one of the great triumphs of molecular biology. The human set of DNA sequences (genome) consists of about 3 billion nucleotide bases, just counting one chromosome from each pair of chromosomes. The 3 billion bases contain about 100,000 genes, which range in size from about 1000 to 2 million base

BOX 4.1

The "Central Dogma" of Molecular Genetics

Genetic information flows from DNA to messenger RNA (mRNA) to protein. Genes are DNA segments that are a few thousand to several million DNA base pairs in length. The DNA molecule contains a linear message with four repeating bases (adenine, thymine, guanine, and cytosine); in this two-stranded molecule, A always pairs with T and G with C. The message is decoded in two basic steps, shown in the figure: (a) transcription of DNA into a different sort of nucleic acid called ribonucleic acid, or RNA, and (b) translation of RNA into proteins.

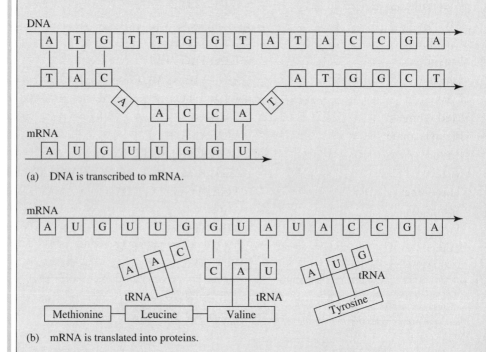

(a) DNA is transcribed to mRNA.

(b) mRNA is translated into proteins.

In the transcription process, the sequence of bases in one strand of the DNA double helix is copied to RNA, specifically a type of RNA called messenger RNA (mRNA) because it relays the DNA code. mRNA is single stranded and is formed by a process of base pairing similar to the replication of DNA,

pairs. About a third of our genes are expressed only in the brain; these are likely to be most important for behavior. Finding all our genes and determining the sequence of the 3 billion bases of DNA is the grand goal of the Human Genome Project. Less than 1 percent of the human genome has been sequenced except that uracil substitutes for thymine (so that A pairs with U instead of T). In the figure, one DNA strand is being transcribed—the DNA bases ACCA have just been copied as UGGU in mRNA. mRNA leaves the nucleus of the cell and enters the cell body (cytoplasm), where it connects with ribosomes, which are the factories where proteins are built.

The second step involves translation of the mRNA into amino acid sequences that form proteins. Another form of RNA, called transfer RNA (tRNA), transfers amino acids to the ribosomes. Each tRNA is specific to 1 of the 20 amino acids. The tRNA molecules, with their attached specific amino acids, pair up with the mRNA in a sequence dictated by the base pairs of the mRNA as the ribosome moves along the mRNA strand. Each of the 20 amino acids found in proteins is specified by a "codon" made up of three sequential mRNA bases. In the figure, the mRNA code has begun to dictate a protein that includes the amino acid sequence methionine-leucine-valine-tyrosine. Valine has just been added to the chain that already includes methionine and leucine. The mRNA triplet code GUA attracts tRNA with the complementary code CAU. This tRNA transfers its attached amino acid valine, which is then bonded to the growing chain of amino acids. The next mRNA codon, UAC, is attracting tRNA with the complementary codon, AUG, for tyrosine. Although this process seems very complicated, amino acids are incorporated into chains at the incredible rate of about 100 per second. Proteins consist of particular sequences of about 100 to 1000 amino acids. The sequence of amino acids determines the shape and function of proteins. Protein shape is altered subsequently in other ways that change its function, but these changes are not controlled by the genetic code and are called *posttranslational changes.*

Surprisingly, DNA that is transcribed and translated like this represents only a small percentage (perhaps as little as 5 percent) of DNA. The rest includes DNA that is not transcribed into RNA. It also includes parts of genes, called *introns,* that are transcribed into RNA but are spliced out before the RNA leaves the nucleus. The parts of genes that are spliced back together are called *exons.* Exons exit the nucleus and are expressed as amino acid sequences. Nearly all eukaryotic genes have introns. Exons usually consist of only a few hundred base pairs, but introns vary widely in length, from 50 to 20,000 base pairs. Only exons are translated into amino acid sequences that make up proteins. The function of introns is not known, but in some cases introns regulate the transcription of other genes.

so far. If 3000 bases are listed on a page and each volume has 1000 pages, a list of one person's DNA sequence will require 1000 volumes. This encyclopedia does not begin to consider the variation in DNA sequences among individuals that is at the heart of behavioral genetics.

For behavioral genetics, the most important thing to understand about the DNA basis of heredity is that the process by which genes affect behavior is not mystical. Genes code for sequences of amino acids that form the thousands of proteins of which organisms are made. Proteins create the skeletal system, muscles, the endocrine system, the immune system, the digestive system, and, most important for behavior, the nervous system. Genes do not code for behavior directly, but DNA variations that create differences in these physiological systems can affect behavior.

SUMMING UP

DNA is a double helix that includes four different bases. Its structure allows DNA to replicate itself and to synthesize proteins. DNA codes for the synthesis of the 20 amino acids by means of sequences of three base pairs, or triplets. The triplets make up the genetic code. The human genome consists of 3 billion nucleotide bases and has about 100,000 genes, a third of which are expressed only in the brain.

Gene Regulation

Genes do not blindly pump out their protein products. When the gene product is needed, many copies of its mRNA will be present, but otherwise very few copies of the mRNA are transcribed. You are changing the rates of transcription of genes for neurotransmitters by reading this sentence. Because mRNA exists for only a couple of minutes and then is no longer translated into protein, changes in the rate of transcription of mRNA are used to control the rate at which genes produce proteins.

In some cases, introns—parts of genes that are transcribed into RNA but spliced out before mRNA leaves the nucleus (Box 4.1)—regulate gene transcription. The sole function of many genes is to regulate the transcription of other genes rather than to code for proteins. Some gene regulation is short term and responsive to the environment. Other gene regulation leads to long-term changes in development. Figure 4.4 shows how regulation often works. Many genes include regulatory sequences that normally block the gene from being transcribed. If a particular molecule binds with the regulatory sequence, it will free the gene for transcription. Most gene regulation involves several mechanisms that act like a committee voting on increases or decreases in transcription. That is, several transcription factors act in concert to regulate the rate of specific mRNA transcription.

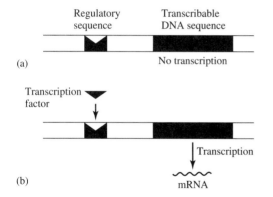

Figure 4.4 Transcription factors regulate genes by controlling mRNA synthesis. (a) A regulatory sequence normally shuts down transcription of its gene; (b) but when a particular transcription factor comes along and binds to the regulatory sequence, the gene is freed for transcription.

Similar mechanisms also lead to long-term developmental changes. The key question about development is how differentiation occurs, how we start life as a single cell and end up with trillions of cells, all of which have the same DNA but many different functions. Some basic aspects of development are programmed in genes. For example, we have 38 homeobox (*Hox*) genes similar to genes in most animals that act as master switches and control the timing of development of different parts of the body. However, for the most part, development is not hard wired in the genes. For example, a thousand different molecules must be synthesized in a specific sequence during the half-hour life cycle of bacteria. It used to be assumed that this sequential synthesis was programmed genetically, with genes programmed to turn on at the right moment. However, the sequence of steps is not efficiently programmed in DNA. Instead, transcription rates of DNA depend on the products of earlier DNA transcription and on experiences. Consider another example. When songbirds are first exposed to their species' songs, the experience causes changes in expression of a set of genes in the brain encoding proteins that regulate the transcription of other brain genes (Mello, Vicario, & Clayton, 1992). Tweaking a rat's whiskers causes changes in gene expression in the sensory cortex (Mack & Mack, 1992).

The point is that many genes regulate the transcription of other genes in response to both the internal and the external environments. It is now thought that much more DNA is involved in regulatory genes than in structural genes that code for proteins (Lawrence, 1992). Although behavioral genetics has been concerned primarily with structural genes, its methods are just as appropriate for detecting genetic variation that arises from gene regulation. For example, identical twins will have identical genes for genetic regulatory processes that are coded in DNA at conception. Changes in gene regulation in response to the environment can differ for identical twins and would properly be attributed to the environment.

SUMMING UP

Many genes are involved in regulating the transcription of other genes rather than in synthesizing proteins. Gene regulation is also responsible for long-term developmental changes.

Mutations

Behavioral genetics asks why people are different behaviorally—for example, why some are mentally ill or retarded. For this reason, it focuses on genetic and environmental differences among people that can account for these observed differences. New DNA differences occur when mistakes, called *mutations*, are made in copying DNA. These mutations result in different alleles (called *polymorphisms*) such as the alleles responsible for the variations that Mendel found in pea plants, for Huntington's disease and PKU, and for complex behavioral traits such as schizophrenia and cognitive abilities. Mutations that occur in the creation of eggs and sperm will be transmitted faithfully unless natural selection intervenes (Chapter 13). The effects that count in terms of natural selection are effects on survival and reproduction. Because evolution has so finely tuned the genetic system, most new mutations in regions of DNA that are translated into amino acid sequences have deleterious effects. However, once in a great while a mutation will make the system function a bit better. In evolutionary terms, this means that individuals with the mutation are more likely to survive and reproduce.

A single-base mutation can result in the insertion of a different amino acid into a protein. Such a mutation can alter the function of the protein. For example, in the figure in Box 4.1, if the first DNA codon TAC is miscopied as TCC, the amino acid arginine will be substituted for methionine. (Table 4.1 indicates that TAC codes for methionine and TCC codes for arginine.) This single amino acid substitution in the hundreds of amino acids that make up a protein might have no noticeable effect on the protein's functioning; then again, it might have a small effect, or it might have a major, even lethal, effect. A mutation that leads to the loss of a single base is likely to be more damaging than a mutation causing a substitution, because the loss of a base shifts the *reading frame* of the triplet code. For example, if the second base in the box figure were deleted, TAC-AAC-CAT- becomes TCA-ACC-AT. Instead of the amino acid chain containing methionine (TAC) and leucine (AAC), the mutation would result in a chain containing serine (TCA) and tryptophan (ACC).

Mutations are often not so simple. For example, a particular gene can have mutations at several locations. As an extreme example, over 60 different mutations have been found in the gene responsible for PKU, some of which yield milder forms of the disorder (Eisensmith & Woo, 1992). Another example of current interest involves triplet repeats, mentioned in Chapter 3. Most cases of Huntington's disease are caused by three repeating base pairs (CAG). Normal alleles have from 11 to 34 CAG repeats at the end of a gene for a protein that is found throughout the brain. For the many individuals with Huntington's disease, the number of CAG repeats varies from 37 to more than 100. Because triplet repeats involve three base pairs, the presence of any number of repeats does not shift the reading frame of transcription. However, the CAG repeat responsible for Huntington's disease is transcribed into mRNA and translated into protein, which means that multiple repeats of an amino acid are inserted into the protein. Which amino acid? CAG is the mRNA code, which means that the DNA code is GTC. Table 4.1 shows that GTC codes for the amino acid glutamine. It is reasonable to expect that having a protein lumbered with many extra copies of glutamine would reduce the protein's normal activity; in other words, the lengthened protein would show loss of function. However, although Huntington's disease is a dominant disorder, the other allele should be operating normally, producing enough of the normal protein to avoid trouble. This possibility suggests that the Huntington's allele, which adds dozens of glutamines to the protein, might confer a new property (gain of function) that creates the problems of Huntington's disease.

About 3 million of our 3 billion nucleotide bases differ among people. Most mutations do not occur in exons that are translated into proteins (Box 4.1). Mutations primarily occur in introns and in regions of DNA that are not transcribed into mRNA and thus have no apparent effect.

Detecting Polymorphisms

Much of the success of molecular genetics comes from the availability of thousands of markers that are DNA polymorphisms. Previously, genetic markers were limited to the products of single genes, such as the red blood cell proteins that define the blood groups. In 1980, new genetic markers that are the actual polymorphisms in the DNA were discovered. Because millions of DNA base pairs are polymorphic, these DNA polymorphisms can be used in linkage studies to track the chromosomal location of genes, as described in Chapter 6. As noted earlier, in 1983, such DNA markers were first used to localize the gene for Huntington's disease at the tip of the short arm of chromosome 4.

The first type of DNA marker has a long but descriptive name: *restriction fragment length polymorphism (RFLP)*. DNA extracted from blood or from cells scraped from the inside of the cheek is chopped up by a type of enzyme called a

BOX 4.2

DNA Markers

The RFLP (restriction fragment length polymorphism) and the SSR (simple sequence repeat) markers are genetic polymorphisms in DNA. They can be identified in the DNA itself rather than indirectly in a gene product such as the red blood cell proteins responsible for blood types. Both types of direct markers can be detected by using a technique called polymerase chain reaction (PCR). PCR makes many copies of small sequences of DNA, a few hundred base pairs in length, and requires that the sequence of DNA surrounding the RFLP or SSR be known. From this DNA sequence, 20 base pairs (bp) on both sides of the polymorphism are synthesized. These 20-bp DNA sequences, called primers, are unique in the genome and identify the precise location of the polymorphism.

Polymerase is an enzyme that begins the process of copying DNA. It begins to copy DNA on each strand of DNA at the point of the primer. That is, one strand is copied from the primer on the left in the right direction and the other strand is copied from the primer on the right in the left direction. In this way, PCR results in a copy of the DNA between the two primers. When this process is repeated many times, even the copies are copied and millions of copies of the DNA between the two primers are produced.

The millions of copies of a PCR-amplified sequence of DNA can easily be distinguished from background DNA. The DNA is put in a lane of a gel and an electrical current is applied to separate DNA fragments according to their size. This technique is called *electrophoresis*, which literally means "carried by electricity." The PCR-amplified sequence will appear as a dark band of the proper size, as shown in the illustration.

To detect an RFLP, PCR-amplified DNA is cut by the particular restriction enzyme that yields the RFLP. If the DNA contains the cutting site, PCR-amplified DNA will be cut into two smaller fragments. If not, the full-length PCR fragment will be seen. In the illustration, the restriction enzyme recognizes a DNA sequence 100 bp in from one end of a 300-bp sequence of DNA. After electrophoresis, DNA with the cutting site will produce two bands, one

restriction enzyme. Restriction enzymes were found in bacteria, which produce them to defend against infecting viruses. There are hundreds of varieties of restriction enzymes, each of which cuts the DNA at a particular sequence, usually six base pairs in length, wherever that sequence is found. For example, one commonly used restriction enzyme, *EcoRI*, recognizes the sequence GAATTC and severs the DNA molecule between the G and A bases. This sequence occurs thousands of times throughout the genome. If this DNA sequence differs at a

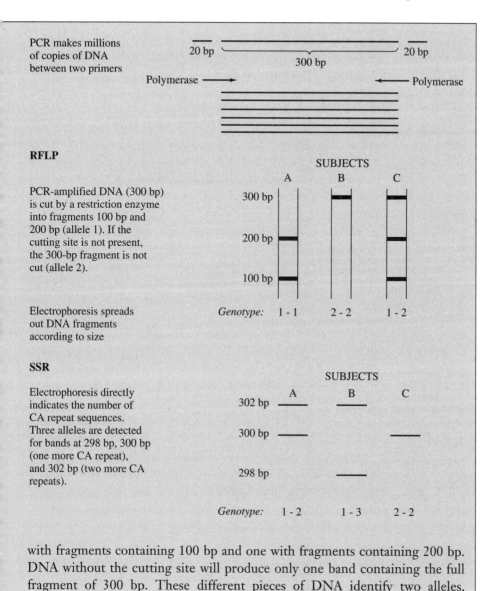

PCR makes millions of copies of DNA between two primers

20 bp   300 bp   20 bp

Polymerase ⟶   ⟵ Polymerase

RFLP

PCR-amplified DNA (300 bp) is cut by a restriction enzyme into fragments 100 bp and 200 bp (allele 1). If the cutting site is not present, the 300-bp fragment is not cut (allele 2).

Electrophoresis spreads out DNA fragments according to size

SUBJECTS
A   B   C

300 bp
200 bp
100 bp

Genotype: 1 - 1   2 - 2   1 - 2

SSR

Electrophoresis directly indicates the number of CA repeat sequences. Three alleles are detected for bands at 298 bp, 300 bp (one more CA repeat), and 302 bp (two more CA repeats).

SUBJECTS
A   B   C

302 bp
300 bp
298 bp

Genotype: 1 - 2   1 - 3   2 - 2

with fragments containing 100 bp and one with fragments containing 200 bp. DNA without the cutting site will produce only one band containing the full fragment of 300 bp. These different pieces of DNA identify two alleles,

(continued on page 50)

particular locus for some individuals, their DNA will not be cut at that point. This failure to cut results in one long DNA fragment for these individuals; this long fragment can be identified when it is compared with fragments from individuals whose DNA is cut at that point into two shorter fragments. In other words, this is a polymorphism that can be detected when a restriction enzyme creates DNA fragments of different lengths from the DNA of two individuals. Box 4.2 describes the process by which RFLPs are detected.

representing the presence or absence of the cutting site. Individuals homozygous for the cutting site will have bands for fragments of 100 and 200 bp (subject A in the illustration). Individuals homozygous for DNA without the cutting site will have only one band of fragments, which contain 300 bp (subject B). Heterozygous individuals will have all three bands (subject C).

SSRs are identified in a similar way. Unlike RFLPs, which are the result of the presence or absence of a restriction enzyme cutting site, SSRs are polymorphisms resulting from various numbers of repeats of short sequences of DNA, such as the two-base sequence CA. An RFLP usually consists of just two alleles, but an SSR usually includes more than two alleles. For example, for a particular SSR, there might be three alleles, with CA repeating 14 times, 15 times, or 16 times. After PCR amplification and electrophoresis, the number of repeats can be detected directly by using a sensitive gel that distinguishes fragments (seen as bands) that differ in size by as few as two base pairs (i.e., one CA repeat). Individuals can have any combination of two of the alleles. As shown in the figure, subject A has a band containing fragments of 300 bp and a band with fragments of 302 bp (i.e., one additional CA repeat). Subject B has fragments containing 298 and 302 bp. Subject C happens to be homozygous for the 300-bp allele. A band containing fragments of 300 bp does not mean that there are 150 CA repeats. There are usually no more than 25 repeats. The rest of each fragment consists of DNA, amplified by PCR, that surrounds the CA repeat sequence.

Many other ways to detect DNA markers are in use. For example, *allele-specific PCR* (AS-PCR) involves a PCR primer that encompasses the actual site of a base-pair difference between two alleles. PCR will amplify the sequence only if the exact primer site is present. Another method for detecting polymorphisms is called *single-strand conformational polymorphisms* (SSCPs). If a short sequence of DNA is polymorphic, it affects the way the DNA folds around itself when the normal double-stranded DNA is made single-stranded. These folding or conformational differences can be detected with the same methods used to detect an SSR. Ultimately, all polymorphisms can be detected by DNA sequencing.

Although there are hundreds of restriction enzymes that detect different DNA sequences, in fact, they can detect only about one-fifth of all DNA polymorphisms. Several other types of DNA markers have been developed to detect other DNA polymorphisms. The most widely used DNA marker, developed in 1987, is called a simple sequence repeat (SSR) marker, previously known as a microsatellite repeat marker. As mentioned in Chapter 3, two, three, or four nucleotide base pairs are repeated dozens of times at as many as 50,000 loci throughout the genome. The number of repeats at each locus differs among individuals and is inherited in a Mendelian manner. For example, an SSR might have three alleles, in which the nucleotide base pairs C-G and

A-T repeat 14, 15, or 16 times. Box 4.2 also describes how SSRs are detected. The human genome has been mapped with several thousands of very closely spaced SSRs to guide the search for genes. SSRs are especially useful because they can be detected by using machines called automated DNA sequencers, which can complete several thousand assays per day.

Several other techniques are available to detect polymorphisms that cannot be identified by RFLP or SSR methods. The ultimate way to detect all polymorphisms for a particular gene is to determine the DNA sequences for individuals. This task is becoming increasingly possible with automated DNA sequencers.

Now that the appropriate techniques have been developed, DNA markers can be used to find genes associated with behavioral traits, as mentioned in Chapter 2 and described in detail in Chapter 6.

SUMMING UP

Mutations are the source of genetic variability. About 3 million of our 3 billion nucleotide bases differ from one individual to the next. Detecting these DNA polymorphisms has been the key to success in molecular genetics. New types of markers in DNA itself include restriction fragment length polymorphisms (RFLPs) and simple sequence repeat polymorphisms (SSRs). Sequencing DNA is the ultimate way to detect all polymorphisms.

Chromosomes

As discussed in Chapter 2, Mendel did not know that genes are grouped together on chromosomes, so he assumed that all genes are inherited independently. However, Mendel's second law of independent assortment is violated when two genes are close together on the same chromosome. In this case, the two genes are *not* inherited independently; and, on the basis of this nonindependent assortment, linkages between DNA markers have been identified and used to produce a map of the genome. With the same technique, mapped DNA markers are used to identify linkages with disorders and dimensions, including behavior, as described in Chapter 6.

Our species has 23 pairs of chromosomes, for a total of 46 chromosomes. The number of chromosome pairs varies widely from species to species. Fruit flies have 4, mice have 20, dogs have 39, and carp have 52. Our chromosomes are very similar to those of the great apes (chimpanzee, gorilla, and orangutan). Although the great apes have 24 pairs, two of their short chromosomes have been fused to form one of our large chromosomes.

One pair of our chromosomes is the *sex chromosomes* X and Y. Females are XX and males are XY. All the other chromosomes are called *autosomes*. As shown in Figure 4.5, chromosomes have characteristic banding patterns when stained with a particular chemical. The *bands*, whose function is not known, are used

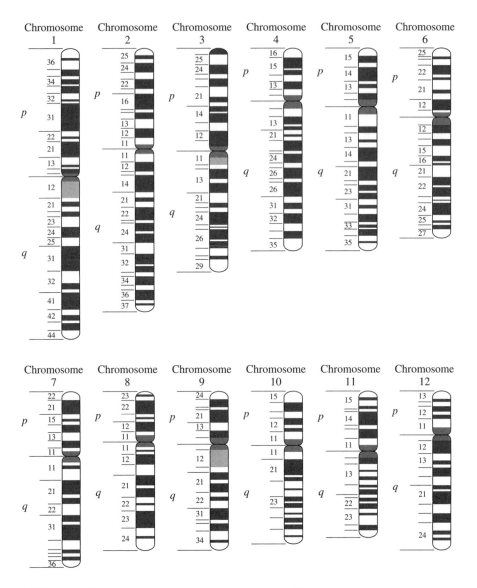

Figure 4.5 The 23 pairs of human chromosomes. The short arm above the centromere is called *p*, and the long arm below the centromere is called *q*. The bands, created by staining, are used to identify the chromosomes and to describe the location of genes. Chromosomal regions are referred to by chromosome number, arm of chromosome, and band. Thus, 1*p*36 refers to band 6 in region 3 of the *p* arm of chromosome 1.

to identify the chromosomes. At some point in each chromosome, there is a *centromere*, a region of the chromosome without genes, where the chromosome is attached to its new copy when cells reproduce. The short arm of the chromosome above the centromere is called *p* and the long arm below the cen-

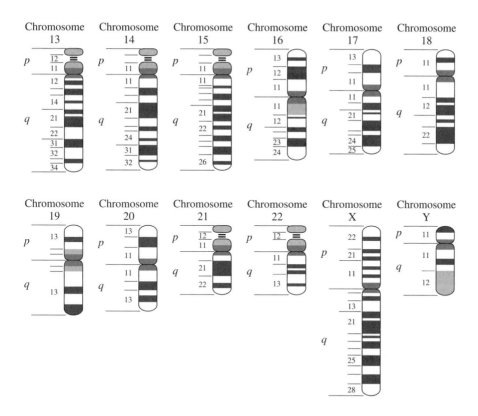

tromere is called *q*. The location of genes is described in relation to the bands. For example, the gene for Huntington's disease is at 4*p*16, which means the short arm of chromosome 4 at a particular band, number 6 in region 1.

In addition to providing the basis for gene mapping, chromosomes are important in behavioral genetics because mistakes in copying chromosomes during cell division affect behavior. There are two kinds of cell division. Normal cell division, called *mitosis*, occurs in all cells, called *somatic cells*, except the sex cells that produce eggs and sperm (*gametes*). In mitosis, each chromosome duplicates and divides to produce two identical cells. A special type of cell division called *meiosis* occurs in ovaries and testes to produce eggs and sperm, which have only one member of each chromosome pair. Each egg and sperm has 1 of over 8 million (2^{23}) possible combinations of the 23 pairs of chromosomes. Moreover, crossover (recombination) of members of each chromosome pair (see Figure 2.7) occurs about once per meiosis and creates even more genetic variability. When a sperm fertilizes an egg to produce a zygote, one chromosome of each pair comes from the mother's egg and the other from the father's sperm, thereby reconstituting the full complement of 23 pairs of chromosomes.

As indicated in Chapter 3, a common copying error for chromosomes is an uneven split of the pairs of chromosomes during meiosis, called nondisjunction (Figure 3.4). The most common form of mental retardation, Down syndrome, is caused by nondisjunction of one of the smallest chromosomes, chromosome 21. Many other chromosomal problems occur, such as breaks in chromosomes that lead to inversion, deletion, duplication, and translocation (see Figure 4.6). About half of all fertilized human eggs have a chromosomal abnormality. Most of these abnormalities result in early spontaneous abortions (miscarriages). At birth, about 1 in 250 babies have an obvious chromosomal abnormality. (Small abnormalities such as deletions are difficult to detect.) Although chromosomal abnormalities occur for all chromosomes, only fetuses with the least severe abnormalities survive to birth. Some of these babies die soon after they are born. For example, most babies with three chromosomes (trisomy) of chromosome 13 die in the first month and most of those with trisomy-18 die within the first year. Other chromosomal abnormalities are less lethal but result in behavioral and physical problems.

Missing a whole chromosome is lethal, except for the X and Y chromosomes. Having an entire extra chromosome is also lethal except for the smallest chromosomes and the X chromosome, which is one of the largest. The reason why the X chromosome is the exception is also the reason why half of all chro-

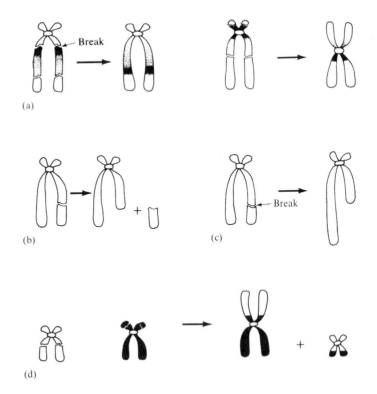

Figure 4.6 Common types of chromosomal abnormalities: (a) inversion; (b) deletion; (c) duplication; (d) translocation between different chromosomes.

mosomal abnormalities that exist in newborns involve the sex chromosomes. In females, one of the two X chromosomes is inactivated, in the sense that its genes are not transcribed. In males and females with extra X chromosomes, the extra X chromosomes also are inactivated. For this reason, even though X is a large chromosome with many genes, having an extra X in males or females or only one X in females is not lethal. The most common sex chromosome abnormalities are XXY (males with an extra X), XXX (females with an extra X), and XYY (males with an extra Y), each with an incidence of about 1 in 1000. The incidence of XO (females with just one X) is 1 in 3000 at birth, because the vast majority of such conceptuses abort.

SUMMING UP

Our species has 23 pairs of chromosomes, including the sex chromosomes X and Y. During meiosis, when egg and sperm are produced, copying errors are occasionally made, leading to an uneven split of pairs of chromosomes, called nondisjunction, and other errors. Such chromosomal abnormalities contribute importantly to behavioral, especially cognitive, disorders. About 1 in 250 births have a major chromosomal abnormality and about half of these involve the sex chromosomes.

The effects of most chromosomal abnormalities are manifold, often involving diverse behavioral and physical traits, which is not surprising because so many genes are involved. Nearly all major chromosomal abnormalities influence cognitive ability, as expected if cognitive ability is affected by many genes. Because the behavioral effects of chromosomal abnormalities often involve mental retardation, they are discussed in Chapter 7.

Summary

One of the most exciting advances in biology in this century has been understanding Mendel's "elements" of heredity. The double helix structure of DNA relates to its dual functions of self-replication and protein synthesis. The genetic code consists of a sequence of three base pairs of DNA that codes for amino acids. DNA is transcribed to mRNA, which is translated into amino acid sequences. Many genes are involved in regulating the transcription of other genes. Gene regulation is responsible for long-term developmental changes as well as short-term responses to environmental conditions.

Mutations are the source of genetic variability. Much of the success of molecular genetics comes from the availability of thousands of markers that detect DNA polymorphisms. Restriction fragment length polymorphisms (RFLPs) and simple sequence repeat polymorphisms (SSRs) are the most widely used DNA markers.

Genes are inherited on chromosomes. Linkage between DNA markers and behavior can be detected by looking for violations of Mendel's law of independent assortment, because a DNA marker and a gene for behavior are not inherited independently if they are close together on the same chromosome. Our species has 23 pairs of chromosomes. Mistakes in duplicating chromosomes often directly affect behavior. About 1 in 250 newborns have a major chromosomal abnormality and about half of these involve the sex chromosomes.

CHAPTER FIVE

Nature, Nurture, and Behavior

Most behavioral traits are much more complex than single-gene disorders such as Huntington's disease and PKU (Chapter 2). Complex dimensions and disorders are influenced by heredity, but not by one gene alone. Multiple genes are usually involved, as well as multiple environmental influences. The purpose of this chapter is to describe ways in which we can study genetic effects on complex behavioral traits.

The first question that needs to be asked about behavioral traits is whether heredity is at all important. For single-gene disorders, this is not an issue, because it is usually obvious that heredity is important. For example, for dominant genes such as the gene for Huntington's disease, you do not need to be a geneticist to notice that every affected individual has an affected parent. Recessive genes are not as easy to observe, but the expected pattern of inheritance is clear. For complex behavioral traits in the human species, two experiments of nature—adoption and twinning—are widely used to assess the net effect of genes and environment. More direct genetic experiments are available to investigate animal behavior.

These methods and the theory underlying them (Appendix B) are called *quantitative genetics*. When heredity is important, it is now possible to identify specific genes, which is the topic of Chapter 6. Methods to identify specific genes are referred to as *molecular genetics*. Behavioral genetics uses both quantitative genetic and molecular genetic methods to study behavior.

Genetic Experiments to Investigate Animal Behavior

Dogs provide a dramatic yet familiar example of genetic variability within species (see Figure 5.1). Despite their great variability in size and physical

Irish Setter   Boxer   Beagle

Dalmation   Afghan

Figure 5.1 Dog breeds illustrate genetic diversity within species for behavior as well as physical appearance.

appearance, they are all members of the same species. Dogs also illustrate within-species genetic effects on behavior. Although physical differences are most obvious, dogs have been bred for centuries as much for their behavior as for their looks. In 1576, the earliest English-language book on dogs classified breeds primarily on the basis of behavior. For example, terriers (from *terra*, which is Latin for "earth") were bred to creep into burrows to drive out small animals. Another book, published in 1686, described the behavior for which spaniels were originally selected. They were bred to creep up on birds and then spring to frighten the birds into the hunter's net. With the advent of the shotgun, different spaniels were bred to point rather than to spring. The author of the 1686 work was especially interested in temperament: "Spaniels by Nature are very loveing, surpassing all other Creatures, for in Heat and Cold, Wet and Dry, Day and Night, they will not forsake their Master" (cited by Scott & Fuller, 1965, p. 47).

Behavioral classification of dogs continues today. Sheepdogs herd, retrievers retrieve, trackers track, and pointers point with minimal training. Breeds also

Standard Poodle   Cocker Spaniel   Miniature Schnauzer

Fox Terrier   Pomeranian   Collie

differ strikingly in intelligence and in temperamental traits such as emotionality, activity, and aggressiveness. The selection process can be quite fine tuned. For example, in France, where dogs are used chiefly for farm work, there are 17 breeds of shepherd and stock dogs specializing in aspects of this work. In England, dogs have been bred primarily for hunting, and there are 26 recognized breeds of hunting dogs. Dogs are not unusual in their genetic diversity, although they are unusual in the extent to which different breeds have been intentionally bred to accentuate genetic differences.

An extensive behavioral genetics research program on breeds of dogs was conducted over two decades by J. Paul Scott and John Fuller (1965). They studied the development of pure breeds and hybrids of the five breeds pictured in Figure 5.2: wire-haired fox terriers, cocker spaniels, basenjis, Shetland sheepdogs, and beagles. These breeds are all about the same size, but they differ markedly in behavior. Although considerable genetic variability remains within each breed, average behavioral differences among the breeds reflect their breeding history. For example, as their history described above would suggest, terriers are aggressive scrappers and spaniels are nonaggressive and people oriented. Unlike the other breeds, Shetland sheepdogs have been bred, not for hunting, but for performing complex tasks under close supervision from their masters.

Figure 5.2 J. P. Scott with the five breeds of dogs used in his experiments with J. L. Fuller. Left to right: wire-haired fox terrier, American cocker spaniel, African basenji, Shetland sheepdog, and beagle. (From *Genetics and the Social Behavior of the Dog* by J. P. Scott and J. L. Fuller. Copyright © 1965 by The University of Chicago Press. All rights reserved.)

They are very responsive to training. In short, Scott and Fuller found behavioral breed differences just about wherever they looked—in the development of social relationships, emotionality, and trainability, as well as many other behaviors.

Selection Studies

Laboratory experiments that select for behavior provide the clearest evidence for genetic influence on behavior. As dog breeders and other animal breeders have known for centuries, if a trait is heritable, you can breed selectively for it. Laboratory experiments typically select high and low lines in addition to maintaining an unselected control line. For example, in one of the largest and longest selection studies of behavior, mice were selected for activity in a brightly lit box called an open field, a measure of fearfulness that was invented more than 60 years ago (see Figure 5.3). In the open field, some animals freeze, defecate, and urinate, whereas others actively explore it. Lower activity scores are presumed to index fearfulness.

Figure 5.3 Mouse in an open field. The holes near the floor transmit light beams that electronically record the mouse's activity.

The most active mice were selected and mated with other high-active mice. The least active mice were also mated with each other. From the offspring of the high-active and low-active mice, the most and least active mice were again selected and mated in a similar manner. This selection process was repeated for 30 generations. (In mice, a generation takes only about three months.)

The results are shown in Figures 5.4 and 5.5 for replicated high, low, and control lines. Over the generations, selection was successful: The high lines became increasingly more active and the low lines less active (Figure 5.4). Successful selection can occur only if heredity is important. After 30 generations of such selective breeding, a thirtyfold average difference in activity has been achieved. There is no overlap between the activity of the low and high lines (Figure 5.5). Mice from the high-active line now boldly run the equivalent total distance of the length of a football field during the six-minute test period, whereas the low-active mice quiver in the corners.

Another important finding is that the difference between the high and low lines steadily increases each generation. This outcome is a typical finding from selection studies of behavioral traits and strongly suggests that many genes contribute to variation in behavior. If just one or two genes were responsible for

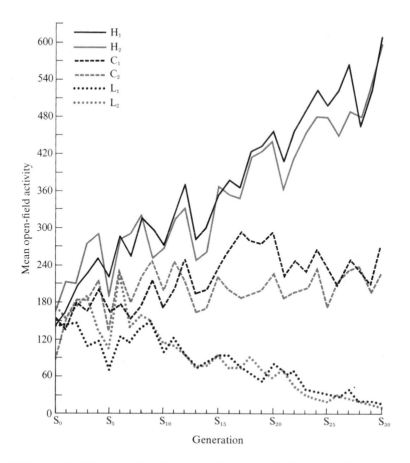

Figure 5.4 Results of a selection study of open-field activity. Two lines were selected for high open-field activity (H$_1$ and H$_2$), two lines were selected for low open-field activity (L$_1$ and L$_2$), and two lines were randomly mated within each line to serve as controls (C$_1$ and C$_2$). (From "Response to 30 generations of selection for open-field activity in laboratory mice" by J. C. DeFries, M. C. Gervais, and E. A. Thomas. *Behavior Genetics, 8,* 3–13. Copyright © 1978 by Plenum Publishing Corporation. All rights reserved.)

open-field activity, the two lines would separate after a few generations and they would not diverge any further in later generations.

Inbred Strain Studies

The other major quantitative genetic design for animal behavior compares *inbred strains*, in which brothers have been mated with sisters for at least 20 generations. This intensive inbreeding makes each animal within the inbred strain virtually a genetic clone of all other members of the strain. Because inbred strains differ genetically from each other, genetically influenced traits will show average differences between inbred strains reared in the same laboratory

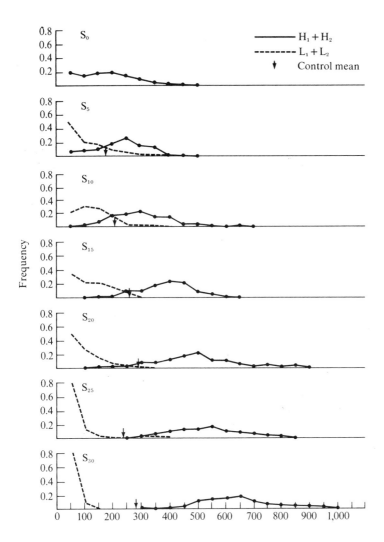

Figure 5.5 Distributions of activity scores of lines selected for high and low open-field activity for 30 generations (S_0 to S_{30}). Average activity of control lines in each generation is indicated by an arrow. (From "Response to 30 generations of selection for open-field activity in laboratory mice" by J. C. DeFries, M. C. Gervais, and E. A. Thomas. *Behavior Genetics, 8,* 3–13. Copyright © 1978 by Plenum Publishing Corporation. All rights reserved.)

environment. Differences within strains are due to environmental influences. In animal behavioral genetic research, mice are most often studied, and well over 100 inbred strains of mice are available. Some of the most frequently studied inbred strains are shown in Figure 5.6.

Studies of inbred strains suggest that most mouse behaviors show genetic influence. For example, Figure 5.7 shows the average open-field activity scores

Figure 5.6 Four common inbred strains of mice: (a) BALB/c; (b) DBA/2; (c) C3H/2; (d) C57BL/6.

of two inbred strains called BALB and C57BL. The C57BL mice are much more active than the BALB mice, suggesting that genetics contributes to open-field activity. The mean activity scores of several crosses are also shown:

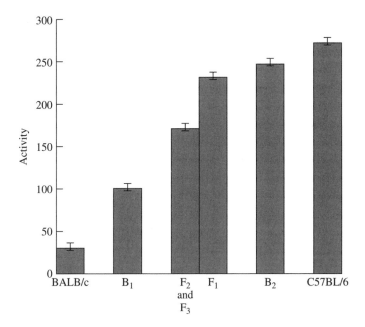

Figure 5.7 Mean open-field activity (± twice the standard error) of BALB and C57BL mice and their derived F$_1$, backcross (B$_1$ and B$_2$), F$_2$, and F$_3$ genera-tions. (From "Response to 30 generations of selection for open-field activity in laboratory mice" by J. C. DeFries, M. C. Gervais, and E. A. Thomas. *Behavior Genetics, 8,* 3–13. Copyright © 1978 by Plenum Publishing Corporation. All rights reserved.)

F$_1$, F$_2$, and F$_3$ crosses (explained in Box 2.1) between the inbred strains, the backcross between the F$_1$ and the BALB strain (B$_1$ in Figure 5.7), and the backcross between the F$_1$ and the C57BL strain (B$_2$ in Figure 5.7). There is a strong relationship between the average open-field scores and the percentage of genes obtained from the C57BL parental strain, which again points to genetic influence.

Rather than just crossing two inbred strains, the *diallel design* compares sev-eral inbred strains and all possible F$_1$ crosses between them. Figure 5.8 shows the open-field results of a diallel cross between BALB, C57BL, and two other inbred strains (C3H and DBA). C3H is even less active than BALB, and DBA is almost as active as C57BL. The F$_1$ crosses tend to correspond to the average scores of their parents. For example, the F$_1$ cross between C3H and BALB is intermediate to the two parents in open-field activity.

Studies of inbred strains are also useful for detecting environmental effects. First, because members of an inbred strain are genetically identical, individ-ual differences within a strain must be due to environmental factors. Large

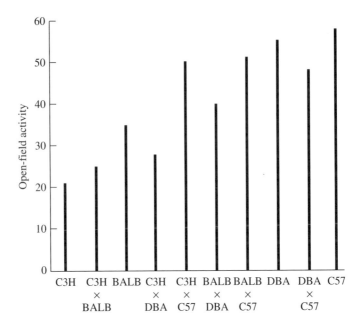

Figure 5.8 Diallel analysis of four inbred mouse strains for open-field activity. The F$_1$ strains are ordered according to the average open-field activity score of their parental inbred strains. (After Henderson, 1967.)

differences within inbred strains are found for open-field activity and most behaviors studied, reminding us of the importance of nurture as well as nature. Second, inbred strains can be used to assess the net effect of mothering by comparing F$_1$ crosses in which the mother is from either one strain or the other. For example, the F$_1$ cross between BALB mothers and C57BL fathers can be compared to the genetically equivalent F$_1$ cross between C57BL mothers and BALB fathers. In the diallel study shown in Figure 5.8, these two hybrids had nearly identical scores, as was the case for comparisons between the other crosses as well. This result suggests that prenatal and postnatal maternal effects do not importantly affect open-field activity. If maternal effects are found, it is possible to separate prenatal and postnatal effects by cross-fostering pups of one strain with mothers of the other strain. Third, the environments of inbred strains can be manipulated in the laboratory in order to investigate interactions between genotype and environment.

Over a thousand behavioral investigations involving genetically defined mouse strains were published between 1922 and 1973 (Sprott & Staats, 1975), and the pace accelerated into the 1980s. Studies such as these played an important role in demonstrating that genetics contributes to most behaviors. Although inbred strain studies now tend to be overshadowed by more sophisticated genetic analyses, inbred strains still provide a simple and highly efficient test for the presence of genetic influence.

SUMMING UP

Differences among breeds of dogs and selection studies of mice and rats in the laboratory provide powerful evidence for the importance of genetic influence on behavior. Behavioral differences between inbred strains of mice, inbred by brother-sister matings for at least 20 generations, demonstrate the widespread contribution of genes to behavior. Differences within an inbred strain indicate the importance of environmental factors.

Investigating the Genetics of Human Behavior

Quantitative genetic methods to study human behavior are not as powerful or direct as selection studies or studies of inbred strains. Rather than using genetically defined populations such as inbred strains for mice or manipulating environments experimentally, human research is limited to studying naturally occurring genetic and environmental variation. Nonetheless, adoption and twinning provide experimental situations that can be used to test the relative influence of nature and nurture. As mentioned in Chapter 1, increasing recognition of the importance of genetics during the past two decades is one of the most dramatic shifts in psychology. This shift is in large part due to the accumulation of adoption and twin research consistently pointing to the important role played by genetics even for complex psychological traits.

Adoption Designs

The most direct way to disentangle genetic and environmental sources of family resemblance involves adoption. Adoption creates pairs of genetically related individuals who do not share a common family environment. Their resemblance estimates the contribution of genetics to family resemblance. Adoption also produces family members who share family environment but are not genetically related. Their resemblance estimates the contribution of family environment to family resemblance. It should be noted that quantitative genetic analyses do not directly assess either genes or specific environmental factors. The effects of nature and nurture are inferred from experiments of nature such as the adoption design. An important direction for future behavioral genetic research is to incorporate direct measures of genes (Chapter 6) and of environment (Chapter 14) in quantitative genetic designs.

For example, consider parents and offspring. Parents in a family study are "genetic-plus-environmental" parents in that they share both heredity and environment with their offspring. The process of adoption results in "genetic" parents and "environmental" parents (Figure 5.9). "Genetic" parents are birth parents who relinquish their child for adoption shortly after birth. Resemblance between birth parents and their adopted-away offspring directly assesses the genetic contribution to parent-offspring resemblance. "Environmental" parents

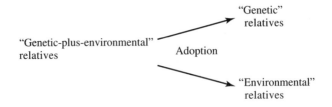

Figure 5.9 Adoption is an experiment of nature that creates "genetic" relatives (biological parents and their adopted-away offspring and siblings adopted-apart) and "environmental" relatives (adoptive parents and their adopted children; genetically unrelated children adopted into the same adoptive family). Resemblance for these "genetic" and "environmental" relatives can be used to test the extent to which resemblance between the usual "genetic-plus-environmental" relatives is due to nature or nurture.

are adoptive parents who adopt children genetically unrelated to them. In the absence of selective placement, resemblance between adoptive parents and their adopted children directly assesses the environmental contribution to parent-offspring resemblance. Because data on birth parents are rare, genetic influence can also be assessed by comparing "genetic-plus-environmental" families with adoptive families who share only family environment.

"Genetic" siblings and "environmental" siblings as well as parents can be studied. "Genetic" siblings are full siblings adopted apart early in life and reared in different homes. "Environmental" siblings are pairs of genetically unrelated children adopted early in life into the same adoptive home. These adoption designs can be depicted more precisely as path models that are used in model-fitting analyses to test the fit of the model, to compare alternative models, and to estimate genetic and environmental influences (Appendix B).

For most psychological traits that have been assessed in adoption studies, genetic factors appear to be important. For example, Figure 5.10 summarizes adoption results for general cognitive ability (see Chapter 8 for details). "Genetic" parents and offspring and "genetic" siblings significantly resemble each other even though they are adopted apart and do not share family environment. You can see that genetics accounts for about half of the resemblance for "genetic-plus-environmental" parents and siblings. The other half of familial resemblance appears to be explained by shared family environment, assessed directly by the resemblance between adoptive parents and adopted children and between adoptive siblings. Chapter 8 describes a recent important finding that the influence of shared environment on cognitive ability decreases dramatically from childhood to adolescence.

One of the most surprising results from genetic research is that, for most psychological traits other than cognitive ability, resemblance between relatives is accounted for by shared heredity rather than by shared environment. For example, the risk of schizophrenia is just as great for offspring of schizophrenic parents whether they are reared by their biological parents or adopted away at birth and reared by adoptive parents. This finding implies that sharing a family environment does not contribute importantly to family resemblance. It does not mean that the environment or even the family environment is unimportant. As discussed in Chapter 14, quantitative genetic research such as adoption studies provides the best available evidence for the importance of environmental influence. The risk for first-degree relatives of schizophrenic probands who are 50 percent similar genetically is only about 10 percent, not 50 percent. Furthermore, although family environment does not contribute to the resemblance of family members, such factors could contribute to *differences* among family members, called *nonshared environment* (Chapter 14).

The first adoption study of schizophrenia, reported by Leonard Heston in 1966, is a classic study that single-handedly turned the tide from assuming that schizophrenia was completely caused by early family experiences to recognizing the importance of genetics (see Box 5.1). Box 5.2 considers some additional issues in adoption studies.

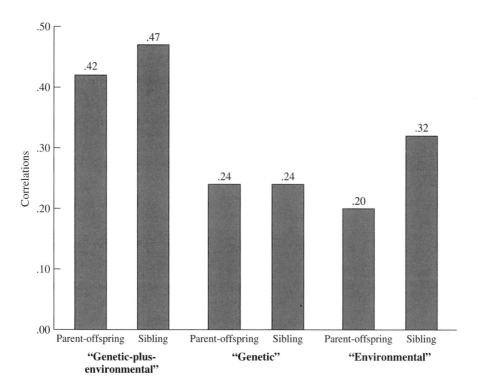

Figure 5.10 Adoption data indicate that family resemblance for cognitive ability is due both to genetic resemblance and to environmental resemblance. "Genetic" relatives refer to genetically related relatives adopted apart. "Environmental" relatives refer to genetically unrelated individuals adopted together. (Data adapted from Loehlin, 1989.)

BOX 5.1

The First Adoption Study of Schizophrenia

Environmentalism, which assumes that we are what we learn, dominated psychology until the 1960s, when a more balanced view that recognized the importance of nature as well as nurture emerged. One reason for this major shift was an adoption study of schizophrenia reported by Leonard Heston in 1966. Although twin studies had for decades suggested genetic influence, schizophrenia was generally assumed to be environmental in origin, caused by early interactions with parents. Heston interviewed 47 adult adopted-away offspring of hospitalized schizophrenic women. He compared their incidence of schizophrenia with that of matched adoptees whose birth parents had no known mental illness. Of the 47 adoptees whose biological mothers were schizophrenic, five had been hospitalized for schizophrenia. Three were chronic schizophrenics hospitalized for several years. None of the adoptees in the control group was schizophrenic.

The incidence of schizophrenia in these adopted-away offspring of schizophrenic biological mothers was 10 percent. This risk is similar to the risk for schizophrenia found when children are reared by their schizophrenic parents. Not only do these findings indicate that heredity makes a major contribution to schizophrenia, they also suggest that shared rearing environment has little effect. When a biological parent is schizophrenic, the risk for schizophrenia is just as great for the offspring when they are adopted away at birth as it is when the offspring are reared by their schizophrenic parents.

Several other adoption studies have confirmed the results of Heston's study. Heston's study is an example of what is called the *adoptees' study method* because the incidence of schizophrenia was investigated in the adopted-away offspring of schizophrenic biological mothers. A second major strategy is called the *adoptees' family method*. Rather than beginning with parents, this method begins with adoptees who are affected (probands) and adoptees who are unaffected. The incidence of the disorder in the biological and adoptive families of the adoptees is assessed. Genetic influence is suggested if the incidence of the disorder is greater for the biological relatives of the affected adoptees than for the biological relatives of the unaffected control adoptees. Environmental influence is indicated if the incidence is greater for the adoptive relatives of the affected adoptees than for the adoptive relatives of the control adoptees.

These adoption methods and their results for schizophrenia are described in Chapter 10.

Twin Design

The other major method used to disentangle genetic from environmental sources of resemblance between relatives involves twins. Identical twins, sometimes called *monozygotic (MZ)* twins because they derive from one zygote, are genetically identical (Figure 5.11). If genetic factors are important for a trait, these genetically identical pairs of individuals must be more behaviorally similar than first-degree relatives, who are only 50 percent similar genetically. Rather than comparing identical twins with nontwin siblings or other relatives, nature has provided a better comparison group: fraternal (*dizygotic*, or *DZ*) twins. Unlike identical twins, fraternal twins develop from separately fertilized eggs. They are first-degree relatives, 50 percent genetically related like other siblings. Half of fraternal twin pairs are same-sex pairs and half are opposite-sex pairs. Twin studies usually use same-sex fraternal twin pairs because they are a better comparison group for identical twin pairs, who are always same-sex pairs. If genetic factors are important for a trait, identical twins must be more similar than fraternal twins. (See Box 5.3, on page 81, for more details about the twin method.)

How can you tell if same-sex twins are identical or fraternal? DNA markers can tell. If a pair of twins differs for any DNA marker (excluding laboratory error), they must be fraternal because identical twins are identical genetically. If many markers are examined and no differences are found, the twin pair must be identical. Physical traits such as eye color and hair color and texture can be used in a similar way to diagnose whether twins are identical or fraternal. Such traits are highly heritable and are affected by many genes. If a twin pair differs

CLOSE UP

Leonard L. Heston is professor emeritus in the Department of Psychiatry, University of Washington. He was raised in Oregon and educated there, graduating from the University of Oregon Medical School. During his residency in psychiatry, he became interested in genetics and in a patient who had schizophrenia, as did the patient's father and two of the patient's three brothers. However, his professors assured him that genes could not be involved with this family's illness because, as was well known, defective mothering caused schizophrenia. But the mother of his patient was a decent, caring woman trying her best to cope with a desperate situation. That case led to Heston's study of adopted-away children of schizophrenics and to ongoing schizophrenia research with collaborators at the Mayo Clinic, which today is being done at the DNA level. Another patient with a severely affected family led to a second research program in Alzheimer's disease; this program is also ongoing at the DNA level with collaborators at the Universities of Minnesota and Washington.

Figure 5.11 Twinning is an experiment of nature that produces identical twins, who are genetically identical, and fraternal twins, who are only 50 percent similar genetically. If genetic factors are important for a trait, identical twins must be more similar than fraternal twins. DNA markers can be used to test whether twins are identical or fraternal, although for most pairs it is easy to tell because identical twins (top photo) are usually much more similar physically than fraternal twins (bottom photo).

for one of these traits, they are likely to be fraternal; if they are the same for many such traits, they are probably identical. In fact, a single question works pretty well because it sums up many such physical traits: When the twins were young, how difficult were they to tell apart? To be mistaken for another person requires that many heritable physical characteristics be identical. Using physical similarity to determine whether twins are identical or fraternal is more than 90 percent accurate when compared with the results of DNA markers. In most cases, it is not difficult to tell whether twins are identical or fraternal (see Figure 5.11).

If a trait is influenced genetically, identical twins must be more similar than fraternal twins. However, when greater similarity of MZ twins is found, it is possible that the greater similarity is caused environmentally rather than genetically. The *equal environments assumption* of the twin method assumes that environmental similarity is roughly the same for both types of twins reared in the same family. If the assumption were violated because identical twins experience more similar environments than fraternal twins, this violation would inflate estimates of genetic influence. The equal environments assumption has been tested in several ways and appears reasonable for most traits (Bouchard & Propping, 1993).

Prenatally, identical twins may experience *greater* environmental differences than fraternal twins. For example, identical twins show greater birth weight differences than fraternal twins do. The difference may be due to greater prenatal competition, especially for the majority of identical twins who share the same chorion. To the extent that identical twins experience less similar environments, the twin method will underestimate heritability.

Postnatally, the effect of labeling a twin pair as identical or fraternal has been studied by using twins who were misclassified by their parents or by themselves (e.g., Kendler et al., 1993a; Scarr & Carter-Saltzman, 1979). When parents think that twins are fraternal but they really are identical, these mislabeled twins are as similar behaviorally as correctly labeled identical twins.

Another way in which the equal environments assumption has been tested takes advantage of the fact that differences within pairs of identical twins can only be due to environmental influences. The equal environments assumption is supported if identical twins who are treated more individually than others do not behave more differently. This is what has been found (e.g., Loehlin & Nichols, 1976; Morris-Yates et al., 1990).

A subtle, but important, issue is that identical twins might have more similar experiences than fraternal twins *because* identical twins are more similar genetically. That is, some experiences may be driven genetically. Such differences between identical and fraternal twins in experience are not a violation of the equal environments assumption because the differences are not caused environmentally. This topic is discussed in Chapter 14.

BOX 5.2

Issues in Adoption Studies

The adoption design is like an experiment of nature that untangles nature and nurture as causes of family resemblance. The first adoption study, which investigated IQ, was reported in 1924 (Theis, 1924). The first adoption study of schizophrenia was reported in 1966 (Box 5.1). Adoption studies have become more difficult to conduct as the number of adoptions has declined. In the 1960s, as many as 1 percent of all children were adopted. Adoption became much less frequent as contraception and abortion increased and more unmarried mothers kept their infants.

One issue about adoption studies is representativeness. If biological parents, adoptive parents, or adopted children are not representative of the rest of the population, the generalizability of adoption results could be affected. However, means are more likely to be affected than variances, and genetic estimates rely primarily on variance. In the population-based Colorado Adoption Project (DeFries et al., 1994), for example, biological and adoptive parents appear to be quite representative of nonadoptive parents, and adopted children seem to be reasonably representative of nonadopted children. Other adoption studies, however, have sometimes shown less representativeness.

Another issue concerns prenatal environment. Because birth mothers provide the prenatal environment for their adopted-away children, the resemblance between them might reflect prenatal environmental influences. A strength

As in any experiment, generalizability is an issue for the twin method. Are twins representative of the general population? One way in which twins are different is that twins are generally born three to four weeks premature. Newborn twins are also about 30 percent lighter at birth than the average singleton newborn, a difference that disappears by middle childhood (MacGillivray, Campbell, & Thompson, 1988). In childhood, language develops somewhat more slowly in twins and twins also perform slightly less well on tests of verbal ability. Both these traits appear to be due to postnatal environment rather than prematurity (Rutter & Redshaw, 1991). Most of this verbal deficit is recovered in the early school years (Wilson, 1983). Twins do not appear to be importantly different from singletons for personality or psychopathology.

In summary, the twin method is a valuable tool for screening behavioral dimensions and disorders for genetic influence (Bouchard & Propping, 1993). The assumptions underlying the twin method are different from those of the adoption method, yet both methods converge on the conclusion that genetics is important in psychology.

of adoption studies is that prenatal effects can be tested independently from postnatal environment by comparing correlations for birth mothers and birth fathers. Although it is more difficult to study birth fathers, results for small samples of birth fathers show results similar to those for birth mothers for IQ and for schizophrenia. Another approach to this issue is to compare adoptees' biological half siblings related through the mother (maternal half siblings) with those related through the father (paternal half siblings). For schizophrenia, paternal half siblings of schizophrenic adoptees show the same risk for schizophrenia as maternal half siblings, an observation suggesting that prenatal factors may not be of great importance (Kety, 1987).

Finally, *selective placement* could cloud the separation of nature and nurture by placing adopted-apart "genetic" relatives into correlated environments. For example, selective placement would occur if the adopted-away children of the brightest biological parents are placed with the brightest adoptive parents. If selective placement matches biological and adoptive parents, genetic influence could inflate the correlation between adoptive parents and their adopted children, and environmental influence could inflate the correlation between biological parents and their adopted-away children. If data are available on biological parents as well as adoptive parents, selective placement can be assessed directly. If selective placement is found in an adoption study, its effects need to be considered in interpreting genetic and environmental results. Although some adoption studies show selective placement for IQ, other psychological dimensions and disorders show little evidence for selective placement.

Recall that for schizophrenia, the risk for a fraternal twin whose co-twin is schizophrenic is about 17 percent; the risk is 48 percent for identical twins (see Figure 3.6). For general cognitive ability, the correlation for fraternal twins is about .60 and .85 for identical twins (see Figure 3.7). The fact that identical twins are so much more similar than fraternal twins strongly suggests genetic influence. For both schizophrenia and general cognitive ability, fraternal twins are somewhat more similar than nontwin siblings, perhaps because twins shared the same uterus at the same time and are exactly the same age.

Combination Designs

During the past two decades, behavioral geneticists have begun to use designs that combine the family, adoption, and twin methods in order to bring more power to bear on these analyses. For example, it is useful to include nontwin siblings in twin studies to test whether twins differ statistically from singletons and whether fraternal twins are more similar than nontwin siblings.

Two major combination designs bring the adoption design together with the family design and with the twin design. The adoption design comparing

CLOSE UP

Lindon Eaves majored in genetics as an undergraduate at the University of Birmingham, England. He obtained his Ph.D. in human behavioral genetics in 1970. He taught at Oxford University for two years before moving to the United States in 1981, where he is now Distinguished Professor of Human Genetics and a professor of psychiatry at the Virginia Commonwealth University School of Medicine in Richmond. With Kenneth Kendler, he directs the Virginia Institute for Psychiatric and Behavioral Genetics. His research includes the study of genetic and environmental effects on personality and social attitudes, the genetic analysis of multiple variables, mate selection, genotype-environment interaction, segregation and linkage analysis, and the genetic analysis of developmental change. With Hans Eysenck and Nick Martin, he wrote *Genes, Culture and Personality: An Empirical Approach*. Eaves holds the James Shields award for twin research and the Paul Hoch award from the American Psychopathological Association. He is a past president of the Behavior Genetics Association and the president of the International Society for Twin Studies. Currently he directs the Virginia Twin Study of Adolescent Behavioral Development, which is analyzing the interaction of genetic and environmental effects in the development of adolescent behavioral problems.

"genetic" and "environmental" relatives is made much more powerful by including the "genetic-plus-environmental" relatives of a family design. This is the design of one of the largest and longest ongoing genetic studies of behavioral development, the Colorado Adoption Project (DeFries, Plomin, & Fulker, 1994). This project has shown, for example, that genetic influence on general cognitive ability increases during infancy and childhood.

The adoption-twin combination involves twins adopted apart and compares them with twins reared together. Two major studies of this type have been conducted, one in Minnesota (Bouchard et al., 1990a) and one in Sweden (Pedersen et al., 1992). These studies have found, for example, that identical twins reared apart from early in life are almost as similar in terms of general cognitive ability as are identical twins reared together, suggesting strong genetic influence and little environmental influence caused by growing up together in the same family (shared family environmental influence).

An interesting combination of the twin and family methods comes from the study of families of identical twins, which has come to be known as the families-of-twins method. When identical twins become adults and have their own children, interesting family relationships emerge. For example, in families of male identical twins, nephews are as related genetically to their twin uncle as they are to their own father. That is, in terms of their genetic relatedness, it is as if they have the same father. Similarly, cousins are as closely related as half siblings are.

Although not as powerful as standard adoption or twin designs, a design that has recently been used takes advantage of the increasing number of stepfamilies created as a result of divorce and remarriage (Neiderhiser et al., 1996). Half siblings typically occur in stepfamilies because a woman brings a child from a former marriage to her new marriage and then has another child with her new husband. These children have only one parent (the mother) in common and are 25 percent similar genetically, unlike full siblings, who have both parents in common and are 50 percent similar genetically. Half siblings can be compared with full siblings in stepfamilies to assess genetic influence. Full siblings in stepfamilies occur when the mother brings full siblings from her former marriage or when she and her new husband have more than one child together. A useful test of whether stepfamilies differ from never-divorced families is the comparison between full siblings in the two types of families.

SUMMING UP

Adoption and twinning are experiments of nature that can be used to assess the relative contributions of nature and nurture to familial resemblance. For schizophrenia and cognitive ability, family members resemble one another even when they are adopted apart. Twin studies show that identical twins are more similar than fraternal twins. Results of family, adoption, and twin studies and of combinations of these designs converge on the conclusion that genetic factors contribute to schizophrenia and cognitive ability.

Heritability

For the complex traits that interest psychologists, it is possible to ask not only *whether* genetic influence is important but also *how much* genetics contributes to the trait. The question about whether genetic influence is important involves *statistical significance*, the reliability of the effect. For example, we can ask whether the resemblance between "genetic" parents and their adopted-away offspring is significant, or whether identical twins are significantly more similar than fraternal twins. Statistical significance depends on the size of the effect and the size of the sample. For example, a "genetic" parent-offspring correlation of .25 will be statistically significant if the adoption study includes at least 45 parent-offspring pairs. Such a result would indicate that it is highly likely (95 percent probability) that the true correlation is greater than zero.

CLOSE UP

Nancy Pedersen is a docent (associate professor) in psychology: specialty behavior genetics at the Karolinska Institute in Stockholm, where she is also head of the Division of Genetic Epidemiology, in the Institute of Environmental Medicine. She holds affiliated associate professorships at Pennsylvania State University and the University of Southern California. As a graduate student at the Institute for Behavioral Genetics in Boulder, she was sent to Sweden to administer a twin-family study of smoking behavior. While there, she "rediscovered," with the help of Robert Plomin and Gerald McClearn, a substantial sample of twins reared apart, in the Swedish Twin Registry. After completing her doctorate in 1980, she returned to Sweden, where she has continued to work with the Swedish Twin Registry, of which she is now the director. Pedersen's research has focused on using the combined adoption-twin design to address issues concerning individual differences in aging. She is particularly interested in applications of multivariate and longitudinal designs, consequences of various ascertainment techniques, and interactions between genotype and environment. Her present dream is to develop the Swedish Twin Registry as an international resource for behavioral, quantitative, and molecular genetic studies of complex disorders and dimensions.

The question about how much genetics contributes to a trait refers to *effect size*, the extent to which individual differences for the trait in the population can be accounted for by genetic differences among individuals. As reiterated later in this chapter, effect size in this sense refers to individual differences for a trait in the entire population, not to certain individuals. For example, if PKU were left untreated, it would have a huge effect on the cognitive development of individuals homozygous for the recessive allele. However, because such individuals represent only 1 in 10,000 individuals in the population, this huge effect for these few individuals would have little effect overall on the variation in cognitive ability in the entire population.

Many statistically significant environmental effects in psychology involve very small effects in the population. For example, birth order is significantly related to intelligence test (IQ) scores (first-born children have higher IQs). This is a small effect in that the mean difference between first- and second-born siblings is less than two IQ points and their IQ distributions almost completely overlap. Birth order accounts for about 1 percent of the variance of IQ scores when other factors are controlled. In other words, if all you know about two siblings is their birth order, then you know practically nothing about their IQs.

In contrast, genetic effect sizes are often very large, among the largest effects found in psychology, accounting for as much as half of the variance. The statistic that estimates the genetic effect size is called *heritability*. Heritability is the proportion of phenotypic variance that can be accounted for by genetic differences among individuals. As explained in Appendix B, heritability can be estimated from the correlations for relatives. For example, if the correlation for "genetic" (adopted-apart) relatives is zero, then heritability is zero. For first-degree "genetic" relatives, their correlation reflects half of the effect of genes because they are only 50 percent similar genetically. That is, if heritability is 100 percent, their correlation would be .50. In Figure 5.10, the correlation for "genetic" (adopted-apart) siblings is .24 for IQ scores. Doubling this correlation yields a heritability estimate of 48 percent, which suggests that about half of the variance in IQ scores can be explained by genetic differences among individuals.

Heritability estimates, like all statistics, include error of estimation, which is a function of the effect size and the sample size. In the case of the IQ correlation of .24 for adopted-apart siblings, the number of sibling pairs is 203. There is a 95 percent chance that the true correlation is between .10 and .38, which means that the true heritability is likely to be between 20 and 76 percent, a very wide range. For this reason, heritability estimates based on a single study need to be taken as very rough estimates surrounded by a large confidence interval, unless the study is very large. For example, if the correlation of .24 were based on a sample of 2000 instead of 200, there would be a 95 percent chance that the true heritability is between 40 and 56 percent. Replication across studies and across designs is also important in order to home in on more precise estimates.

If identical and fraternal twin correlations are the same, heritability is estimated as zero. If identical twins correlate 1.0 and fraternal twins correlate .50, a heritability of 100 percent is implied. In other words, genetic differences among individuals completely account for their phenotypic differences. A rough estimate of heritability in a twin study can be made by doubling the difference between the identical and fraternal twin correlations. As explained in Appendix B, because identical twins are identical genetically and fraternal twins are 50 percent similar genetically, the difference in their correlations reflects half of the genetic effect and is doubled to estimate heritability. For example, in Figure 3.7, IQ correlations for identical and fraternal twins are .85 and .60, respectively. Doubling the difference between these correlations results in a heritability estimate of 50 percent, which also suggests that about half of the variance of IQ scores can be accounted for by genetic factors. Because these studies include more than 10,000 pairs of twins, the error of estimation is small. There is a 95 percent chance that the true heritability is between .48 and .52.

CLOSE UP

Nick Martin is senior principal research fellow at the Queensland Institute of Medical Research, Brisbane, Australia. He graduated with a degree in genetics from the University of Adelaide in 1972. His doctoral dissertation in 1976 was entitled "The classical twin study in human behaviour genetics" and was supervised by Lindon Eaves at the University of Birmingham. In 1978 he moved to the Australian National University in Canberra, where he cofounded the Australia Twin Registry, which has been the focus of most of his research. From 1983 to 1986, he worked again with Lindon Eaves, as well as with Andrew Heath and Ken Kendler at the Medical College of Virginia, Richmond, before returning to his present position in Brisbane. Martin's interest has been in the use of twins to investigate the causes of individual differences in behavior. With Eaves, he developed the genetic analysis of covariance structures, which is the basis of most current multivariate behavioral genetic analysis, and formalized the investigation of statistical power in behavioral genetic applications. His current interest is to use linkage and association studies to detect quantitative trait loci for human behavioral traits.

Because disorders are diagnosed as either-or dichotomies, familial resemblance is assessed by concordances rather than by correlations. As explained in Appendix A, concordance is an index of risk. For example, if sibling concordance is 10 percent for a disorder, we say that siblings of probands have a 10 percent risk for the disorder. If identical and fraternal twin concordances are the same, heritability must be zero. To the extent that identical twin concordances are greater than fraternal twin concordances, genetic influence is implied. For schizophrenia (Figure 3.6), the identical twin concordance of .48 is much greater than the fraternal twin concordance of .17, a difference suggesting substantial heritability. The fact that in 52 percent of the cases identical twins are *dis*cordant for schizophrenia, even though they are genetically identical, implies that heritability is much less than 100 percent.

One way to estimate heritability for disorders is to use the liability-threshold model (see Box 3.1) to translate concordances into correlations on the assumption that a continuum of genetic risk underlies the dichotomous diagnosis. For schizophrenia, the identical and fraternal twin concordances of .48 and .17 translate into liability correlations of .86 and .57, respectively.

BOX 5.3

The Twin Method

Francis Galton (1876) studied developmental changes in twins' similarity, but the first real twin study in which identical and fraternal twins were compared in order to estimate genetic influence was conducted in 1924 (Merriman, 1924). This first twin study assessed IQ and found that identical twins were markedly more similar than fraternal twins, a result suggesting genetic influence. Dozens of subsequent twin studies of IQ confirmed this finding. Twin studies have also been reported for many other psychological dimensions and disorders and provide the bulk of the evidence for the widespread influence of genetics in behavioral traits.

Although most mammals have large litters, primates, including our species, tend to have single offspring. However, primates occasionally have multiple births. Human twins are more common than people usually realize—about 1 in 85 births are twins. Surprisingly, as many as 20 percent of fetuses are twins, but because of the hazards associated with twin pregnancies, often one member of the pair dies very early in pregnancy. Until recently in the United States, among live births, the numbers of identical and same-sex fraternal twins are approximately equal. That is, of all twin pairs, about one-third are identical twins, one third are same-sex fraternal twins, and the other third are opposite-sex fraternal twins.

Identical twins result from a single fertilized egg (called a *zygote*) that divides for unknown reasons, producing two (or sometimes more) genetically identical individuals. For about a third of identical twins, the zygote splits during the first five days after fertilization as it makes it way down to the womb. In this case, the identical twins have different sacs (called *chorions*) within the placenta. Two-thirds of the time, the zygote splits after it implants in the placenta and the twins share the same chorion. Identical twins who share the same chorion may be more similar for some psychological traits than identical twins who do not share the same chorion, although the evidence on this hypothesis is mixed (Sokol et al., 1995). "Siamese" twins occur when the zygote splits after about two weeks, resulting in twins whose bodies are partially fused. Fraternal twins occur when two eggs are separately fertilized; they have different chorions. Like other siblings, they are 50 percent similar genetically.

The rate of fraternal twinning differs across countries, increases with maternal age, and may be inherited in some families. Increased use of fertility drugs results in greater numbers of fraternal twins because these drugs make it likely that more than one egg will ovulate. The rate of identical twinning is not affected by any of these factors.

Doubling the difference between these liability correlations suggests a heritability of about 60 percent. As explained in Box 3.1, this statistic refers to a hypothetical construct of continuous liability as derived from a dichotomous diagnosis of schizophrenia rather than to the diagnosis of schizophrenia itself.

For combination designs that compare several groups, and even for simple adoption and twin designs, modern genetic studies are typically analyzed by using an approach called *model fitting*. Model fitting tests the significance of the fit between a model of genetic and environmental relatedness against the observed data. Different models can be compared, and the best-fitting model is used to estimate the effect size of genetic and environmental effects. Model fitting is described in Appendix B.

Genetic designs can also be used to assess genetic and environmental contributions to the covariance between two traits, called *multivariate genetic analysis* (Martin & Eaves, 1977). That is, the correlation between two traits could be due to genetic factors that affect both traits or to overlapping environmental factors. Multivariate genetic analysis is explained in Appendix B. Many examples of multivariate genetic analyses are described in later chapters.

Interpreting Heritability

Heritability refers to the genetic contribution to individual differences (variance), *not* to the phenotype of a single individual. For a single individual, both genotype and environment are indispensable—a person would not exist without both genes and environment. As noted by Theodosius Dobzhansky, the first president of the Behavior Genetics Association:

> The nature-nurture problem is nevertheless far from meaningless. Asking right questions is, in science, often a large step toward obtaining right answers. The question about the roles of genotype and the environment in human development must be posed thus: To what extent are the *differences* observed among people conditioned by the differences of their genotypes and by the differences between the environments in which people were born, grew and were brought up? (Dobzhansky, 1964, p. 55)

This issue is critical for the interpretation of heritability. You can still read in introductory psychology textbooks that genetic and environmental effects on behavior cannot be disentangled because behavior is the product of genes and environment. An example sometimes given is the area of a rectangle. It is nonsensical to ask about the separate contributions of length and width to the area of a single rectangle because area is the product of length and width. Area does not exist without both length and width. However, if we ask, not about a single rectangle, but about a population of rectangles (Figure 5.12), the variance in areas could be due entirely to length (b), entirely to width (c), or to both (d). Obviously, there can be no behavior without both an organism and an

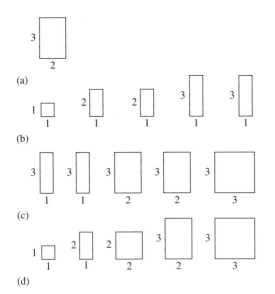

Figure 5.12 Individuals and individual differences. Genetic and environmental contributions to behavior do not refer to a single individual, just as the area of a single rectangle (a) cannot be attributed to the relative contributions of length and width, because area is the product of length and width. However, in a population of rectangles, the relative contribution of length and width to differences in area can be investigated. It is possible that length alone (b), width alone (c), or both (d) account for differences in area among rectangles.

environment. The scientifically useful question is the origins of differences among individuals.

For example, the heritability of height is about 90 percent, but this does not mean that you grew to 90 percent of your height for reasons of heredity and that the other inches were added by the environment. What it means is that most of the height differences among individuals are due to the genetic differences among them. Heritability is a statistic that describes the contribution of genetic differences to observed differences among individuals in a particular population at a particular time. In different populations or at different times, environmental or genetic influences might differ, and heritability estimates in such populations could differ.

A counterintuitive example concerns the effects of equalizing environments. If environments were made the same for everyone in a particular population, heritability would be high in that population because individual differences that remained in the population would be due exclusively to genetic differences.

It should be emphasized that heritability refers to the contribution of genetic differences to observed differences among individuals for a particular trait in a particular population at a particular time. Much of our DNA does not vary from person to person, or even from our species to other primates or other mammals. Because these genes are the same for everyone, they obviously cannot contribute to differences among individuals. If these nonvarying genes were disrupted by mutation, they could have a devastating, even lethal, effect on development, even though they do not normally contribute at all to variation in the population. Similarly, many environmental factors do not vary substantially;

for example, the air we breath and the essential nutrients we eat. Again, although at this level of analysis such nonvarying environmental factors do not contribute to differences among individuals, disruption of these essential environments could have devastating effects.

A related issue concerns average differences between groups, such as average differences between males and females, between social classes, or between ethnic groups. It should be emphasized that the causes of individual differences within groups have no necessary implications for the causes of average differences between groups. Specifically, heritability refers to the genetic contribution to differences among individuals within a group. High heritability within a group does not necessarily imply that average differences between groups are due to genetic differences between groups. The average differences between groups could be due solely to environmental differences even when heritability within both groups is very high.

This point extends beyond the politically sensitive issues of gender, social class, and ethnic differences. As discussed in Chapter 10, a key issue in psychopathology concerns the links between the normal and the abnormal. Finding heritability for individual differences within the normal range of variation does not necessarily imply that the average difference between an extreme group and the rest of the population is also due to genetic factors. For example, if individual differences in depressive symptoms for an unselected sample are highly heritable, this finding does not necessarily imply that severe depression is also due to genetic factors. This point is worth repeating: The causes of average differences between groups are not necessarily related to the causes of individual differences within groups.

A related point is that heritability describes *what is* in a particular population at a particular time rather than *what could be*. That is, if either genetic influences change (e.g., changes due to migration) or environmental influences change (e.g., changes in educational opportunity), then the relative impact of genes and environment will change. Even for a highly heritable trait such as height, changes in the environment *could* make a big difference, for example, if an epidemic struck or if children's diets were altered. Indeed, the huge increase in children's heights during this century is almost certainly a consequence of improved diet. Conversely, a trait that is largely influenced by environmental factors *could* show a big genetic effect. For example, genetic engineering can knock out a gene or insert a new gene that greatly alters the trait's development, something that can now be done in laboratory animals, as discussed in the next chapter.

Although it is useful to think about what could be, it is important to begin with what is: the genetic and environmental sources of variance in existing populations. Knowledge about what is can sometimes help to guide research concerning what could be, as in the example of PKU. Most important, heritability has nothing to say about *what should be*. Evidence of genetic influence for a behavior is compatible with a wide range of social and political views, most of which depend on values, not facts.

A related point is that heritability does not imply genetic determinism. Just because a trait shows genetic influence does not mean that nothing can be done to change it. Environmental change is possible even for single-gene disorders. For example, when PKU was found to be a single-gene cause of mental retardation, it was not treated by means of eugenic (breeding) intervention or genetic engineering. An environmental intervention was successful in bypassing the genetic problem of high levels of phenylalanine: Administer a diet low in phenylalanine. This important environmental intervention was made possible by recognition of the genetic basis for this type of mental retardation.

For behavioral disorders and dimensions, the links between specific genes and behavior are weaker because behavioral traits are generally influenced by multiple genes and environmental factors. For this reason, genetic influence on behavior involves probabilistic propensities rather than predetermined programming. In other words, the complexity of most behavioral systems means that genes are not destiny. Although specific genes that contribute to complex disorders such as late-onset Alzheimer's disease are beginning to be identified, these genes only represent genetic risk factors in that they increase the probability of occurrence of the disorder but do not guarantee that the disorder will occur. An important corollary of the point that heritability does not imply genetic determinism is that heritability does not constrain environmental interventions such as psychotherapy.

We hasten to note that finding a gene that is associated with a disorder does not mean that the gene is "bad" and should be eliminated. For example, the gene might involve protection rather than risk. A gene associated with novelty seeking (Chapter 11) may be a risk factor for antisocial behavior, but it could also predispose individuals to scientific creativity. The gene that causes the flushing response to alcohol in Asian individuals protects them against becoming alcoholics (Chapter 12). The classic evolutionary example is a gene that causes sickle-cell anemia in the recessive condition but protects carriers against malaria (Chapter 13). As we shall see, most complex traits are influenced by multiple genes, which means that we are all likely to be carrying one or more genes that contribute to risk for some disorders.

Finally, finding genetic influence on complex traits does not mean that the environment is unimportant. For simple single-gene disorders, environmental factors may have little effect. In contrast, for complex traits, environmental influences are usually as important as genetic influences. When one member of an identical twin pair is schizophrenic, for example, the other twin is not schizophrenic in about half the cases, even though members of identical twin pairs are identical genetically. Such differences within pairs of identical twins can only be caused by nongenetic factors. Despite its name, behavioral genetics is as useful in the study of environment as it is in the study of genetics. In providing a "bottom line" estimate of all genetic influence on behavior, genetic research also provides a "bottom line" estimate of environmental influence. Indeed, genetic research provides the best available evidence for the importance of the environment. Moreover, genetic research has made some of the most important discoveries in recent years about how the environment works in psychological development (Chapter 14).

In the field of quantitative genetics, the word *environment* includes all influences other than inherited factors. This use of the word *environment* is much broader than is usual in psychology. In addition to environmental influences traditionally studied in psychology, such as parenting, environment includes prenatal events and nongenetic biological events after birth, such as illnesses and nutrition. As mentioned in Chapter 3, environment even includes changes in DNA that are not inherited because they occur in cells other than testes and ovaries, where sperm and eggs are formed. For example, identical twins are not identical for such environmentally induced changes in DNA.

SUMMING UP

The size of the genetic effect can be quantified by the statistic called heritability. Heritability estimates the proportion of observed (phenotypic) differences among individuals that can statistically be attributed to genetic differences. For schizophrenia and cognitive ability, genetic influence is not only significant but also substantial. Heritability describes *what is* in a particular population at a particular time, not *what could be* or *what should be*. Phenotypic differences not explained by genetic differences can be attributed to the environment. In this way, genetic studies provide the best available evidence for the importance of environment.

Equality

It was a self-evident truth to the signers of the American Declaration of Independence that all men are created equal. Does this not mean that democracy rests on the absence of genetic differences among people? Absolutely not! The founding fathers of America were not so naive as to think that all people are created *identical*. The essence of a democracy is that all people should have legal equality *despite* their genetic differences.

The central message from behavioral genetics involves genetic individuality. With the exception of identical twins, each one of us is a unique genetic experiment, never to be repeated again. Here is the conceptualization on which to build a philosophy of the dignity of the individual! Human variability is not simply imprecision in a process that, if perfect, would generate unvarying representatives of an ideal person. Genetic diversity is the essence of life.

Summary

Quantitative genetic methods can detect genetic influence for complex traits. For animal behavior, selection studies and studies of inbred strains provide powerful tests of genetic influence. Adoption and twin studies are the workhorses for human quantitative genetics. They capitalize on the quasi-experimental situations caused by adoption and twinning to assess the relative contributions of nature and nurture. For schizophrenia and cognitive ability, resemblance of relatives increases with genetic relatedness, an observation suggesting genetic influence. Adoption studies show family resemblance even when family members are adopted apart. Twin studies show that identical twins are more similar than fraternal twins. Results of such family, adoption, and twin studies converge on the conclusion that genetic factors contribute substantially to complex human behavioral traits, among other traits.

The size of the genetic effect is quantified by heritability, a statistic that describes the contribution of genetic differences to observed differences in a particular population at a particular time. For most behavioral dimensions and disorders, including cognitive ability and schizophrenia, genetic influence is not only detectable but also substantial, often accounting for as much as half of the variance in the population. Genetic influence in psychology has been controversial in part because of misunderstandings about heritability.

Genetic influence on behavior is just that—an influence or contributing factor, not preprogrammed and deterministic. Environmental influences are usually as important as genetic influences. Behavioral genetics focuses on why people differ, that is, the genetic and environmental origins of individual differences that exist at a particular time in a particular population. Recognition of individual differences, regardless of their environmental or genetic origins, does not vitiate the value of equality.

CHAPTER SIX

Identifying Genes

M uch more quantitative genetic research of the kind described in the previous chapter is needed to identify the most heritable components and constellations of behavior, to investigate environmental contributions in genetically sensitive designs, and, especially, to explore the developmental interplay between nature and nurture. However, one of the most exciting directions for research in behavioral genetics is the coming together of quantitative genetics and molecular genetics in attempts to identify specific genes responsible for genetic influence on behavior, even for complex behaviors for which many genes as well as many environmental factors are at work.

As illustrated in Figure 6.1, quantitative genetics and molecular genetics both began around the end of the nineteenth century. The two groups, biometricians (Galtonians) and Mendelians, quickly came into conflict, as described in Chapter 3. Their ideas grew apart as quantitative geneticists focused on naturally occurring genetic variation and complex quantitative traits and molecular geneticists analyzed single-gene mutations. During the past decade, however, quantitative genetics and molecular genetics have begun to come together to identify genes for complex, quantitative traits. Such a gene in multiple-gene systems is called a *quantitative trait locus (QTL)*.

In addition to producing indisputable evidence of genetic influence, the identification of specific genes will revolutionize behavioral genetics by providing measured genotypes for investigating genetic links between behaviors, for tracking the developmental course of genetic effects, and for identifying interactions and correlations between genotype and environment. Moreover, once a gene is identified, it is possible to determine what protein the gene produces and to investigate how the gene's product affects behavior.

This new knowledge about specific genes associated with behavioral disorders is creating new problems as well, as discussed at the end of this chapter and also in Chapter 15.

Animal Behavior

The previous chapter indicated that inbred strain and selection studies with animals provide direct experiments to investigate genetic influence. In contrast, quantitative genetic research on human behavior is limited to the experiments

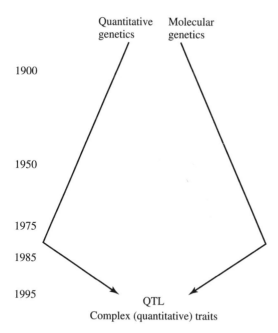

1900

1950

1975

1985

1995

Quantitative   Molecular
genetics   genetics

QTL
Complex (quantitative) traits

Figure 6.1 Quantitative genetics and molecular genetics are coming together in the study of complex quantitative traits and quantitative trait loci (QTLs).

of nature, adoption and twinning. Similarly, animal models provide more powerful means to identify genes than are available for our species.

Long before DNA markers became available in the 1980s, associations were found between single genes and behavior. The first example was discovered in 1915 by A. H. Sturtevant, inventor of the chromosome map. He found that a single-gene mutation that alters eye color in the fruit fly *Drosophila* also affects their mating behavior. Another example involves the single recessive gene that causes albinism and also affects open-field activity in mice. Albino mice are less active in the open field. It turns out that this effect is largely due to the fact that albinos are more sensitive to the bright light of the open field. With a red light that reduces visual stimulation, albino mice are almost as active as pigmented mice. These relationships are examples of what is called *allelic association*, the association between a particular allele and a phenotype. Rather than using genes like eye color and albinism that are known by their phenotypic effect, it is now possible to use thousands of polymorphisms in DNA itself.

Induced Mutations

In addition to studying naturally occurring genetic variation, geneticists have used chemicals to create mutations that enable them to dissect the effects of genes. This mutational strategy has been applied to the study of behavior. During the past 30 years, hundreds of behavioral mutants have been selected in organisms as diverse as single-celled organisms, worms, fruit flies, and mice. This work illustrates the point that most normal behavior is influenced by many genes. Although any one of many single-gene mutations can seriously disrupt behavior, normal development is orchestrated by many genes working together. An analogy is an automobile, which requires thousands of parts for its normal functioning. If any one part breaks down, the automobile may not run properly. In the same way, single genes can drastically affect behavior that is normally influenced by many genes.

Bacteria Although the behavior of bacteria is by no means attention grabbing, they do behave. They move toward or away from many kinds of chemicals by rotating their propellerlike flagella. Since the first behavioral mutant in bacteria was isolated in 1966, the dozens of mutants that have been created emphasize the genetic complexity of an apparently simple behavior in a simple organism. For example, many genes are involved in rotating the flagella and controlling the duration of the rotation.

Paramecia Like bacteria, paramecia are one-celled organisms, but they are larger and their movements are more obvious. Propelled by cilia, paramecia avoid certain chemicals and heat by backing up and then swimming forward in a new direction. Hundreds of induced behavioral mutants have been isolated. For example, at least 20 genes are involved in the avoidance behavior of paramecia. Some mutants that cannot swim backward are called *pawn* mutants (after the chess piece that can only move forward). Mutations at any of three loci may result in this behavior. Other mutants include *paranoiac* (prolonged backward movement), which involves mutations at five loci, and *sluggish* (involving very slow movement), which is due to a single locus. Figure 6.2 shows tracks of the movements of the normal (wild) type and these three mutants.

Roundworms The nematode (roundworm) *Caenorhabditis elegans* has 959 cells, of which 302 are nerve cells. Its behavior is more complex than that of single-celled organisms like bacteria and paramecia, and many behavioral mutants have been identified. For example, several mutations that affect locomotion and foraging behavior have recently been identified (McIntire, Jorgensen, & Horvitz, 1993). *C. elegans* is especially important for behavioral genetic analysis because the developmental fate of each of its cells, the wiring diagram of its 302 nerve cells, and most of its several thousand genes are known. The complete description of its DNA sequence of 100 million base pairs (3 percent of the size of the human genome) is expected soon.

Fruit flies The fruit fly *Drosophila* is the star organism in terms of behavioral mutants, with hundreds known. The earliest behavioral research involved response to light (phototaxis) and to gravity (geotaxis). Normal *Drosophila* move toward light (positive phototaxis) and away from gravity (negative geotaxis). Many mutants that were either negatively phototaxic or positively geotaxic were created. The hundreds of other behavioral mutants included *sluggish* (generally slow), *hyperkinetic* (generally fast), *easily shocked* (jarring produces a seizure), and *paralyzed* (collapses when the temperature goes above 28° C). A *drop dead* mutant walks and flies normally for a couple of days and then sud-

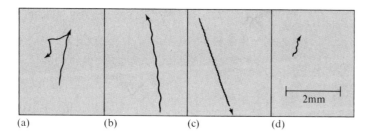

(a)   (b)   (c)   (d)

Figure 6.2 Traces of movements of paramecia during 10-second periods. (a) This wild type individual shows a spontaneous avoidance reaction, which leads to turns. (b) The *pawn* mutant moves forward but shows no avoidance reactions. (c) The *paranoiac* mutant shows rapid and sustained backward motion. (d) The *sluggish* mutant moves very slowly. (Adapted from "Genetic dissection of behavior in *Paramecium*" by C. Kung et al. *Science, 188,* 898–904. Copyright © 1975 by the American Association for the Advancement of Science.)

denly falls on its back and dies. More complex behaviors have also been studied, especially courtship and learning. Behavioral mutants for various aspects of courtship and copulation have been found. One male mutant, called *fruitless,* courts males as well as females and does not copulate. Another male mutant cannot disengage from the female after copulation and is given the dubious title *stuck.* The first learning behavior mutant was called *dunce* and could not learn to avoid an odor associated with shock even though it had normal sensory and motor behavior. Other learning mutants have been found as well as a memory mutant called *amnesiac.*

Drosophila also offer the possibility of creating genetic mosaics in which individuals have different genes in various cells of the body. This method has been applied to many behavioral mutants to analyze which body parts must be altered to allow the mutant behavior to occur. In an early study, mosaics were created with one eye with a normal gene and positive phototaxis and one eye with a mutant gene and nonphototaxis. As illustrated in Figure 6.3, in a dark tube these mutants climb straight up, because their negative geotaxis is not affected. However, when a light is placed at the top of the tube, the mosaic flies attempt to equalize the light coming into the two eyes and thus keep their defective eye toward the light. As a result, the mosaic flies climb up the tube in a spiral, keeping their mutant eye closer to the light.

The most interesting mosaic mutant studies involved sexual behavior and the X chromosome. *Drosophila* were made mosaic for the X chromosome so that some body parts have two X chromosomes and are female and other body parts have only one X chromosome and are male. As long as a small region toward the back of the brain is male, courtship behavior is male. Of course, sex is not all in the head. Different parts of the nervous system are involved in aspects of courtship behavior such as tapping, "singing," and licking. Successful copulation also requires a male thorax (containing the fly's version of a

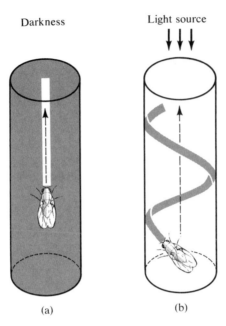

Darkness   Light source

(a)   (b)

Figure 6.3 Mosaic eye mutant in *Drosophila* with normal vision in one eye and a mutant gene for a diminished response to light. (a) In a dark tube, the fly climbs straight up because its negative geotaxis is unaffected. (b) When a light is shone down the tube, the same fly turns its defective eye toward the light in an attempt to equalize the light coming into the two eyes and climbs in a spiral path. (From Hotta & Benzer, 1970.)

spinal cord between the head and abdomen) and, of course, male genitals (Greenspan, 1995).

Mice: Gene Targeting and Transgenics

With mice, random chemically induced mutations have not been used to detect associations with behavior, perhaps because so many naturally occurring polymorphisms are known. An important development in recent years is targeted mutation (*gene targeting*), a technique in which mutations are created in specific genes (Capecchi, 1994). The mutated gene is then transferred to mouse embryos (called *transgenics* when they contain "foreign" DNA from another species). It is possible to alter parts of a gene that increase or decrease rates of its own transcription. Most amazing is the ability to add bits to a gene that act as a switch, turning the gene on or off (i.e., creating inducibility), for example, when a particular hormone is added or even when the temperature changes. As described in Chapter 8, this has been done in fruit flies for genes that affect their ability to learn to discriminate odors.

For mice, most gene targeting research is cruder. A gene is "knocked out" entirely by deleting some of its key DNA sequences. Once mice homozygous for the knock-out gene are bred, the effect of the knock-out gene on behavior can be investigated. One of the first experiments of this kind for behavior produced mice with a knock-out version of a gene (α-CaMKII) for a protein enriched in the hippocampus and thought to be involved in long-term memory. Mutant mice homozygous for the knock-out gene learned a spatial task significantly more poorly than control mice, although otherwise their behavior

Jeanne Wehner is a professor of pharmacoge-netics in the Institute for Behavioral Genetics and the School of Pharmacy at the University of Colorado, Boulder. She received her doctor-ate in biochemistry from the University of Minnesota in 1976. From 1976 to 1981, she was a postdoctoral fellow at the Dight Institute for Human Genetics and was a faculty mem-ber at the University of Minnesota in 1981 to 1982. The goal of Wehner's research is to elu-cidate the neurochemical substrates that reg-ulate genetic differences in complex learning ability. She also conducts studies on genetic regulation of sensitivity to alcohol and has been a member of the Colorado Alcohol Research Center. In both areas of research, she uses traditional behavioral genetic strategies, including inbred strain and recombinant inbred strain studies, as well as newer knock-out mouse models. Behavioral differences are examined with a wide variety of techniques, including pharmacology, neurochemistry, and molecular biology.

seemed normal (Silva et al., 1992). Knock-out alleles of several other genes have been shown to affect learning and memory (Wehner, Bowers, & Paylor, 1996). Another example relevant to behavior involved aggression. Increased aggres-sion was shown by male mice with a knock-out version of a gene that codes for a receptor for an important neurotransmitter, serotonin (Saudou et al., 1994). Increased aggression was also found for another knock-out of a gene for an enzyme (neuronal nitric oxide synthase) that normally plays a basic role in neuro-transmission (Nelson et al., 1995).

Hundreds of knock-out studies are in progress and many associations with behavior are likely to emerge. The significance of these associations between single-gene mutations and behavior is that knocking out a single gene can dras-tically affect behavior that is normally influenced by many genes.

SUMMING UP

Many behavioral mutants have been identified in organisms as diverse as single-celled bacteria and paramecia, roundworms, fruit flies, and mice. In mice, genes can be transferred from other organisms (transgenics), including a mutated form of a gene that knocks out the functioning of the normal allele (gene targeting).

QTL Linkage

Complex quantitative traits, whether medical or behavioral, are likely to be influenced by multiple genes as well as by multiple environmental factors. Finding such QTLs has only become possible in recent years. Animal models have been particularly useful. For example, researchers have identified 23 QTLs that account for most of the genetic variation in diabetes in the mouse (Ghosh et al., 1993; Risch, Ghosh, & Todd, 1993). The QTLs have various effect sizes. More than half of the genetic variation is due to interactive effects between QTLs. Work on rat models of hypertension shows a similar complex picture (Hilbert et al., 1991).

Allelic association between candidate genes and particular phenotypes is a simple and powerful approach to QTL identification. However, it is usually difficult to know which of the thousands of genes to investigate as possible candidate genes affecting behavior. Finding allelic association requires that a DNA marker must itself be the gene that causes the association or that the marker must be so close to the gene that the marker and the gene are not separated by recombination. In contrast to allelic association, linkage can be used to scan the entire genome with just a few hundred markers. As discussed in Chapter 2, linkage is an exception to Mendel's second law of independent assortment. If a DNA marker is on the same chromosome and within the general chromosomal region of a gene that affects a behavioral disorder, the marker and disorder will not assort independently, thus indicating linkage.

Linkage can be identified by using Mendelian crosses to trace the cotransmission of a marker whose chromosomal location is known and a single-gene disorder, as illustrated in Figure 2.6. However, as emphasized in previous chapters, behavioral dimensions and disorders are likely to be influenced by many genes. If many genes contribute to behavior, behavioral traits will be distributed quantitatively. The goal is to find some of the many genes (QTLs) that affect these quantitative traits. Allelic association is able to detect small QTL effects, but only recently have linkage approaches been able to identify QTLs.

For example, QTL linkage strategies have been applied to open-field activity in mice. One QTL strategy involves F_2 mice derived from a cross between the high and low lines selected for open-field activity and subsequently inbred by using brother-sister matings for over 30 generations (Flint et al., 1995a). Each F_2 mouse has a unique combination of alleles from the parental strains. The most active and the least active F_2 mice were examined for 84 DNA markers spread throughout the mouse chromosomes in order to identify chromosomal regions that are associated with open-field activity. The analysis simply asks whether the frequencies of various alleles differ between the most active and least active groups. Although this analysis is like allelic association, F_2 animals have an average of only one recombination per chromosome, so a marker can detect linkage with a gene millions of base pairs away on a chromosome.

Figure 6.4 shows that regions of chromosomes 1, 12, and 15 harbor QTLs for open-field activity. A QTL on chromosome 15 is related primarily to open-field activity and not to other measures of fearfulness, an observation suggesting the possibility of a gene specific to open-field activity. The QTL regions on chromosomes 1 and 12, on the other hand, are related to other measures of fearfulness, an association suggesting that the same QTL affects these diverse measures of fearfulness. The exception is exploration in an enclosed arm of a maze (Figure 6.4), which was included in the study as a control, because other research suggests that this measure is not genetically correlated with measures of fearfulness.

Another method used to identify QTLs for behavior involves special inbred strains called *recombinant inbred (RI) strains*. As explained in Box 6.1, RI strains are inbred strains derived from an F_2 cross between two inbred strains; this process leads to recombination of parts of chromosomes from the parental strains. The chromosomes of RI strains have been examined and mapped for more than a thousand markers, thus enabling investigators to determine whether any of these markers are linked to behavior. For example, an RI QTL analysis of open-field activity suggests a QTL on chromosome 6 and possible QTLs on other chromosomes (Phillips et al., 1995). These QTL results may be different from those shown in Figure 6.4, because different progenitor strains were used, as well as a different test protocol and apparatus.

Various QTL approaches are being applied to many domains of behavior, especially in the area of psychopharmacogenetics (Crabbe, Belknap, & Buck, 1994). For statistical reasons, only the largest QTL effects can be detected by these methods. Although many QTLs are likely to have much smaller effect sizes, it makes sense to begin the quest for QTLs with the biggest QTL effects.

Synteny Homology

Especially exciting is the possibility of using QTLs found in mice as candidate QTLs for human research. Specific genes that are associated with behavior in mice can be used in human studies because nearly all mouse genes are similar to human genes. Moreover, chromosomal regions linked to behavior in mice can be used as candidate regions in human studies because parts of mouse chromosomes have the same genes in the same order as parts of human chromosomes, a relationship called *synteny homology*. For example, the region of mouse chromosome 1 shown in Figure 6.4 to be linked with open-field activity has the same order of genes as part of the long arm of human chromosome 1. This region of human chromosome 1 can be considered a candidate QTL region for human fearfulness or anxiety.

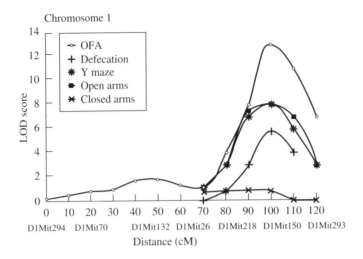

Chromosome 1

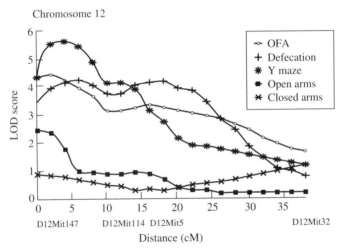

Chromosome 12

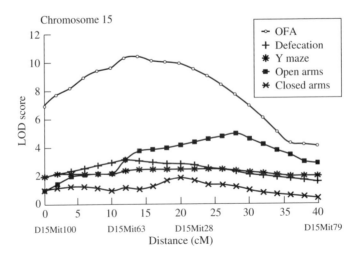

Chromosome 15

SUMMING UP

Linkage studies scan the genome for linkage between behavior and DNA markers, even when many genes are involved for complex, quantitative traits (quantitative trait loci, or QTLs). Recombinant inbred strains can be used to identify QTL linkage. Mouse QTL research can nominate candidate QTLs for human behavior because of the extensive synteny homology between mouse and human chromosomes.

Human Behavior

In our species, experimental crosses such as those producing recombinant inbred strains cannot be used to identify genes for behavior. However, as in research on animal behavior, allelic association studies can be used to examine the effects of candidate genes on behavior. For example, as mentioned in Chapter 1, a particular allele of a gene (apolipoprotein E on chromosome 19) involved in cholesterol transport is associated with late-onset Alzheimer's disease (Corder et al., 1993). This allele quadruples the risk for Alzheimer's disease. Although the association remains controversial, a gene for one of the neuronal receptor sites for dopamine (dopamine receptor D_2 on chromosome 11) has been associated with alcoholism and other drug abuse (Uhl et al., 1993). Allelic associations are beginning to be reported for psychopathology, personality, and cognitive abilities, as mentioned in later chapters. Allelic association is especially useful for detecting small QTL effects.

As mentioned earlier, in relation to mouse behavior, a systematic search of the genome for genes for behavior requires linkage rather than allelic association. For single-gene disorders, linkage can be identified by using large family

Figure 6.4 QTLs for open-field activity and other measures of fearfulness in an F_2 cross between high and low lines selected for open-field activity. The five measures are (1) open-field activity (OFA), (2) defecation in the open field, (3) activity in the Y maze, (4) entry in the open arms of the elevated plus maze, and (5) entry in the closed arms of the elevated plus maze, which is not a measure of fearfulness. LOD (*logarithm* to the base 10 of the *od*ds) scores indicate the strength of the linkage, and a LOD score of 3 or greater is generally accepted as significant. Distance in cM (centimorgans) indicates position on the chromosome, with each cM referring roughly to 1 million base pairs. Below the cM scale are listed the specific short-sequence repeat markers for which the mice were examined and mapped. (Reprinted with permission from "A simple genetic basis for a complex psychological trait in laboratory mice" by J. Flint et al. *Science, 269,* 1432–1435. Copyright © 1995 American Association for the Advancement of Science. All rights reserved.)

BOX 6.1

Recombinant Inbred Strains and QTLs

Recombinant inbred (RI) strains were developed for mapping genes to chromosomes. They are called recombinant because they are derived from the F_2 generation resulting from an initial mating between two parental inbred strains. Because of recombination in subsequent inbred generations, RI strains will show combinations of alleles that are different from those of the two parental strains. As shown in the figure below, when the parental strains have different alleles for a locus, each RI strain will become homozygous for one or the other allele during the subsequent inbreeding process. About half of the RI strains will be homozygous for one allele and half for the other. For the hypothetical *A* locus in the figure, RI-1 is A_2A_2 and RI-2 is A_1A_1. The set of RI strains illustrated in the figure is named C×B because it resulted from an initial cross between *BALB/c* and *C57BL* inbred strains of mice.

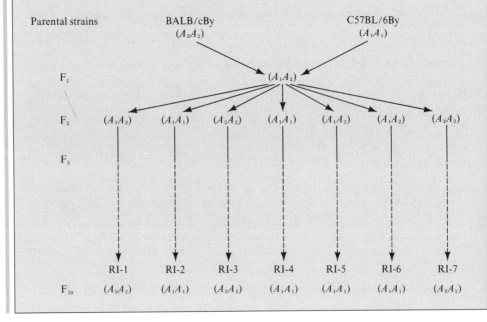

pedigrees in which cotransmission of a DNA marker allele and a disorder can be traced. Because recombination occurs an average of once per chromosome in the formation of gametes passed from parent to offspring, a marker allele within millions of base pairs and a gene for a disorder on the same chromosome will be inherited together within a family. Linkage research has been boosted by the construction of systematic maps of landmarks, called sequence-tagged sites (STSs), that cover the entire human genome. The human genome map now includes more than 15,000 STSs, at an average interval of less than 200,000 base pairs (Hudson et al., 1995).

Another RI series is called B×D because it was derived from a cross between the C57BL and DBA inbred strains. The B×D RI series includes
26 RI strains, not just the 7 strains shown in the figure. If a major gene
affected a trait, we would expect a bimodal (two-hump) distribution of RI
strain means because half of the RI strains would have the allele for the trait
and half would not. However, behavioral traits usually show continuous distributions in RI strains, indicating that more than one gene is involved.

B×D RIs can also be used to localize quantitative trait loci (QTLs) for
behavioral traits. There are now more than a thousand DNA markers that have
been mapped in the B×D RI series, with thousands more on the way. DNA
markers that have similar patterns for all the 26 B×D RI strains must be close
together on a chromosome. For any marker, each RI strain is either BB (i.e.,
homozygous for the allele from the C57BL progenitor inbred strain) or DD. By
chance, any RI strain will be BB half of the time for any particular marker. The
chance that a particular pattern for a specific marker is seen in three
RI strains—for example, BB for RI-1, DD for RI-2, and BB for RI-3—is about
12 percent ($.5^3$). For all 26 RI strains, the pattern of genotypes for each marker
is unique ($.5^{26}$) and will not be the same as that of another marker unless the
two DNA markers are closely linked on the same chromosome. The number of
RI strains that have different alleles for the two markers is a measure of how far
apart the markers are on the chromosome.

RI QTL analysis compares quantitative trait scores for strains with one
allele and those with the other allele for each DNA marker. For example,
open-field activity scores of B×D RI strains suggest linkage to a region of
chromosome 6 (Phillips et al., 1995). The RI QTL approach has been applied
especially to drugs and behavior, a field called psychopharmacogenetics
(Crabbe et al., 1994). The special value of the RI QTL approach is that
it enables all investigators to study essentially the same animals, because the
RI strains are extensively inbred. This feature of RI QTL analysis means that
each RI strain needs to be genotyped only once and that genetic correlations
can be assessed across measures, across studies, and across laboratories.

In 1983, the first DNA marker linkage was found for Huntington's disease
in the five-generation pedigree shown in Figure 6.5. In this family, the gene for
Huntington's disease is linked to the allele labeled C. All but one person with
Huntington's disease has inherited a chromosome that happens to have the
C allele in this family. This marker is not the Huntington's gene itself, because a
recombination was found between the marker allele and Huntington's disease
for one individual; the leftmost woman shown in generation IV had
Huntington's disease but did not inherit the C allele for the marker. That is, this
woman received that part of her affected father's chromosome carrying the gene

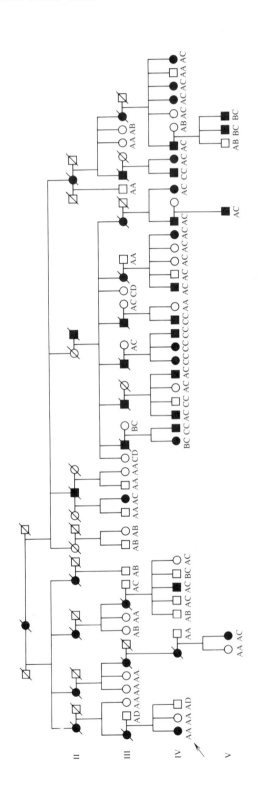

for Huntington's disease, which is normally linked in this family with the *C* allele but in this woman is recombined with the *A* allele from the father's other chromosome. The farther the marker is from the disease gene, the more recombinations will be found within a family. Markers even closer to the Huntington's gene were later found. Finally, in 1993, a genetic defect was identified as the CAG repeat sequence associated with most cases of Huntington's disease, as described in Chapter 3. Precisely how the CAG repeat causes Huntington's disease is not yet known. A similar approach was used to locate the genes responsible for other single-gene disorders such as PKU on chromosome 12 and fragile X mental retardation on the X chromosome.

Although linkage analysis of large pedigrees has been very effective for locating genes for single-gene disorders, it is less powerful when several genes are involved. Early attempts to use this approach to investigate linkage for major psychiatric disorders led in the late 1980s to well-publicized reports of linkage, but these reports were later retracted. Even for complex disorders like schizophrenia, for which many genes may be involved in the population, the traditional large-pedigree design might be successful in identifying some families in which a particular gene has a large effect. For example, schizophrenia researchers are currently focusing on the end of the short arm of chromosome 6 (6*p*24-22), because a pedigree study reported linkage there for about a quarter of families with multiple schizophrenic members (Straub et al., 1995).

Another linkage approach is less restrictive than the traditional large-pedigree design. It examines *allele sharing* for pairs of affected relatives, usually siblings, in many different families. For a particular DNA marker, a sibling pair can share 0, 1, or 2 of their parents' alleles. Half of sibling pairs are expected to share a particular marker allele. If both members of a sibling pair are affected, they should show greater than 50 percent allele sharing for markers in a region that harbors a gene that contributes to risk for schizophrenia. In other words, linkage is indicated in an affected sib-pair linkage study when markers in a region show allele sharing that exceeds the expected 50 percent. For example, a study of schizophrenic sibling pairs found 58 percent allele sharing in the chromosome 6 region (6*p*24-22) mentioned earlier (Schwab et al., 1995), although other affected sib-pair studies have not found evidence for linkage in this region.

Figure 6.5 Linkage between the Huntington's disease gene and a DNA marker at the tip of the short arm of chromosome 4. In this pedigree, Huntington's disease occurs for individuals who inherit a chromosome bearing the *C* allele for the DNA marker. A single individual shows a recombination (marked with an arrow) in which Huntington's disease occurred in the absence of the *C* allele. (From "DNA markers for nervous system disease" by J. F. Gusella et al. *Science, 225*, 1320–1326. Copyright © 1984. Used with permission of the American Association for the Advancement of Science.)

The affected sib-pair linkage design also was used to identify and replicate a linkage on the X chromosome for homosexuality inherited through the mother (Hamer et al., 1993; Hu et al., 1995). Allele sharing for markers in the $q28$ region of the X chromosome was 67 percent for homosexual brothers. In contrast, allele sharing for the heterosexual brothers of these homosexual men was only 22 percent, significantly less than the expected 50 percent allele sharing, again suggesting linkage between sexual preference and the X chromosome (see Chapter 11).

Linkage based on allele sharing can also be investigated for quantitative traits by correlating allele sharing for DNA markers with sibling differences on a quantitative trait. That is, a marker linked to a quantitative trait will show greater than expected allele sharing for siblings who are more similar for the trait. The sib-pair QTL linkage design was first used to identify and replicate a linkage for reading disability on chromosome 6 (6p21; Cardon et al., 1994).

SUMMING UP

Allelic association, which has been found for late-onset Alzheimer's disease, can also detect small QTL effects. Linkage studies of large family pedigrees can localize genes for single-gene disorders. Other linkage designs such as the affected sib-pair linkage design can identify linkages for more complex disorders, as has been reported for schizophrenia and homosexuality. Techniques are also available to detect QTL linkage for quantitative traits and have been used to show linkage with reading disability.

It is clear that the field of psychology is at the dawn of a new era in which behavioral genetic research is moving beyond the demonstration of the importance of heredity to the identification of specific genes (Plomin, 1995a). In clinics and research laboratories, psychologists of the future will routinely collect cells from the inside of the cheek and send them to a laboratory to extract DNA

CLOSE UP

Peter Propping is a professor of human genetics at the University of Bonn, Germany. He is also director of the Institute of Human Genetics in the Medical Faculty. He studied medicine at the Free University of Berlin, where he wrote a thesis on pharmacology and received an M.D. degree in 1970. In 1970 Propping also studied human genetics at the University of Heidelberg. From 1980 to 1983, he was a recipient of a Heisenberg fellowship for psychiatric genetics, for which he had a joint appointment at the Institute of Human Genetics in Heidelberg and the Central Institute of Mental Health in Mannheim. In 1984 he was appointed chair of human genetics at the University of Bonn. Propping has contributed to various fields of medical genetics and the genetic analysis of human neural function and mental disorders. He examined the genetic control of ethanol action on the electroencephalogram and its implications for the etiology of alcoholism, the evolution of neuroreceptors, and variation in human neuroreceptor genes. Since 1990, his group has conducted association and linkage studies in complex disorders, particularly in manic-depressive disorder.

BOX 6.2

Genetic Counseling

Genetic counseling is an important interface between the behavioral sciences and genetics. Genetic counseling goes well beyond simply conveying information about genetic risks and burdens. It helps individuals come to terms with the information by dispelling mistaken beliefs and allaying anxiety in a nondirective manner that aims to inform rather than to advise.

Until recently, most genetic counseling was requested by parents who had an affected child and were concerned about risk for other children. Now, genetic risk is often assessed directly by means of DNA testing. As more genes are identified for disorders, genetic counseling is increasingly involved in issues related to prenatal diagnoses, prediction, and intervention. This new information will create new ethical dilemmas. Huntington's disease provides a good example. If you had a parent with the disease, you would have a 50 percent chance of developing the disease. However, with the discovery of the gene responsible for Huntington's disease, it is now possible to diagnose in almost all cases whether a fetus or an adult will have the disease. Would you want to take the test? It turns out that the majority of people at risk choose *not* to take the test, largely because there is as yet no cure (Tyler, Ball, & Crawford, 1992). If you did take the test, the results are likely to affect knowledge of risk for your relatives. Do your relatives have the right to know, or is their right not to know more important?

These are difficult questions that are only now being considered. One generally accepted rule is that informed consent is required for testing; moreover, children should not be tested before they become adults, unless a treatment is available (Morris & Harper, 1991). Another increasingly important problem concerns the availability of genetic information to employers and insurance companies (Harper, 1992).

These issues are most pressing for single-gene disorders like Huntington's disease, in which a single gene is necessary and sufficient to develop the disorder. In some behavioral disorders, single genes are important, for example, in fragile X mental retardation. For these disorders, the same issues will emerge. For most behavioral disorders, however, genetic risks will involve QTLs that are probabilistic risk factors rather than certain causes of the disorder. Even for big QTL effects, such as apolipoprotein E and late-onset Alzheimer's disease, the issues are somewhat less acute because of the lack of predictive certainty.

It is most important that genetic counseling be nondirective and emphasize the rights of individuals to make their own decisions. Counseling must avoid the type of eugenics practiced by the German Third Reich (Müller-Hill, 1988). Despite the ethical dilemmas that arise with the new genetic information, it should also be emphasized that these findings have the potential for profound improvements in the prediction, prevention, and treatment of diseases.

and to genotype specific genes associated with particular psychological traits. This is already happening for apolipoprotein E and late-onset dementia and for fragile X mental retardation in males.

As is the case with most important advances, identifying genes for behavior will raise new ethical issues (e.g., Wright, 1990) that are already affecting genetic counseling (see Box 6.2). Identifying genes for single-gene disorders has led to concerns about employment and insurance discrimination (e.g., Bishop & Waldholz, 1990; Nelkin & Tancredi, 1989). The benefits of identifying genes for understanding the etiology of behavioral disorders and dimensions seem likely to far outweigh the potential abuses. Moreover, forewarned of problems and solutions that have arisen with single-gene disorders, we should be fore-armed as well to prevent abuses as genes that influence complex behavioral traits are discovered.

Summary

Although much more quantitative genetic research is needed, one of the most exciting directions for genetic research in psychology involves harnessing the power of molecular genetics to identify specific genes responsible for the wide-spread influence of genetics on behavior. Two major strategies are allelic associa-tion and linkage. Allelic association is simply a correlation between an allele and a trait for individuals in a population. Linkage is like an association within fami-lies, tracing the coinheritance of a DNA marker and a disorder within families.

Animal studies provide powerful designs to identify specific genes. Many associations between single genes and behavior have been identified. In addi-tion to naturally occurring genetic variation, many behavioral mutants have been identified from studies of chemically induced mutations in organisms as diverse as single-celled organisms, roundworms, fruit flies, and mice. In mice, mutations that knock out the gene's functioning can be targeted to a particular gene. Associations between such single-gene mutations and behavior generally underline the point that disruption of a single gene can drastically affect behav-ior that is normally influenced by many genes.

Because it is often difficult to know which of thousands of genes to investi-gate for a possible association with behavior, linkage studies are used to scan the genome for linkage with markers. Experimental crosses of inbred strains are powerful tools for identifying linkages, even for complex, quantitative traits for which many genes are involved. Such quantitative trait loci (QTLs) have been identified for several behaviors in mice, such as fearfulness and responses to drugs.

Genes associated with behavior in mice can be used in human studies, because nearly all mouse genes are similar to human genes. Moreover, chro-mosomal regions linked to behavior in mice can be used as candidate regions in human studies because corresponding regions can be identified in the two

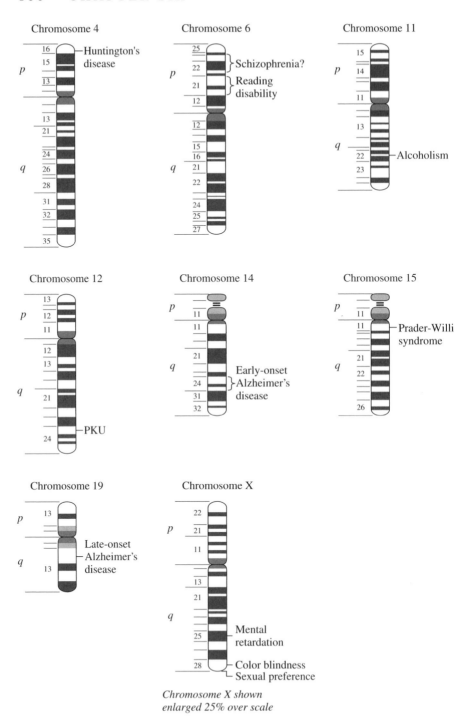

Figure 6.6 Some associations and linkages for human behavioral disorders and dimensions.

species. For complex human behaviors, several associations and linkages have been reported. Those mentioned in this chapter are illustrated in Figure 6.6.

Psychology is at the dawn of a new era. Many genes responsible for the widespread genetic influence on behavior will be identified and routinely used to assess genetic risk within the next decade.

The rest of the text presents what is known about quantitative genetic and molecular genetic research in psychology: cognitive disabilities (Chapter 7); general cognitive ability (Chapter 8); specific cognitive abilities (Chapter 9); psychopathology (Chapter 10); personality and personality disorders (Chapter 11); health psychology and aging (Chapter 12); evolutionary psychology (Chapter 13); and experience (Chapter 14). Chapter 15 predicts behavioral genetics in the twenty-first century.

CHAPTER SEVEN

Cognitive Disabilities

In an increasingly technological world, cognitive disabilities—such as mental retardation, learning disabilities, and dementia—are important liabilities. More is known about specific genetic causes of cognitive disabilities than about any other area of behavioral genetics. Many single genes and chromosomal abnormalities that contribute to mental retardation are known. Although most of these are rare, together they account for a substantial amount of mental retardation, especially severe retardation (often defined as IQ scores below 50). (The average IQ in the population is 100, with a standard deviation of 15, which means that about 95 percent of the population has IQ scores between 70 and 130.) Less is known about mild mental retardation (IQs from 50 to 70), even though it is much more common. Specific types of cognitive disabilities, especially reading disability and dementia, are foci of current research because genes linked to these disabilities have recently been identified.

In this chapter and later chapters on disorders, we follow the terminology of the American Psychiatric Association's *Diagnostic and Statistical Manual of Mental Disorders-IV* (DSM-IV), which is consistent with the *International Classification of Diseases-10* (ICD-10). For example, DSM-IV defines mental retardation in terms of subaverage intellectual functioning, onset before 18 years, and related limitations in adaptive skills. Four levels of retardation are considered: mild (IQ 50 to 70), moderate (IQ 35 to 50), severe (IQ 20 to 35), and profound (IQ below 20). About 85 percent of all individuals with retardation are classified as mildly retarded, and most can live independently and hold a job. Moderately retarded people usually have good self-care skills and can carry on simple conversations. Although they generally do not live independently and, in the past, were usually institutionalized, today they often live in the community in special residences or with their families. Severely retarded individuals can learn some self-care skills and understand language but have trouble speaking and require considerable supervision. Profoundly retarded people may understand a simple communication but usually cannot speak; they remain institutionalized.

Although these distinctions in levels of retardation are useful, there are two problems. First, the American Association on Mental Retardation argues that these DSM-IV distinctions rely too heavily on IQ and not sufficiently on adaptive skills. Because genetic research on mental retardation has also focused on IQ, the origins of differences in adaptive skills among retarded people have yet to be investigated. Second, genetic research provides little support for the existence of four levels. For example, as described below, most single-gene and chromosomal causes of mental retardation depress IQ, but a wide range of IQs that encompasses several of these levels remains (Rutter, Simonoff, & Plomin, 1996).

Mental Retardation: Quantitative Genetics

In psychology, it is now widely accepted that genetics substantially influences general cognitive ability; this belief is based on evidence presented in Chapter 8. Although one might expect that low IQ scores are also due to genetic factors, this conclusion does not necessarily follow. For example, mental retardation can be caused by environmental trauma, such as birth problems, nutritional deficiencies, and head injuries. Given the importance of mental retardation, it is surprising that no twin or adoption studies of mental retardation have been reported. Nonetheless, one sibling study suggests that moderate and severe mental retardation may be due largely to nonheritable factors. In a study of over 17,000 white children, 0.5 percent were moderately to severely retarded (Nichols, 1984). As shown in Figure 7.1, the siblings of these retarded children were not retarded. The siblings' average IQ was 103 and ranged from 85 to 125. In other words, moderate to severe mental retardation showed no familial resemblance, a finding implying that mental retardation is not heritable. Although most moderate and severe mental retardation may not be inherited from generation to generation, it is often caused by noninherited DNA events, such as new gene mutations and new chromosomal abnormalities, as discussed in the following sections.

In contrast, siblings of mildly retarded children tend to have lower than average IQ scores (Figure 7.1). The average IQ for these siblings of mildly retarded children (1.2 percent of the sample) was only 85. This important finding, that mild mental retardation is familial and moderate and severe retardation are not familial, also emerged from the largest family study of mild mental retardation, which considered 80,000 relatives of 289 mentally retarded individuals (Reed & Reed, 1965). This family study showed that mild mental retardation is very strongly familial. If one parent is mildly retarded, the risk for retardation in their children is about 20 percent. If both parents are retarded, the risk is nearly 50 percent. Familial resemblance for moderate retardation is much lower (Johnson, Ahern, & Johnson, 1976).

Although mild mental retardation runs in families, it could do so for reasons of nurture rather than nature. Twin and adoption studies of mild mental

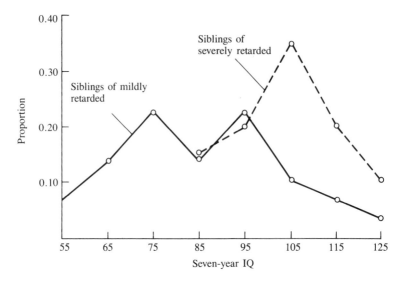

Figure 7.1 Siblings of mildly retarded children tend to have lower than average IQs. In contrast, siblings of severely retarded children tend to have normal IQs. These trends suggest that mild retardation is familial but severe retardation is not. (From Nichols, 1984.)

retardation are needed to disentangle the relative roles of nature and nurture. Two twin studies of unselected samples of twins have been used to investigate the origins of low IQ (Petrill et al., in press-c; Saudino et al., 1994). Both studies found that low IQ is at least as heritable as IQ in the normal range, suggesting that heritable factors might contribute to the familial resemblance found for mild mental retardation. However, because few subjects in these studies were actually retarded, with IQs below 70, the jury is still out on this basic question about the origins of mild mental retardation.

In studies considering disorders, the co-occurrence of several disorders in the same person is a common finding. For example, mental retardation co-occurs with medical problems and with behavioral problems. About a third of children with mild mental retardation have medical problems. Although it has been assumed that the medical problems caused the retardation, another possibility is that genetic factors account for both the medical problems and the retardation. That this might be so is suggested by studies that show that children with both mild mental retardation and medical problems are more likely to have parents who have cognitive disabilities (Bregman & Hodapp, 1991). Similarly, about half of children with mild mental retardation also have behavioral problems. Nothing is known as yet about whether the behavioral problems follow from the retardation or whether there are genetic factors at work that affect both retardation and behavioral problems.

Mental Retardation: Single-Gene Disorders

More than 100 genetic disorders, most extremely rare, include mental retardation among their symptoms (Wahlström, 1990). The classic disorder is PKU, discussed in Chapter 2; and the newest discovery is fragile X mental retardation, mentioned in Chapter 3. We will first discuss these two single-gene disorders, which are known for their effect on mental retardation. Then we will mention three single-gene disorders that also contribute to mental retardation, even though their primary defect is something other than cognitive ability.

Until recently, much of what was known about these disorders, as well as the chromosomal disorders described in the next section, came from studies of patients in institutions. These earlier studies painted a gloomy picture. But more recent systematic surveys of entire populations show a wide range of individual differences, including individuals whose cognitive functioning is in the normal range. These genetic disorders shift the IQ distribution downward, but a wide range of individual differences remains.

Phenylketonuria

The most well known inherited form of moderate mental retardation is phenylketonuria (PKU), which occurs in about 1 in 10,000 births. In the untreated condition, IQ scores are often below 50, although the range includes some near-normal IQs. As mentioned in Chapter 2, PKU is a single-gene recessive disorder that previously accounted for about 1 percent of severely retarded individuals in institutions. PKU is the best example of the usefulness of finding genes for behavior. Knowledge that PKU is caused by a single gene led to an understanding of how the genetic defect causes mental retardation. Mutations in the gene that produces the enzyme phenylalanine hydroxylase (PAH) create an enzyme that does not work properly, that is, it cannot break down phenylalanine. Phenylalanine comes from food, especially red meats; and if it cannot be broken down, its metabolic products build up and damage the developing brain.

Although PKU is inherited as a simple single-gene recessive disorder, the molecular genetics of PKU is not so simple. The PAH gene, which is on chromosome 12, shows dozens of different mutations, some of which cause milder forms of retardation. Similar findings have emerged for many classic single-gene disorders. Different mutations can do different things to the gene's product, and this variability makes understanding the disease process more difficult. It also makes DNA diagnosis more difficult, because DNA markers that identify all the mutations have to be used.

Because of fears about how genetic information will be used, it is important to note that knowledge about the single-gene cause of PKU did not lead to

sterilization programs or genetic engineering. Instead, an environmental intervention—a diet low in phenylalanine—successfully prevented the development
of retardation. Nevertheless, PKU individuals still tend to have a slightly lower
IQ, especially when the low phenylalanine diet has not been strictly followed
(Smith et al., 1991). It is generally recommended that the diet be maintained as
long as possible, at least through adolescence. PKU women must return to a
strict low phenylalanine diet before becoming pregnant in order to prevent
their high levels of phenylalanine from damaging the fetus (Medical Research
Council, 1993).

Fragile X Syndrome

As mentioned in Chapters 1 and 3, fragile X is the second most common cause
of mental retardation after Down syndrome. It is twice as common in males as
in females. The frequency of fragile X is usually given as 1 in 1250 males and
1 in 2500 females, although a recent study suggests a much lower frequency
of 1 in 5000 males (Murray et al., 1996). At least 2 percent of the male residents
of schools for mentally retarded persons have the fragile X syndrome. Most cases
of fragile X males are moderately retarded; but many are only mildly retarded,
and some have normal intelligence. Only about one-half of girls with fragile X
are affected, because one of the two X chromosomes for girls inactivates, as
mentioned in Chapter 4.

For fragile X males, IQ declines after childhood. In addition to generally
lower IQ, about three-quarters of fragile X males show large, often protruding
ears, a long face with a prominent jaw, and, after adolescence, enlarged testicles.
They also often show unusual speech, poor eye contact (gaze aversion), and
flapping movements of the hand. Language difficulties range from an absence
of speech to mild communication difficulties. Often observed is a speech pattern called "cluttering" in which talk is fast, with occasional garbled, repetitive,
and disorganized speech. Comprehension of language is often better than
expression and better than expected on the basis of IQ scores (Dykens, Hodapp,
& Leckman, 1994; Hagerman, 1995). Parents frequently report overactivity,
impulsivity, and inattention.

Until the gene for fragile X was found in 1991, its inheritance was puzzling.
It did not conform to a simple X-linkage pattern because its risk increased
across generations (a phenomenon called *anticipation*). The fragile X syndrome
is caused by an expanded triplet repeat (CGG) on the X chromosome ($Xq27.3$).
The disorder is called fragile X because the many repeats cause the chromosome to be fragile at that point and to break during laboratory preparation of
chromosomes. As mentioned in Chapter 3, parents who inherit X chromosomes
with a normal number of repeats (6 to 54 repeats) can produce eggs or sperm
with an expanded number of repeats (up to 200 repeats), called a *premutation*.
This premutation does not cause retardation in their offspring, but it is unstable

and often leads to much greater expansions (more than 200 repeats) in later generations, especially when the premutated X chromosome is inherited through the mother. The full mutation causes fragile X in almost all males but in only half of the females. The risk that a premutation will expand to a full mutation increases over four generations from 5 to 50 percent. A recent discovery is that pure CGG repeats are more likely to expand than CGG repeats that are interrupted by an occasional T base. This finding accounts for some of the variability of transmission of the disorder.

The triplet repeat is adjacent to a gene (*FMR-1*) and prevents that gene from being transcribed. It is not yet known what *FMR-1* does, although it is expressed in the brain. Its protein product appears to bind RNA, which means that the gene product probably regulates expression of other genes (Siomi et al., 1994). A mouse with a knock-out version of *FMR-1* shows learning deficits; and studies of this experimental model may shed light on the functioning of *FMR-1* (Dutch-Belgian Fragile X Consortium, 1994). There is hope that once the protein is understood it can be replaced, because there is evidence that just a little of the protein goes a long way toward ameliorating the effect (Hagerman et al., 1994).

Two other, much rarer, triplet repeats also have been found to cause fragile X syndrome. Molecular genetic advances in this area are fast paced (Warren & Nelson, 1994).

Other Single-Gene Disorders

Many other single-gene disorders, whose primary defect is something other than retardation, also show effects on IQ. Three of the most common disorders are Duchenne's muscular dystrophy, Lesch-Nyhan syndrome, and neurofibromatosis.

Duchenne's muscular dystrophy   This syndrome is an X-linked ($Xp21$) recessive disorder that occurs in 1 in 3500 males, with about a third of the cases being due to new mutations. It is a neuromuscular disorder that causes progressive wasting of muscle tissue and usually leads to death by age 20 years, as a result of respiratory or cardiac failure. Mutations in the dystrophin gene also affect neurons in the brain. The average IQ of males with Duchenne's muscular dystrophy is 85. Verbal abilities are more severely impaired than nonverbal abilities, although effects on cognitive ability are highly variable (Emery, 1993).

Lesch-Nyhan syndrome   This syndrome is another X-linked ($Xq26$–27) recessive disorder, with an incidence of about 1 in 20,000 male births. The gene codes for an enzyme (hypoxanthine phosphoribosyltransferase, HPRT) involved in the production of nucleic acids. The most striking feature of this disorder is compulsive self-injurious behavior, reported in over 85 percent of cases (Anderson & Ernst, 1994). Most typical is lip and finger biting, which is often so severe that it leads to extensive loss of tissue. The self-injurious behavior begins as early as infancy or as late as adolescence. The behavior is painful to the individual, yet uncontrollable. In terms of cognitive disability, most individuals have moderate or severe learning difficulties, and speech is usually impaired. Memory for both recent and past events appears to be unaffected. One of the first transgenic knock-out mouse strains created involved the HPRT gene responsible for Lesch-Nyhan (Kuehn et al., 1987). Knocking out the HPRT gene seemed to have no effect on the brain or behavior of the mice, until it was found that another gene compensated for the missing HPRT enzyme. When the other gene was inhibited with drugs, the HPRT knock-out mice showed self-mutilation similar to that associated with the Lesch-Nyhan syndrome (Wu & Melton, 1993). It is not known whether the mice also have learning deficits.

Neurofibromatosis Type 1 (NF1) First described more than a century ago, NF1 involves skin tumors and tumors in nerve tissue. Its symptoms are highly variable, beginning with chocolate-colored spots that appear in early childhood. The tumorous lumps are not cancerous and primarily cause cosmetic disfigurement, although the tumors can cause more serious problems if they compress a nerve. NF1 is caused by a dominant allele on chromosome 17 ($17q11.2$) and is quite common (about 1 in 3000 births). The allele, which is thought to be involved in tumor suppression, is inherited from the father in more than 90 percent of cases. However, the gene has a very high mutation rate; approximately half of all cases are new mutations. Only recently have cognitive abilities of NF1 individuals been examined. It was found that the majority of affected individuals have IQ scores in the low to average range (Ferner, 1994). About half have learning difficulties, although verbal abilities tend to be higher than nonverbal abilities. Another medical disorder with effects on cognitive abilities is tuberous sclerosis.

SUMMING UP

Although little is known about the quantitative genetics of mental retardation, many single-gene disorders that cause mental retardation have been discovered. The classic disorder is PKU, caused by a gene on chromosome 12. The newest discovery is fragile X mental retardation, caused by a triplet repeat on the X chromosome that expands over several generations and prevents a nearby gene from being transcribed. More than 100 other single-gene disorders contribute to mental retardation, such as Duchenne's muscular dystrophy, Lesch-Nyhan syndrome, and neurofibromatosis Type 1.

Some of the most common single-gene causes of mental retardation are summarized in Figure 7.2. The figure shows the average IQ of individuals with these disorders, but it should be remembered that the range of cognitive functioning is very wide for these disorders. The defective allele shifts the IQ distribution downward, but a wide range of individual IQs remains.

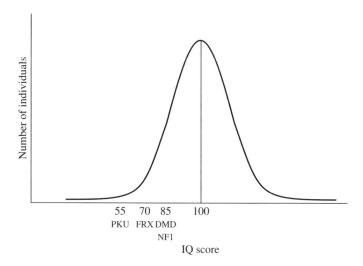

Figure 7.2 The most common single-gene causes of mental retardation. Despite the lower average IQs, a wide range of cognitive functioning is found.

Mental Retardation: Chromosomal Abnormalities

Even more common than single-gene causes of mental retardation are chromosomal abnormalities. Most well known are abnormalities that involve an entire extra chromosome like *Down syndrome* (three copies of chromosome 21) or a missing chromosome like *Turner's syndrome* (females with only one X chromosome). As the resolution of chromosomal analysis becomes finer, more minor abnormalities, especially deletions, are likely to be found.

Angelman syndrome, mentioned in Chapter 3 as an example of gametic imprinting, involves a small deletion in chromosome 15 (Zori et al., 1992). When inherited from the mother, this rare disorder (1 in 25,000 births) causes moderate retardation. When inherited from the father (1 in 15,000 births), the same (or very similar) chromosomal deletion causes *Prader-Willi syndrome*, which most noticeably involves overeating and temper outbursts but also leads to multiple learning difficulties and an IQ in the low normal range (Donaldson et al., 1994). Most of the time, this chromosomal abnormality occurs spontaneously. For example, the risk to siblings of probands with the disorder is low, less than 1 percent.

Another example is *Williams syndrome*, with an incidence of about 1 in 25,000 births. It involves a small deletion from chromosome 7. Most cases are spontaneous. Williams syndrome involves disorders of connective tissue that lead to growth retardation and multiple medical problems. Mental retardation is common, and most affected individuals have learning difficulties that require special schooling. As adults, most are unable to live independently. The disorder is of special interest to psychologists because expressive language skills have been reported to be superior to other cognitive abilities (Bellugi, Wang, & Jernigan, 1994).

New techniques have been developed to identify micro-deletions of chromosomes. Results of studies using these techniques suggest that some cases of mental retardation may be due to micro-deletions (Flint et al., 1995b).

The next sections have descriptions of the classic chromosomal abnormalities that involve mental retardation. Chromosomes and chromosomal abnormalities, such as nondisjunction and the special case of abnormalities involving the X chromosome, were introduced in Chapters 3 and 4.

Down Syndrome

As described in Chapter 3, Down syndrome, caused by a trisomy of chromosome 21, is the single most important cause of mental retardation, occurring in about 1 in 1000 births. Down syndrome accounts for about 10 percent of institutionalized mentally retarded individuals. It is so common that its general features are probably familiar to everyone (see Figure 7.3). Although more than

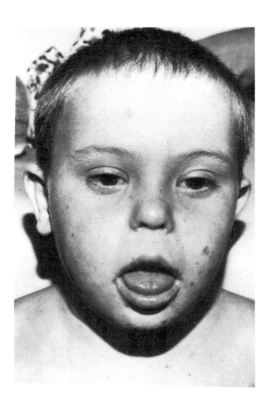

Figure 7.3 Child with Down syndrome.

300 abnormal features have been reported for Down syndrome children, a handful of specific physical disorders are diagnostic because they occur so frequently. These features include increased neck tissue, muscle weakness, speckled iris of eye, open mouth, and protruding tongue. Some symptoms, such as increased neck tissue, become less prominent as the child grows. Other symptoms, such as mental retardation and short stature, are noted only as the child grows. About two-thirds of affected individuals have hearing deficits, and one-third have heart defects. As first noted by Langdon Down, who identified the disorder in 1866, children with Down syndrome appear to be obstinate but otherwise many are generally amiable. Although it might be assumed that these diverse effects come from overexpressing genes on chromosome 21 because there are three copies of the chromosome, it is possible that having so much extra genetic material creates a general instability in development (Shapiro, 1994).

The most striking feature of Down syndrome is mental retardation (Cicchetti & Beeghly, 1990). The average IQ among children with Down syndrome is 55, with only the top 10 percent falling within the lower end of the normal range of IQs. By adolescence, language skills are generally at about the level of a 3-year-old child. Most individuals with Down syndrome who reach the age of 45 years suffer from the cognitive decline of dementia, which was an early clue suggesting that a gene related to dementia might be on chromosome 21 (see later).

As is the case for all single-gene and chromosomal effects on mental retardation, affected individuals show a wide range of IQs. This variation can come about in two ways. First, the variability in IQs may be due to the same biological processes involved in the average lowering of IQ that is associated with the genetic or chromosomal disorder, which seems to be the case for Down syndrome (Carr, 1995) and for autism (Le Couteur et al., 1996). It is well demonstrated that medical disorders show substantial variable expression of genetic effects. For example, tuberous sclerosis may manifest either as subtle skin changes that require an expert to detect them or as gross brain swellings associated with epilepsy and mental handicap. The reasons for this variable expression remain poorly understood, but they are likely to apply to intelligence as well. Alternatively, it is possible that the variability in IQ among individuals with a genetic or chromosomal disorder may be due to the same genetic and environmental factors that contribute to variability of IQ in unselected samples. The presence of a major handicapping disorder does not make individuals immune from effects that apply in the general population, although, in the case of severe disorders, the effects may be attenuated because of the major biological abnormality.

In Chapter 3, Down syndrome was used as an example of an exception to Mendel's laws because it does not run in families. Because individuals with Down syndrome do not reproduce, most cases are created anew each generation by nondisjunction of chromosome 21. Another important feature of Down syndrome is that it occurs much more often in women giving birth later in life, for reasons explained in Chapter 3.

Sex Chromosome Abnormalities

Extra X chromosomes also cause cognitive disabilities, although the effect is highly variable. In males, an extra X chromosome (XXY) causes *Klinefelter's syndrome*. It occurs in about 1 in 750 male births. The major problems involve low testosterone levels after adolescence, leading to infertility, small testes, and breast development (see Figure 7.4). Early detection and hormonal therapy are important to alleviate the condition, although infertility remains. Males with Klinefelter's syndrome also have a somewhat lower than average IQ, and most have speech and language problems and poor school performance (Mandoki et al., 1991). In females, extra X chromosomes (called *triple X syndrome*) occur in about 1 in 1000 female births. Females with triple X show an average IQ of

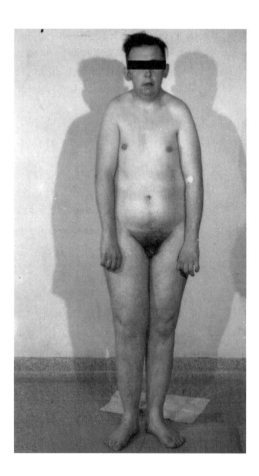

Figure 7.4 Man with XXY Klinefelter's syndrome.

about 85, lower than for Klinefelter's males (Bender, Linden, & Robinson, 1993). Verbal scores are lower than nonverbal scores and many require speech therapy. For both XXY and XXX individuals, head circumference at birth is smaller than average, a feature suggesting that the cognitive deficits may be prenatal in origin (Ratcliffe, 1994).

In addition to having an extra X chromosome, it is possible for males to have an extra Y chromosome (XYY) and for females to have just one X chromosome (XO, called *Turner's syndrome*). XYY males, about 1 in 1000 male births, are taller than average after adolescence and have normal sexual development. Although XYY males have fewer cognitive problems than XXY males, about half have speech difficulties, often requiring speech therapy, and language and reading problems. Juvenile delinquency is also associated with XYY. The XYY syndrome was the center of a furor in the 1970s, when it was suggested that such males are more violent, a suggestion possibly triggered by the notion of a "supermale" with exaggerated masculine characteristics caused by their extra Y chromosome. Although little support has been found for this hypothesis, this uproar made it much more difficult to conduct population surveys on chromosomal abnormalities.

Turner's syndrome females (XO) occur in about 1 in 2500 births, although 99 percent of XO fetuses miscarry. The main problems are short stature and abnormal sexual development, and infertility is common. Puberty rarely occurs without hormone therapy; and even with therapy, infertility remains as a result of nonfunctioning ovaries. Although verbal IQ is about normal, performance IQ is lower, about 90, after adolescence (El Abd, Turk, & Hill, 1995).

SUMMING UP

Small deletions of chromosomes can result in mental retardation, as in Angelman syndrome, Prader-Willi syndrome, and Williams syndrome. The most common cause of mental retardation is Down syndrome, which is due to the presence of three copies of chromosome 21. Mild mental retardation generally occurs in individuals with extra X chromosomes: XXY Klinefelter's males and XXX females. Some cognitive disability appears in XYY males with an extra Y chromosome and in Turner's females with a missing X chromosome. Although the average cognitive ability of these groups is generally lower than the average for the entire population, individuals in these groups show a wide range of cognitive functioning.

Figure 7.5 illustrates the most common chromosomal causes of mental retardation. Again, it should be emphasized that there is a wide range of cognitive functioning around the average IQ scores shown in the figure.

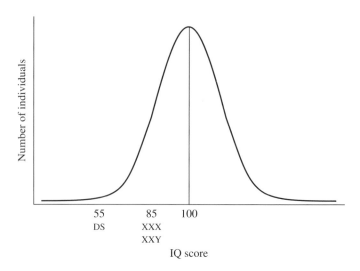

Figure 7.5 The most common chromosomal causes of mental retardation. Despite the low average IQ for Down syndrome, a wide range of cognitive functioning is found.

Learning Disorders

Many children have difficulty learning to read. For some, specific causes can be identified, such as mental retardation, brain damage, sensory problems, and deprivation. However, many children without such problems find it difficult to read. In fact, reading is the primary problem in about 80 percent of children with a diagnosed learning disorder. Children with specific reading disorder (also known as *dyslexia*) read slowly, and often with poor comprehension. When reading aloud, they perform poorly. In addition to reading disorder, DSM-IV recognizes mathematics disorder and disorder of written expression. Although no behavioral genetic research has as yet been reported for the latter disorder, results from a small twin study of mathematics disorder suggest moderate genetic influence (Alarcón et al., in press) and substantial genetic overlap with reading disability (Light & DeFries, 1995). These disorders are diagnosed on the basis of performance substantially below what would be expected for the child's general cognitive ability.

Family studies have shown that reading disability runs in families. For example, the largest family study included 1044 individuals in 125 families with a reading-disabled child and 125 matched control families (DeFries, Vogler, & LaBuda, 1986). Siblings and parents of the reading-disabled children performed significantly worse on reading tests than did siblings and parents of control children. Earlier twin studies suggested that familial resemblance for reading disability involves genetic factors (Bakwin, 1973; Decker & Vandenberg, 1985).

David W. Fulker is a professor in the Institute for Behavioral Genetics at the University of Colorado and currently on leave at the Institute of Psychiatry in London. He is principal investigator on several grants that support the Colorado Adoption Project and the Colorado Twin Study. These are large collaborative longitudinal studies of behavioral development of children followed since birth. He also leads a family study of young males engaged in severe substance abuse. Other collaborations include genetic studies of mice and rats that are designed to identify quantitative trait loci (QTLs) for alcohol sensitivity, drug responses, and memory. He did his undergraduate work in psychology at the University of London and received a master's degree in statistical genetics at the University of Birmingham, where he also completed his doctorate. Fulker has taught at the Universities of Birmingham, London, and Colorado. His research has focused on the development of statistical methods for analyzing behavioral genetic data on humans and animals. Most recently he has developed novel methods for detecting QTLs in humans and has applied these methods to identify QTLs for reading disability and sexual orientation.

Even though one twin study showed little evidence of genetic influence (Stevenson et al., 1987), the largest twin study confirmed genetic influence on reading disability (DeFries, Fulker, & LaBuda, 1987). For more than 200 twin pairs in which at least one member of the pair was reading disabled, twin concordances were 66 percent for identical twins and 43 percent for fraternal twins, a result suggesting moderate genetic influence.

As part of this twin study, a new method was developed to estimate the genetic contribution to the mean difference between the reading-disabled probands and the mean reading ability of the population. This type of analysis is called DF extremes analysis after its creators (DeFries & Fulker, 1985, 1988). As described in Box 7.1, DF extremes analysis for reading disability estimates that about half of the mean difference between the probands and the population is heritable. The analysis also suggests that the heritability of reading disability may differ from that of individual differences in reading ability, a conclusion that would imply that different genetic factors affect reading disability and reading ability.

Various modes of transmission have been proposed for reading disability, especially autosomal dominant transmission and X-linked recessive transmission.

BOX 7.1

DF Extremes Analysis

The genetic and environmental causes of individual differences throughout the range of variability in a population can differ from the causes of the average difference between an extreme group and the rest of the population. For example, finding genetic influence on individual differences in reading ability in an unselected sample (Chapter 9) does not mean that the average difference in reading ability between reading disabled individuals and the rest of the population is also influenced by genetic factors. Alternatively, it is possible that reading disability represents the extreme end of a continuum of reading ability, rather than a distinct disorder. That is, reading disability might be quantitatively rather than qualitatively different from the normal range of reading ability. DF extremes analysis, named after its creators (DeFries & Fulker, 1985, 1988), addresses these important issues concerning the links between the normal and abnormal.

DF extremes analysis takes advantage of quantitative scores of the relatives of probands rather than just assigning a dichotomous diagnosis to the relatives and assessing concordance for the disorder. The following figure shows hypothetical distributions of reading performance of an unselected sample of twins and of the identical (MZ) and fraternal (DZ) co-twins of probands (P) with reading disability (DeFries, Fulker, & LaBuda, 1987). The mean score of the probands is $\overline{P}$. The differential regression of both the MZ and the DZ co-twin means ($\overline{C}_{MZ}$ and $\overline{C}_{DZ}$ toward the mean of the unselected population (μ) provides a test of genetic influence. That is, to the extent that reading deficits of probands are heritable, the quantitative reading scores of identical co-twins will be more similar to that of the probands than will the scores of fraternal twins. In other words, the mean reading score of identical co-twins will regress less far back toward the population mean than will that of fraternal co-twins.

The results for reading disability are similar to those illustrated in the figure. The scores of the identical co-twins regress less far back toward the population mean than do those of the fraternal co-twins. This finding suggests that genetics contributes to the mean difference between the reading-disabled probands and the population. Twin *group correlations* provide an index of how far the co-twins regress toward the population mean. For reading disability, the twin group correlations are .90 for identical twins and .65 for fraternal twins. Doubling the difference between these group correlations suggests a *group heritability* of 50 percent, similar to the results of more sophisticated DF extremes analysis (DeFries & Gillis, 1993). In other words, half of the mean difference between the probands and the population is heritable. This is called "group heritability" to distinguish it from the usual heritability estimate, which refers to differences between individuals rather than to mean differences between groups.

DF extremes analysis is conceptually similar to the liability-threshold model described in Box 3.1. The major difference is that the threshold model assumes a

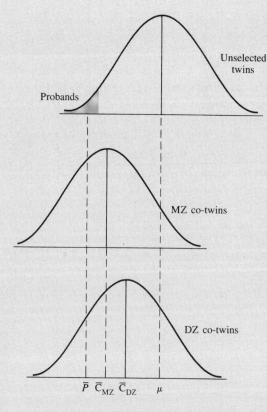

Probands

Unselected twins

MZ co-twins

DZ co-twins

$\bar{P}$ $\bar{C}_{MZ}$ $\bar{C}_{DZ}$ μ

continuous dimension even though it assesses a dichotomous disorder. The liability-threshold analysis converts dichotomous diagnostic data to a hypothetical construct of a threshold with an underlying continuous liability. In contrast, DF extremes analysis assesses rather than assumes a continuum. If all of the assumptions of the liability-threshold model are correct for a particular disorder, it will yield results similar to the DF extremes analysis to the extent that the quantitative dimension assessed underlies the qualitative disorder (Plomin, 1991). In the case of reading disability, a liability-threshold analysis of these twin data yields an estimate of group heritability similar to that of the DF extremes analysis.

DF extremes analysis is also useful for exploring genetic links between the normal and the abnormal. That is, to what extent do individuals with reading disability merely represent the lower tail of a normal distribution of individual differences in reading ability? One way to investigate this question is to compare group heritability for reading disability with individual heritability for reading ability. If these two types of heritability differ, it suggests that the genetics of reading disability are not the same as the genetics of reading ability. Group heritability of reading disability is about 50 percent, similar to the heritability of reading ability within the normal range of variation. However, a recent DF extremes analysis suggests that group heritability is less than that for reading ability within the affected group, a finding that in turn suggests the possibility that different genetic factors—a major gene, for example—may be involved in reading disability (DeFries & Alarcón, 1996).

In addition, DF extremes analysis can be used to examine the genetic and environmental origins of the co-occurrence between disorders. For example, hyperactivity (see Chapter 10) is often found among reading-disabled children. Multivariate DF extremes analysis suggests that genetic factors are largely responsible for this overlap in the two disorders (Light et al., 1995). In other words, the two disorders appear to share some genetic influences.

The autosomal dominant hypothesis takes into account the high rate of familial resemblance but fails to account for the fact that about a fifth of reading-disabled individuals do not have affected relatives. An X-linked recessive hypothesis is suggested when a disorder occurs more often for males than females, as is the case for reading disability. However, the X-linked recessive hypothesis does not work well as an explanation of reading disability. As described in Chapter 3, one of the hallmarks of X-linked recessive transmission is the absence of father-to-son transmission, because sons inherit their X chromosome only from their mother. Contrary to the X-linked recessive hypothesis, reading disability is transmitted from father to son as often as from mother to son. It is generally accepted that, like most complex disorders, reading disability is caused by multiple genes as well as by multiple environmental factors.

One of the most exciting findings in behavioral genetics in recent years is that the first quantitative trait locus (QTL) for a human behavioral disorder has been reported for reading disability (Cardon et al., 1994). As explained in Chapter 6, siblings can share 0, 1, or 2 alleles for a particular DNA marker. If siblings who share more alleles are also more similar for a quantitative trait

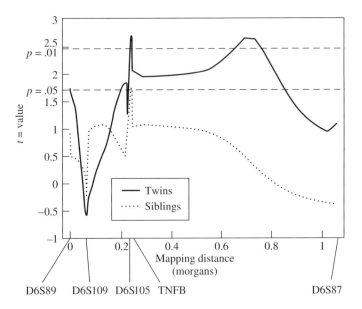

Figure 7.6 QTL linkage for reading disability in two independent samples in which at least one member of the pair is reading disabled: siblings (dotted line) and fraternal twins (solid line). D6S89, D6S109, D6S105, and TNFB are DNA markers in the 6p21 region of chromosome 6. The t-values are an index of statistical significance. The marker D6S105 is significant at the $p = .05$ level for siblings and at the $p = .01$ level for fraternal twins. (After Cardon et al., 1994; modified from DeFries & Alarcón, 1996; courtesy of Javier Gayan.)

such as reading ability, then QTL linkage is likely. QTL linkage analysis is much more powerful when one sibling is selected because of an extreme score on the quantitative trait. When one sibling was selected for reading disability, the reading ability score of the co-sibling was also lower when the two siblings shared alleles for markers on the short arm of chromosome 6 ($6p21$). These QTL linkage results for four DNA markers on $6p$ are depicted by the dotted line in Figure 7.6, showing significant linkage for the D6S105 marker. Significant linkage was also found for markers in this region in an independent sample of fraternal twins (see solid line in Figure 7.6) and in another study of siblings (Grigorenko et al., 1996). Although additional replications are needed before considering this linkage confirmed, the report is exciting because it suggests the possibility of finding QTLs for complex behavioral disorders.

Communication Disorders

Very little is known about the genetics of communication disorders. DSM-IV includes four types of communication disorders: expressive language (putting thoughts into words) disorder, mixed receptive (understanding the language of others) and expressive language disorder, phonological (articulation) disorder, and stuttering (speech interrupted by prolonged or repeated words, syllables, or sounds). Hearing loss, mental retardation, and neurological disorders are excluded.

Several family studies, examining communication disorders broadly, indicate that communication disorders are familial. For children with communication disorders, about a quarter of their first-degree relatives report similar disorders, compared with about 5 percent for the relatives of controls (Felsenfeld, 1994). Two twin studies of communication disorders found very substantial genetic influence, with average concordances of about 90 percent for identical twins and 50 percent for fraternal twins (Bishop, North, & Donlan, 1995; Lewis & Thompson, 1992). Multivariate genetic analysis (Appendix B) suggested that DSM-IV diagnostic categories may not reflect the genetic origins of these disorders. Expressive and receptive language disorders overlap genetically, whereas the genetic factors appear to be different for individuals with expressive disorders who have articulation problems and those who do not (Bishop et al., 1995).

Family studies of stuttering over the past 50 years have shown that about a third of stutterers have other stutterers in their families. Most of our recent knowledge comes from the Yale Family Study of Stuttering, which includes nearly 600 stutterers and more than 2000 of their first-degree relatives (Kidd, 1983). About 15 percent of the first-degree relatives reported that they had stuttered at some point in their life, about five times greater than the base rate of approximately 3 percent in the general population. Moreover, about half of the affected first-degree relatives were considered to be chronic stutterers. Interestingly, stutterers who had affected relatives did not have more severe stuttering. One small twin study of stuttering suggests that familial resemblance is heritable, with concordances of 77 percent for identical twins (17 pairs) and 32 percent for fraternal twins (13 pairs) (Howie, 1981). A large twin study that included a single item about stuttering in a lengthy questionnaire study also found evidence for substantial genetic influence (Andrews et al., 1991).

Dementia

Although aging is a highly variable process, as many as 15 percent of individuals over 80 years of age suffer severe cognitive decline known as dementia (Skoog et al., 1993). Prior to the age of 65 years, the incidence is less than 1 percent. Among the elderly, dementia accounts for more days of hospitalization than any other psychiatric disorder (Cumings & Benson, 1992).

At least half of all cases of dementia involve Alzheimer's disease (AD). AD occurs very gradually over many years, beginning with loss of memory for recent events, which affects many older individuals but is much more severe in individuals with AD. Irritability and difficulty in concentrating are also often noted. Memory gradually worsens to include simple behaviors, such as forgetting to turn off the stove or bath water, and wandering off and getting lost. Eventually— sometimes after 3 years, sometimes after 15 years—individuals with AD become bedridden. Biologically, AD involves extensive changes in brain nerve cells, including plaques and tangles (described later) that build up and result in death of the nerve cells. Although these plaques and tangles occur to some extent in most older people, they are usually restricted to the hippocampus. In individuals with AD, they are much more numerous and widespread.

Another type of dementia is the result of the cumulative effect of multiple small strokes in which blood flow to the brain becomes blocked, thus damaging the brain. This type of dementia is called multiple-infarct dementia (MID). (An infarct is an area damaged as a result of a stroke.) Unlike AD, MID is usually more abrupt and involves focal symptoms such as loss of language rather than general cognitive decline. Co-occurrence of AD and MID is seen in about a third of all cases. DSM-IV recognizes nine other kinds of dementia, such as dementia due to AIDS, to head trauma, and to Huntington's disease.

Like the situation for mental retardation, surprisingly little is known about the quantitative genetics of either AD or MID. Recent family studies of AD probands estimate risk to first-degree relatives of nearly 50 percent by the age of 85, when the data are adjusted for age of the relatives (McGuffin et al., 1994). However, this estimate may be inflated, because families are more likely to be included in such studies when multiple family members are affected. Until recently, the only twin study of dementia was one reported 40 years ago. That twin study, which did not distinguish AD and MID, found concordances of 43 percent for identical twins and 8 percent for fraternal twins, results suggesting moderate genetic influence (Kallmann, 1955). More recent twin studies of AD also found evidence for genetic influence, with identical and fraternal twin concordances of about 60 percent and 30 percent, respectively (Breitner et al., 1993).

In contrast, some of the most important molecular genetic findings for behavioral disorders have come from research on dementia (Pollen, 1993). Research has focused on a rare (1 in 10,000) type of Alzheimer's disease that appears before 65 years of age and shows evidence for autosomal dominant inheritance. Most of these early-onset cases are due to a gene on chromosome 14. In 1992, linkage was discovered simultaneously by three groups (e.g., St. George-Hyslop et al., 1992). In 1995, the offending gene (presenilin-1) was identified (Sherrington et al., 1995), although it is not yet known how the gene causes early-onset AD. A similar gene (presenilin-2) on chromosome 1 is also responsible for some cases and may also be related to late-onset AD (Hardy & Hutton, 1995). As is often the case, dozens of different mutations in the presenilin genes have been found, which will make screening difficult. A small percentage of early-onset cases are linked to chromosome 21.

The great majority of Alzheimer's cases occur after 65 years of age, typically in persons in their seventies and eighties. A major advance toward understanding late-onset Alzheimer's disease is the discovery of a strong allelic association with a gene (apolipoprotein E) on chromosome 19 (Corder et al., 1993). This gene has three alleles (confusingly called alleles 2, 3, and 4). The frequency of allele 4 is about 40 percent in individuals with Alzheimer's disease and 15 percent in control samples. This result translates to about a sixfold increased risk for late-onset Alzheimer's disease for individuals who have one or two of these alleles. There is some evidence that allele 2, the least common allele, may play a protective role (Corder et al., 1994). Finding QTLs that protect rather than increase risk for a disorder is an important direction for genetic research.

Apolipoprotein E is a QTL in the sense that, although it is a risk factor, allele 4 is neither necessary nor sufficient for developing dementia. For instance, nearly half of late-onset Alzheimer's patients do not have that allele. Assuming a liability-threshold model, investigators have estimated that allele 4 accounts for about 15 percent of the variance in liability (Owen, Liddell, & McGuffin, 1994).

SUMMING UP

Twin studies suggest genetic influence for learning disorders, especially reading disability, communication disorders, and dementia. DF extremes analysis suggests that genetic influence on reading disability may differ from genetic influence on the normal range of reading ability. The first QTL linkage for human behavioral disorders has been reported for reading disability. A QTL association has been established between late-onset Alzheimer's disease and allele 4 of the apolipoprotein E gene. Genes that are responsible for a rare early-onset form of Alzheimer's disease have also been identified.

Because apolipoprotein E is known for its role in transporting lipids throughout the body, its association with late-onset AD was puzzling at first. Each of the three polymorphisms are in exons and produce in the protein a structural change that involves a single amino acid substitution. Other roles for the gene became known, such as its increased production following injury to the nervous system, as in head injury, and, most important, its role in plaques (Hardy & Allsop, 1991).

A hypothesis has been proposed to explain the effects of these alleles on the buildup of nerve cell plaques characteristic of Alzheimer's disease (Hardy & Higgins, 1992). The plaques consist of a protein fragment called β-amyloid. When β-amyloid builds up, it somehow kills nerve cells. The type of apolipoprotein E coded by allele 4 binds more readily with β-amyloid, leading to amyloid deposits, which in turn lead to plaques and, eventually, to death of nerve cells. Allele 2 may block this buildup of β-amyloid. Apolipoprotein E is also responsible for the other main feature of the brains of individuals with AD. These are called neurofibrillary tangles and are dense bundles of abnormal fibers that appear in the cytoplasm of certain nerve cells. Allele 3 of apolipoprotein E appears to buffer nerve cells against such tangles.

Summary

More is known about specific genetic causes in cognitive disabilities than in any other area of behavioral genetics, although less is known about basic quantitative genetic issues. Surprisingly, no twin or adoption study has been reported for mental retardation, although family studies suggest that it is familial.

PKU has been known for decades as a single-gene recessive disorder that causes severe mental retardation if untreated, although PKU is rare (1 in 10,000). The recent discovery of fragile X mental retardation is especially important. It is the second most common cause of mental retardation (1 in several thousand males, half as common in females). It is caused by a triplet repeat (CGG) on the X chromosome that expands over several generations until it reaches more than 200 repeats, when it often causes moderate retardation in males. Other single genes known primarily for other effects also contribute to mental retardation, such as genes for Duchenne's muscular dystrophy, Lesch-Nyhan syndrome, and neurofibromatosis.

Chromosomal abnormalities play an important role in mental retardation. The most common cause of mental retardation is Down syndrome, caused by the presence of three copies of chromosome 21. Down syndrome occurs in about 1 in 1000 births and is responsible for about 10 percent of institutionalized mentally retarded individuals. Risk for mental retardation is also increased by having an extra X chromosome (XXY Klinefelter's males, XXX triple X females). An extra Y chromosome (XYY males) and a missing X chromosome (Turner's females) cause less retardation, although XYY males show some speech and language problems and Turner's females generally perform less well on non-verbal tasks such as spatial tasks. There is a wide range of cognitive functioning around the lowered average IQ scores found for all of these genetic causes of mental retardation.

Specific genes have also been localized for reading disability and for dementia. For reading disability, a replicated linkage on chromosome 6 has been reported, the first QTL linkage for human behavioral disorders. Quantitative genetic research indicates moderate genetic influence for reading disability. DF extremes analysis, which assesses genetic links between the normal and the abnormal, suggests that the etiology of reading disability may differ in part from that of individual differences in reading ability. Much less is known about the genetics of other learning disorders. Substantial genetic influence has been found for communication disorders.

For dementia, several genes, especially presenilin-1 on chromosome 14, have been found in most cases of early-onset Alzheimer's disease, which is a rare (1 in 10,000) form of Alzheimer's disease that occurs before 65 years of age and often shows pedigrees consistent with autosomal dominant inheritance. Late-onset Alzheimer's disease is very common, striking as many as 15 percent of individuals over 85 years of age. Apolipoprotein E is a gene strongly associated with late-onset Alzheimer's disease. Allele 4 of this gene increases risk about six-fold. Apolipoprotein E is a QTL in the sense that it is a probabilistic risk factor, not a single gene necessary and sufficient to develop the disorder.

CHAPTER EIGHT

General Cognitive Ability

General cognitive ability is one of the most well studied domains in behavioral genetics. Nearly all this genetic research is based on a model, called a *psychometric model*, that considers cognitive abilities to be organized hierarchically (Carroll, 1993) from specific tests to broad factors to general cognitive ability (often called g; Figure 8.1). There are hundreds of tests of diverse cognitive abilities. These tests measure several broad factors (specific cognitive abilities) such as verbal ability, spatial ability, memory, and speed of processing. Specific cognitive abilities are the focus of the next chapter.

These broad factors intercorrelate modestly. That is, in general, people who do well on tests of verbal ability tend to do well on tests of spatial ability. g, that which is in common among these broad factors, was discovered by Charles Spearman over 90 years ago, about the same time that Mendel's laws of inheritance were rediscovered (Spearman, 1904). The phrase *general cognitive ability*, or g, is preferable to the word *intelligence* because the latter has so many different meanings in psychology and in the general language (Jensen, 1994). A general text on intelligence is available (Brody, 1992).

Most people are familiar with intelligence tests, often called IQ tests (intelligence quotient tests). These tests typically assess several cognitive abilities and yield total scores that are reasonable indices of g. For example, the Wechsler tests of intelligence, widely used clinically, include ten subtests such as vocabulary, picture completion (indicating what is missing in a picture), analogies, and block design (using colored blocks to produce a design that matches a picture). In research contexts, g is usually derived by using a technique that is called factor analysis and weights tests differently, according to how much they contribute to g. This weight can be thought of as the average of a test's correlations with every other test.

A test's contribution to g is related to the complexity of the cognitive operations it assesses. More complex cognitive processes such as abstract reasoning

General cognitive
ability (*g*)

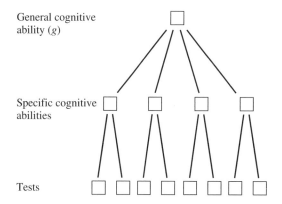

Specific cognitive
abilities

Tests

Figure 8.1 Hierarchical
model of cognitive abilities.

are better indices of *g* than less complex cognitive processes such as simple sensory discriminations. *g* is the best psychological predictor of educational achievement and occupational success (Jensen, 1993; Thorndike, 1985). Its prediction of occupational success increases with the cognitive complexity demanded by the job. Despite the massive data pointing to *g*, considerable controversy continues to surround *g* and IQ tests, especially in the media (Brody, 1992).

There are other ways to study cognitive processes, for example, information-processing approaches. Especially exciting are neuroscience measures that directly assess brain function (Vernon, 1993). However, very few genetic studies have as yet used these other approaches to cognition, in part because the measures are expensive and time-consuming, features that make it difficult to test large samples of twins or adoptees.

Historical Highlights

The relative influence of nature and nurture on *g* has been studied since the beginning of psychology. Indeed, a year before the publication of Gregor Mendel's seminal paper on the laws of heredity, Francis Galton (1865) published a two-article series on high intelligence and other abilities, which he later expanded into the first book on heredity and cognitive ability, *Hereditary Genius: An Inquiry into Its Laws and Consequences* (1869; see Box 8.1). The first twin and adoption studies in the 1920s also focused on *g* (Burks, 1928; Freeman, Holzinger, & Mitchell, 1928; Merriman, 1924; Theis, 1924).

Animal Research

Cognitive ability, at least problem-solving behavior and learning, can also be studied in other species. For example, in a well-known experiment in learning psychology, begun in 1924 by the psychologist Edward Tolman and continued by Robert Tryon, rats were selectively bred for their performance in learning a maze in order to find food. The results of subsequent selective breeding by

BOX 8.1

Francis Galton

Sir Francis Galton

Francis Galton's life (1822–1911) as an inventor and explorer changed as he read the now-famous book on evolution written by Charles Darwin, his half cousin. Galton understood that evolution depends on heredity, and he began to ask whether heredity affects human behavior. He suggested the major methods of human behavioral genetics—family, twin, and adoption designs—and conducted the first systematic family studies that showed that behavioral traits "run in families." Galton invented correlation, one of the fundamental statistics in all of science, in order to quantify degrees of resemblance among family members.

One of Galton's studies on mental ability was reported in an 1869 book, *Hereditary Genius: An Inquiry into Its Laws and Consequences*. Since there was no satisfactory way at the time to measure mental ability, Galton had to rely on reputation as an index. By "reputation," he did not mean notoriety for a single act, nor mere social or official position, but "the reputation of a leader of opinion, or an originator, of a man to whom the world deliberately acknowledges itself largely indebted" (1869, p. 37). Galton identified approximately 1000 "eminent" men and found that they belonged to only 300 families, a finding indicating that the tendency toward eminence is familial.

Taking the most eminent man in each family as a reference point, the other individuals who attained eminence were tabulated with respect to closeness of family relationship. As indicated in the diagram, eminent status was

Robert Tryon for "maze-bright" rats (few errors) and "maze-dull" rats (many errors) are shown in Figure 8.2. Substantial response to selection was achieved after only a few generations of selective breeding. There was practically no overlap between the maze-bright and maze-dull lines; that is, all rats in the maze-bright line were able to learn to run through a maze with fewer errors than any of the rats in the maze-dull line. The difference between the bright and dull lines did not increase after the first half-dozen generations, possibly

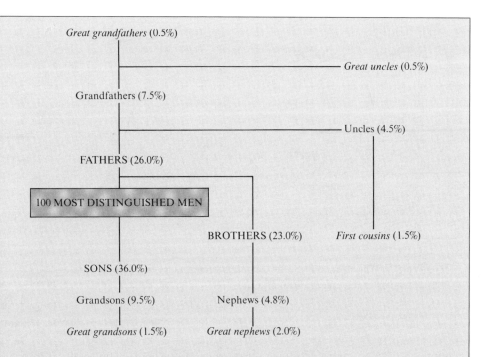

Great grandfathers (0.5%)

_____ *Great uncles* (0.5%)

Grandfathers (7.5%)

_____ Uncles (4.5%)

FATHERS (26.0%)

100 MOST DISTINGUISHED MEN

BROTHERS (23.0%)   *First cousins* (1.5%)

SONS (36.0%)

Grandsons (9.5%)   Nephews (4.8%)

Great grandsons (1.5%)   *Great nephews* (2.0%)

more likely to appear in close relatives, with the likelihood of eminence decreasing as the degree of relationship became more remote.

Galton was aware of the possible objection that relatives of eminent men share social, educational, and financial advantages. One of his counterarguments was that many men had risen to high rank from humble backgrounds. Nonetheless, such counterarguments do not today justify Galton's assertion that genius is solely a matter of nature (heredity) rather than nurture (environment). Family studies by themselves cannot disentangle genetic and environmental influences.

Galton set up a needless battle by pitting nature against nurture, arguing that "there is no escape from the conclusion that nature prevails enormously over nurture" (1883, p. 241). Nonetheless, his work was pivotal in documenting the range of variation in human behavior and in suggesting that heredity underlies behavioral variation. For this reason, Galton can be considered the father of behavioral genetics.

because brothers and sisters were often mated. Such inbreeding greatly reduces the amount of genetic variability within selected lines, which inhibits progress in a selection study.

These maze-bright and maze-dull selected rats were used in one of the best known psychological studies of genotype-environment interaction (Cooper & Zubek, 1958). Rats from the two selected lines were reared under one of three conditions. One condition was "enriched," in that the cages were large and

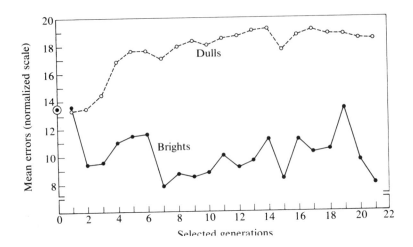

contained many movable toys. For the comparison condition, called "restricted," small gray cages without movable objects were used. In the third condition, rats were reared in a standard laboratory environment.

The results of testing the maze-bright and maze-dull rats reared in these conditions are shown in Figure 8.3. Not surprisingly, in the normal environment in which the rats had been selected, there was a large difference between the two selected lines. A clear genotype-environment interaction emerged for the enriched and restricted environments. The enriched condition had no effect on the maze-bright rats, but it greatly improved the performance of the maze-

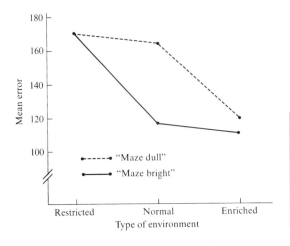

Figure 8.3 Genotype-environment interaction. The effects of rearing in a restricted, normal, or enriched environment on maze-learning errors differ for maze-bright and maze-dull selected rats. (From Cooper & Zubek, 1958.)

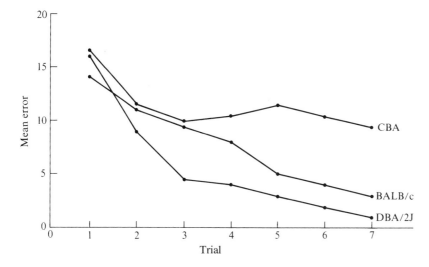

Figure 8.4 Maze-learning errors (Lashley III maze) for three inbred strains of mice. (From "Genetic aspects of learning and memory in mice" by D. Bovet, F. Bovet-Nitti, and A. Oliverio. *Science, 163,* 139–149. Copyright © 1969 by the American Association for the Advancement of Science.)

dull rats. On the other hand, the restricted environment was very detrimental to the maze-bright rats but had little effect on the maze-dull ones. In other words, there is no simple answer concerning the effect of restricted and enriched environments in this study. It depends on the genotype of the animals. This example illustrates genotype-environment interaction, the differential response of genotypes to environments. Despite this persuasive example, other systematic research on learning generally failed to find widespread evidence of genotype-environment interaction (Henderson, 1972).

In the 1950s and 1960s, studies of inbred strains of mice showed the important contribution of genetics to most aspects of learning. Genetic differences have been shown for maze learning as well as for other types of learning, such as active avoidance learning, passive avoidance learning, escape learning, lever pressing for reward, reversal learning, discrimination learning, and heart rate conditioning (Bovet, 1977). For example, differences among widely used inbred strains in maze-learning errors (Figure 8.4) confirm the evidence for genetic influence found in the maze-bright and maze-dull selection experiment. The DBA/2J strain learned quickly, the CBA animals were slow, and the BALB/c strain was intermediate. Similar results were obtained for active avoidance learning, in which mice learn to avoid a shock by moving from one compartment to another whenever a light is flashed on (Figure 8.5), although in this case the CBA strain did not learn at all (Figure 8.6). Genetic influence on learning has also been studied in many other species. For example, fruit flies can learn to avoid an odor associated with shock. As in rats and mice, differences in learning ability among fruit flies show substantial genetic influence (McGuire, 1984).

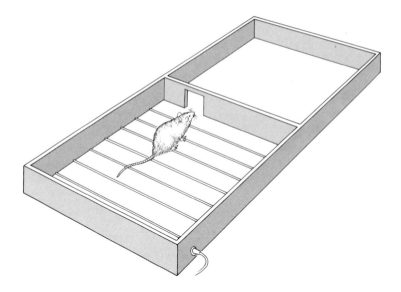

Figure 8.5 Avoidance learning in mice has been investigated by using a "shuttle box," which has two compartments and an electrified floor. The mouse is placed in one compartment, a light is flashed on and is followed by a shock (delivered by an electrified grid on the floor) that continues until the mouse moves to the other compartment. Animals learn to avoid the shock by moving to the other compartment as soon as the light comes on. (From *The Experimental Analysis of Behavior* by Edmund Fantino and Cheryl A. Logan. W. H. Freeman and Company. Copyright © 1979.)

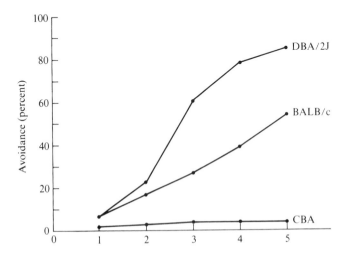

Figure 8.6 Avoidance learning for three inbred strains of mice. (From "Genetic aspects of learning and memory in mice" by D. Bovet, F. Bovet-Nitti, and A. Oliverio. *Science, 163,* 139–149. Copyright © 1969 by the American Association for the Advancement of Science.)

Human Research

Highlights in the history of human research on genetics and intelligence include Leahy's (1935) adoption study, in which she compared IQ resemblance for nonadoptive and adoptive families. This study confirmed an earlier adoption study (Burks, 1928) that showed genetic influence, in that IQ correlations were greater in nonadoptive than in adoptive families. The first adoption study that included IQ data for biological parents of adopted-away offspring also showed significant parent-offspring correlation, suggesting genetic influence, although the increased average IQ score of the adopted-away offspring suggested environmental effects (Skodak & Skeels, 1949). Begun in the early 1960s, the Louisville Twin Study was the first major longitudinal twin study of IQ that charted the developmental course of genetic and environmental influences (Wilson, 1983).

In 1963, a review of genetic research on *g* was influential in showing the convergence of evidence pointing to genetic influence (Erlenmeyer-Kimling & Jarvik, 1963). In 1966, Cyril Burt summarized his decades of research on MZ twins reared apart, which added the dramatic evidence that MZ twins reared apart are nearly as similar as MZ twins reared together. After his death in 1973, Burt's work was attacked, with allegations that some of his data were fraudulent (Hearnshaw, 1979). Two subsequent books have reopened the case (Fletcher, 1990; Joynson, 1989). Although the jury is still out on some of the charges (Mackintosh, 1995), it appears that some of Burt's data are very dubious.

During the 1960s, environmentalism, which had been rampant until then in American psychology, was beginning to wane, and the stage was set for increased acceptance of genetic influence on *g*. Then, in 1969, a monograph on the genetics of intelligence by Arthur Jensen almost brought the field to a halt, because the monograph suggested that ethnic differences in IQ might involve genetic differences. Twenty-five years later, this issue was resurrected in a book called *The Bell Curve* (Herrnstein & Murray, 1994) and caused a similar uproar. The causes of average differences between groups need not be related to the causes of individual differences within groups. The former question is much more difficult to investigate than the latter, which is the focus of the vast majority of genetic research on IQ. The question of the origins of ethnic differences in performance on IQ tests remains unresolved.

The storm raised by Jensen's monograph was appropriate in its emphasis on his misleading inferences about both the possible benefits of intervention and the causes of group differences in IQ but inappropriate in the way it led to intense criticism of all behavioral genetic research, especially in the area of cognitive abilities (e.g., Kamin, 1974). These criticisms of older studies had the positive effect of generating a dozen bigger and better behavioral genetic studies that used family, adoption, and twin designs. These new projects produced much more data on the genetics of *g* than had been obtained in the previous 50 years. The new data contributed in part to a dramatic shift that occurred in the 1980s in psychology toward acceptance of the conclusion that genetic

CLOSE UP

Arthur R. Jensen is professor emeritus of educational psychology at the University of California, Berkeley. After receiving his doctorate in psychology from Teachers College, Columbia University, and completing a clinical internship at the University of Maryland's Psychiatric Institute, he was awarded a two-year postdoctoral fellowship at the University of London's Institute of Psychiatry under Hans J. Eysenck. He then joined the UC Berkeley faculty. After a decade of research on classical problems in verbal learning, Jensen turned to differential psychology and the nature and causes of individual, cultural, and racial group differences in scholastic performance. His 1969 article "How Much Can We Boost IQ and Scholastic Achievement?", which argued that genetic, as well as environmental and cultural, factors should be considered for understanding not only individual differences but also social class and racial group differences in intelligence and scholastic performance, caused a storm of protest from many educators and social scientists. Jensen has since explicated his position in several books: *Genetics and Education*, *Educability and Group Differences*, *Bias in Mental Testing*, and *Straight Talk About Mental Tests*.

differences between individuals are significantly associated with differences in *g* (Snyderman & Rothman, 1988).

The major textbook on intelligence describes the great shift of opinion that has occurred over the past few decades in terms of acceptance of the role of genetics in the origins of individual differences in *g:*

> In 1974 Kamin wrote a book suggesting that there was little or no evidence that intelligence was a heritable trait. I believe that he was able to maintain this position by a distorted and convoluted approach to the literature. It is inconceivable to me that any responsible scholar could write a book taking this position in 1990. In several respects our understanding of the behavior genetics of intelligence has been significantly enhanced in the last 15 years. We have new data on separated twins, large new data sets on twins reared together, better adoption studies, the emergence of developmental behavior genetics and longitudinal data sets permitting an investigation of developmental changes in genetic and environmental influences on intelligence, and the development of new and sophisticated methods of analysis of behavior

genetic data. These developments provide deeper insights into the ways in which genes and the environment influence intelligence. They are, in addition, relevant to an understanding of general issues in the field of intelligence. . . . Our ability to address many of the central issues in contemporary discussions of intelligence is enhanced by a knowledge of the results of behavior genetic research. (Brody, 1992, p. 167)

SUMMING UP

Selection and inbred strain studies indicate genetic influence on animal learning, such as the famous maze-bright and maze-dull selection study of learning in rats. Human twin and adoption studies of general cognitive ability have been conducted for more than 70 years. This research has led to widespread acceptance of the conclusion that genetic factors contribute to individual differences in general cognitive ability.

Overview of Genetic Research

In 1981, a review of genetic research on g was published that summarized results from dozens of studies (Bouchard & McGue, 1981). Figure 8.7 is an expanded version of the summary of the review presented earlier in Chapter 3 (see Figure 3.7).

Genetic Influence

First-degree relatives living together are moderately correlated for g (about .45). As in Galton's original family study on hereditary genius (see Box 8.1), this resemblance could be due to genetic or to environmental influences, because such relatives share both. Adoption designs disentangle these genetic and environmental sources of resemblance. Because adopted-apart parents and offspring and siblings share heredity but not family environment, their similarity indicates that resemblance among family members is due in part to genetic factors. For g, the correlation between children adopted away from their "genetic" parents is .24. The correlation between genetically related siblings reared apart is also .24. Because first-degree relatives are only 50 percent similar genetically, doubling these correlations gives a rough estimate of heritability of 48 percent. As discussed in Chapter 5, this outcome means that about half of the variance in IQ scores in the populations sampled in these studies can be accounted for by genetic differences among individuals.

The twin method supports this conclusion. Identical twins are nearly as similar as the same persons tested twice. (Test-retest correlations for g are

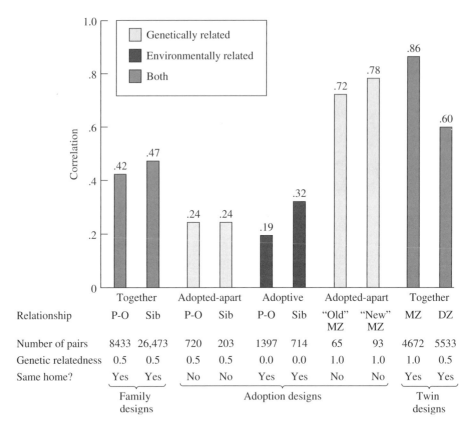

Figure 8.7 Average IQ correlations for family, adoption, and twin designs. (Based on reviews by Bouchard & McGue, 1981, as amended by Loehlin, 1989. "New" data for adopted-apart MZ twins includes Bouchard et al., 1990a, & Pedersen et al., 1992.)

generally between .80 and .90.) The average twin correlations are .86 for identical twins and .60 for fraternal twins. Doubling the difference between MZ and DZ correlations estimates heritability as 52 percent.

The most dramatic adoption design involves MZ twins who were reared apart. Their correlation provides a direct estimate of heritability. For obvious reasons, the number of such twin pairs is small. For several small studies published before 1981, the average correlation for MZ twins reared apart is .72 (excluding the suspect data of Cyril Burt). This outcome suggests higher heritability (72 percent) than the other designs. This high heritability estimate has been confirmed in two new studies of twins reared apart. One of these is the Minnesota Study of Twins Reared Apart, which is being conducted by Thomas J. Bouchard, Jr., and his colleagues at the University of Minnesota. In a report on 45 pairs of MZ twins reared apart, the correlation was .78 (Bouchard et al.,

CLOSE UP

Thomas J. Bouchard, Jr., is a professor of psychology, a member of the Institute of Human Genetics, and director of the Minnesota Center for Twin and Adoption Research at the University of Minnesota. He received his doctorate in psychology from the University of California, Berkeley, in 1966. From 1966 to 1969, he held an appointment as an assistant professor in the Department of Psychology at the University of California, Santa Barbara. In the fall of 1969, he joined the faculty of the Department of Psychology at the University of Minnesota. In 1979, with colleagues in the Departments of Psychology and Psychiatry and the Medical School, he began the Minnesota Study of Twins Reared Apart (MISTRA), a comprehensive medical and psychological study of monozygotic and dizygotic twins separated early in life and reared apart during their formative years. In 1983, Bouchard joined with other colleagues in the Department of Psychology to found the Minnesota Twin Family Registry, a resource for genetic research throughout the University of Minnesota. His research interests include genetic and environmental influences on personality, mental abilities, psychological interests, social values, and psychopathology, as well as the evolution of human behavior. He teaches differential psychology and evolutionary psychology.

1990a). A study of Swedish twins also included 48 pairs of MZ twins reared apart and reported the same correlation of .78 (Pedersen et al., 1992). Possible explanations for this higher heritability estimate for adopted-apart MZ twins are discussed later.

Model-fitting analyses that simultaneously analyze all the family, adoption, and twin data summarized in Figure 8.7 yield heritability estimates of about 50 percent (Chipuer, Rovine, & Plomin, 1990; Loehlin, 1989). It is noteworthy that genetics can account for half of the variance of a trait as complex as general cognitive ability. In addition, the total variance includes error of measurement. Corrected for unreliability of measurement, heritability estimates would be higher. Regardless of the precise estimate of heritability, the point is that genetic influence on g is not only statistically significant, it is also substantial.

Although heritability could differ in different cultures, so far it appears that the level of heritability of g applies not only to American and Western

European countries, where most studies have been conducted. Similar heritabilities have also been found in twin studies in Moscow (Lipovechaja, Kantonistowa, & Chamaganova, 1978) and former East Germany (Weiss, 1982).

Another interesting finding is that heritabilities for cognitive test scores are higher the more a test relates to g (Jensen, 1987). This result has been found in studies of older twins (Pedersen et al., 1992), in retarded individuals (Spitz, 1988), and in a twin study using information-processing tasks (Vernon, 1989). These results suggest that g is the most highly heritable composite of cognitive tests.

Environmental Influence

If half of the variance of g can be accounted for by heredity, the other half is attributed to environment (plus errors of measurement). Some of this environmental influence appears to be shared by family members, making them similar to one another. Direct estimates of the importance of shared environmental influence come from correlations for adoptive parents and children and for adoptive siblings. Particularly impressive is the correlation of .32 for adoptive siblings. Because they are unrelated genetically, what makes adoptive siblings similar is shared rearing, perhaps having the same parents. The adoptive sibling correlation of .32 suggests that about a third of the total variance can be explained by shared environmental influences. The correlation for adoptive parents and their adopted children is lower ($r = .19$) than that for adoptive siblings, a result suggesting that shared environment accounts for less resemblance between parents and offspring than between siblings.

Shared environmental effects are also suggested because correlations for relatives living together are greater than correlations for adopted-apart relatives. Twin studies also suggest shared environmental influence. In addition, shared environmental effects appear to contribute more to the resemblance of twins than to that of nontwin siblings, because the correlation of .60 for DZ twins exceeds the correlation of .47 for nontwin siblings. Twins may be more similar than other siblings because they share the same womb and are exactly the same age. Because they are the same age, twins also tend to be in the same school, if not the same class, and share many of the same peers.

Model-fitting estimates of the role of shared environment for g based on the data in Figure 8.7 are about 20 percent for parents and offspring, about 25 percent for siblings, and about 40 percent for twins (Chipuer et al., 1990). The rest of the variance is attributed to nonshared environment and errors of measurement, which accounts for about 10 percent of the variance.

Assortative Mating

Several factors need to be considered for a more refined estimate of genetic influence. One issue is *assortative mating*, which refers to nonrandom mating. Old adages are sometimes contradictory. Do "birds of a feather flock together" or do "opposites attract"? Research shows that, for some traits, "birds of a feather" do "flock together," in the sense that individuals who mate tend to be similar, although not as similar as you might think. For example, although there is some positive assortative mating for physical characters, the correlations between spouses are relatively low—about .25 for height and about .20 for weight (Spuhler, 1968). Spouse correlations for personality are even lower, in the .10 to .20 range (Vandenberg, 1972). Assortative mating for g is substantial, with average spouse correlations of about .40 (Jensen, 1978). In part, spouses select each other for g on the basis of education. Spouses correlate about .60 for education, which correlates about .60 with g.

Assortative mating is important for genetic research for two reasons. First, assortative mating increases genetic variance in a population. For example, if spouses mated randomly in relation to height, tall women would be just as likely to mate with short men as with tall men. Offspring of the matings of tall women and short men would generally be of moderate height. However, because there is positive assortative mating for height, children with tall mothers are also likely to have tall fathers, and the offspring themselves are likely to be taller than average. The same thing happens for short parents. In this way, positive assortative mating increases variance in that the offspring differ more from the average than they would if mating were random. Even though spouse correlations are modest, assortative mating can greatly increase genetic variability in a population, because its effects accumulate generation after generation.

Assortative mating is also important because it affects estimates of heritability. For example, it increases correlations for first-degree relatives. If assortative mating were not taken into account, it could inflate heritability estimates obtained from studies of parent-offspring (e.g., birth parents and their adopted-apart offspring) or sibling resemblance. For the twin method, however, assortative mating could result in underestimates of heritability. Assortative mating does not affect MZ correlations because MZ twins are identical genetically, but it raises DZ correlations because they are first-degree relatives. In this way, assortative mating lessens the difference between MZ and DZ correlations, and it is this difference that provides estimates of heritability in the twin method. The model-fitting analyses described above took assortative mating into account in estimating the heritability of g to be about 50 percent.

Nonadditive Genetic Variance

Nonadditive genetic variance also affects heritability estimates. For example, when we double the difference between MZ and DZ correlations to estimate heritability, we assume that genetic effects are largely additive. *Additive genetic effects* occur when alleles at a locus and across loci "add up" to affect behavior. However, sometimes the effects of alleles can be different in the presence of other alleles. These interactive effects are called *nonadditive*.

Dominance is a nonadditive genetic effect in which alleles at a locus interact rather than add up to affect behavior. For example, having one PKU allele is not half as bad as having two PKU alleles. Even though many genes operate with a dominant-recessive mode of inheritance, much of the effect of such genes can nonetheless be attributed to the average effect of the alleles. The reason is that, even though heterozygotes are phenotypically similar to the homozygote dominant, there is a substantial linear relationship between genotype and phenotype.

When several genes affect a behavior, the alleles at different loci can add up to affect behavior, or they can interact. This type of interaction between alleles at different loci is called *epistasis*. (See Appendix B for details.)

Additive genetic variance is what makes us resemble our parents, and it is the raw material for natural selection. Our parents' genetic decks of cards are shuffled when our hand is dealt at conception. We and each of our siblings receive a random sampling of half of each parent's genes. We resemble our parents to the extent that each allele that we share with our parents has an average additive effect. Because we do not have exactly the same combination of alleles as our parents (we inherit only one of each of their pairs of alleles), we will differ from our parents for nonadditive interactions as a result of dominance or epistasis. The only relatives who will resemble each other for all dominance and epistatic effects are identical twins, because they are identical for all combinations of genes. For this reason, the hallmark of nonadditive genetic variation is that first-degree relatives are less than half as similar as MZ twins.

In the case of *g*, the correlations in Figure 8.7 suggest that genetic influence is largely additive. For example, first-degree relatives are just about half as similar as MZ twins. However, there is evidence that assortative mating for *g* masks some nonadditive genetic variance. As indicated in the previous section, assortative mating, which is greater for *g* than for any other known trait, inflates correlations for first-degree relatives but does not affect MZ correlations. When assortative mating is taken into account in model-fitting analyses, some evidence appears for nonadditive genetic variance, although most genetic influence on *g* is additive (Chipuer et al., 1990; Fulker, 1979).

The presence of dominance can be seen from studies of inbreeding. (*Inbreeding* is mating between genetically related individuals.) If inbreeding occurs, offspring are more likely to inherit the same alleles at any locus. Thus, inbreeding makes it more likely that two copies of rare recessive alleles will be inherited, including harmful recessive disorders. In this sense, inbreeding reduces heterozygosity by "redistributing" heterozygotes as dominant homozygotes and recessive homozygotes. Therefore, inbreeding also alters the average phenotype of a population. Because the frequency of recessive homozygotes for harmful recessive disorders is increased with inbreeding, the average phenotype will be lowered.

Inbreeding data suggest some dominance for *g*, because inbreeding lowers IQ (Vandenberg, 1971). Children of marriages between first cousins generally perform worse than controls. The risk of mental retardation is more than three times greater for children of a marriage between first cousins than for unrelated controls (Böök, 1957). Children of double first cousins (double first cousins are the children of two siblings who are married to another pair of siblings) perform even worse (Agrawal, Sinha, & Jensen, 1984; Bashi, 1977). Nonetheless, inbreeding does not have an appreciable effect in general in the population because it is rare, with the exception of a few societies and small isolated groups.

An extreme version of epistasis called *emergenesis* has been suggested as a model for unusual abilities (Lykken, 1982). Luck of the draw at conception can result in certain unique combinations of alleles that have extraordinary effects not seen in parents or siblings. For example, the great race horse Secretariat was bred to many fine mares to produce hundreds of offspring. Many of Secretariat's offspring were good horses, thanks to additive genetic effects, but none came even close to the unique combination of strengths responsible for Secretariat's greatness. Such genetic luck of the draw might contribute to human genius as well.

SUMMING UP

Family, twin, and adoption studies converge on the conclusion that about half of the total variance of measures of general cognitive ability can be accounted for by genetic factors. For example, twin correlations for general cognitive ability are about .85 for identical twins and .60 for fraternal twins. Heritability estimates are affected by assortative mating (which is substantial for general cognitive ability) and nonadditive genetic variance (dominance and epistasis). About half of the environmental variance appears to be accounted for by shared environmental factors.

Despite these complications, the general summary of behavioral genetic results for *g* is surprisingly simple (see Figure 8.8). About half of the variance is due to genetic factors. Some, but not much, of this genetic variance might be nonadditive. Of the half of the variance that is due to nongenetic factors, about half is accounted for by shared environmental factors. The other half is due to nonshared environment and errors of measurement. However, during the past decade, it has been discovered that these average results differ dramatically during development, as described in the following section.

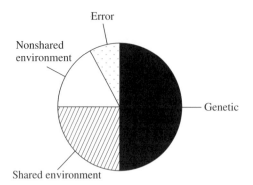

Figure 8.8 About half of the variance of general cognitive ability can be accounted for by genetic factors.

Developmental Research

When Francis Galton first studied twins in 1876, he investigated the extent to which the similarity of twins changed during development. Other early twin studies were also developmental (Merriman, 1924; Thorndike, 1905), but this developmental perspective faded from genetic research until recent years.

Two types of developmental questions can be addressed in genetic research. Does heritability change during development? Do genetic factors contribute to developmental change?

Does Heritability Change During Development?

Try asking people this question: As you go through life, do you think the effects of heredity become more important or less important? Most people will usually guess "less important" for two reasons. First, it seems obvious that life events such as accidents and illnesses, education and occupation, and other experiences accumulate during a lifetime. This fact implies that environmental differences increasingly contribute to phenotypic differences, so heritability necessarily decreases. Second, most people mistakenly believe that genetic effects never change from the moment of conception.

Because it is so reasonable to assume that genetic differences become less important as experiences accumulate during the course of life, one of the most interesting findings about g is that the opposite is closer to the truth. Genetic factors become increasingly important for g throughout the life span (McCartney, Harris, & Bernieri, 1990; McGue et al., 1993a; Plomin, 1986).

For example, an ongoing longitudinal adoption study called the Colorado Adoption Project (DeFries, Plomin, & Fulker, 1994) provides parent-offspring correlations for general cognitive ability from infancy through adolescence. As illustrated in Figure 8.9, correlations between parents and children for control (nonadoptive) families (CP–CC in Figure 8.9) increase from less than .20 in infancy to about .20 in middle childhood and to about .30 in adolescence. The correlations between biological mothers and their adopted-away children (BM–AC in Figure 8.9) follow a similar pattern, thus indicating that parent-

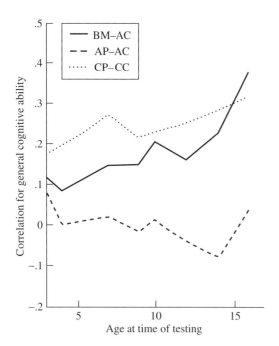

Figure 8.9 Parent-offspring correlations from infancy to adolescence in the Colorado Adoption Project. BM–AC, biological mother–adopted child; AP–AC, adoptive parent–adopted child; CP–CC, control parent–control child. (J. C. DeFries & R. Plomin, unpublished data.)

offspring resemblance for *g* is due to genetic factors. Parent-offspring correlations for adoptive parents and their adopted children hover around zero. This study suggests that family environment shared by parents and offspring does not contribute importantly to parent-offspring resemblance for *g*.

Figure 8.10 summarizes MZ and DZ twin correlations for *g* by age (McGue et al., 1993a). The difference between MZ and DZ twin correlations increases slightly from early to middle childhood and then increases dramatically in adulthood. Because relatively few twin studies of *g* have included adults, summaries of IQ data (Figure 8.7) rest primarily on data from childhood. Heritability in adulthood is higher. This conclusion is supported by five studies of MZ twins reared apart. These studies, unlike studies of twins reared together, almost exclusively include adults. The average heritability estimate from these studies of MZ twins reared apart is 75 percent (McGue et al., 1993a). For example, a recent study included twins reared apart and matched twins reared together tested at the average age of 60 years as part of the Swedish Adoption–Twin Study of Aging (SATSA). These twins are much older than twins in other studies, and the study yielded a heritability estimate of 80 percent for *g* (Pedersen et al., 1992), a result that was replicated when the SATSA twins were retested three years later (Plomin et al., 1994a). This is one of the highest heritabilities reported for any behavioral dimension or disorder.

Why does heritability increase during the life span? Perhaps completely new genes come to affect *g* in adulthood. A more likely possibility is that relatively small genetic effects early in life snowball during development, creating

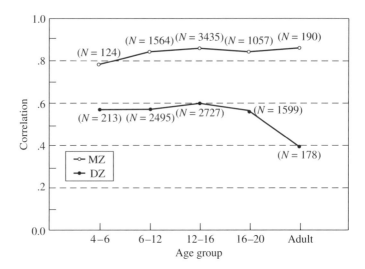

Figure 8.10 The difference between MZ and DZ twin correlations for *g* increases during adolescence and adulthood, a trend suggesting increasing genetic influence. (From McGue et al., 1993a, p. 63.)

larger and larger phenotypic effects. For the young child, parents and teachers contribute importantly to intellectual experience; but for the adult, intellectual experience is more self-directed. For example, it seems likely that adults with a genetic propensity toward high *g* keep active mentally by reading, arguing, and simply thinking more than other people. Such experiences not only reflect but also reinforce genetic differences (Bouchard et al., in press; Scarr, 1992; Scarr & McCartney, 1983).

Another important developmental finding is that the effects of shared environment appear to decrease. Twin study estimates of shared environment are weak because shared environment is estimated indirectly by the twin method; that is, shared environment is estimated as twin resemblance that cannot be explained by genetics. Nonetheless, the world's twin literature indicates that shared environment effects for *g* are negligible in adulthood.

Adoption studies provide two types of evidence for the importance of shared environment in childhood. The classic adoption study of Skodak and Skeels (1949) found that adopted-away offspring had higher IQ scores than expected from the IQ of their biological parents, results confirmed in more recent studies (e.g., Capron & Duyme, 1989, in press). This finding suggests that IQ scores are raised when children whose biological parents have lower than average IQ scores are adopted by adoptive parents whose IQ scores are higher than average.

The most direct evidence for the important effect of shared environment on individual differences in *g* comes from the resemblance of adoptive siblings, pairs of genetically unrelated children adopted into the same adoptive families. Figure 8.7 indicates an average IQ correlation of about .30 for adoptive siblings. However, these studies assessed adoptive siblings when they were children. In 1978, the first study of older adoptive siblings yielded a strikingly different result: The IQ correlation was –.03 for 84 pairs of adoptive siblings who were 16–22 years of age (Scarr & Weinberg, 1978a). Other studies of older siblings have also found similarly low IQ correlations. The most impressive evidence comes from a ten-year longitudinal follow-up study of more than 200 pairs of adoptive siblings. At the average age of 8 years, the IQ correlation was .26. Ten years later, their IQ correlation was near zero (Loehlin, Horn, & Willerman, 1989). Figure 8.11 shows the results of studies of adoptive siblings in childhood and in adulthood (McGue et al., 1993a). In childhood, the average adoptive sibling correlation is .25, but in adulthood, the correlation for adoptive siblings is near zero.

These results represent a dramatic example of the importance of genetic research for understanding the environment. Shared environment is important for g during childhood when children are living at home. However, its importance fades in adulthood as influences outside the family become more salient. What are these mysterious nonshared environmental factors that make siblings growing up in the same family so different? This topic is discussed in Chapter 14.

In summary, from childhood to adulthood, heritability increases and shared environment decreases for g (Figure 8.12).

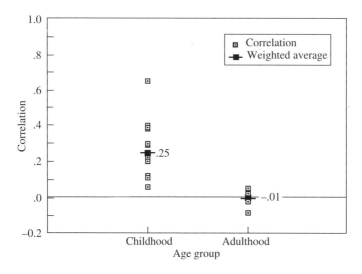

Figure 8.11 The correlation for adoptive siblings provides a direct estimate of the importance of shared environment. For g, the correlation is .25 in childhood and –.01 in adulthood, a trend suggesting that shared environment becomes less important after childhood. (From McGue et al., 1993a, p. 67.)

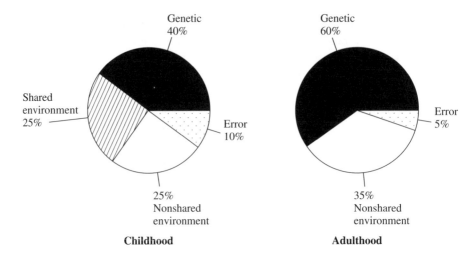

Figure 8.12 From childhood to adulthood, heritability of *g* increases and shared environment declines in importance.

Do Genetic Factors Contribute to Developmental Change?

The second type of genetic change in development refers to age-to-age change seen in longitudinal data in which individuals are assessed several times. It is important to recognize that genetic factors can contribute to change as well as to continuity in development. Change in genetic effects does not necessarily mean that genes are turned on and off during development, although this does happen. Genetic change simply means that genetic effects at one age differ from genetic effects at another age. For example, genes that affect cognitive processes involved in language cannot show their effect until language appears in the second year of life.

The issue of genetic contributions to change and continuity can be addressed by using longitudinal genetic data in which twins or adoptees are tested repeatedly. The simplest way to think about genetic contributions to change is to ask whether change scores from age to age show genetic influence. That is, although *g* is quite stable from year to year, some children's scores increase and some decrease. Genetic factors in part account for such changes, especially in childhood (Fulker, DeFries, & Plomin, 1988), and perhaps even in adulthood (Loehlin et al., 1989). Still, not surprisingly, most genetic effects on *g* contribute to continuity from age to age.

Model-fitting analysis (Appendix B) is especially useful for longitudinal data because of the complexity of having multiple measurements for each subject. Several types of longitudinal genetic models have been proposed (Loehlin et al., 1989). A longitudinal model applied to twin and adoptive sibling data

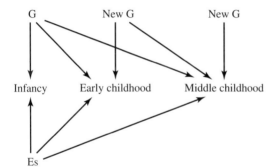

Figure 8.13 Genetic factors (G) contribute to change as well as to continuity in *g* during childhood. Shared environment (Es) contributes only to continuity. (Adapted from Fulker et al., 1993.)

from infancy to middle childhood found evidence for genetic change at two important developmental transitions (Fulker, Cherny, & Cardon, 1993). The first is the transition from infancy to early childhood, an age when cognitive ability rapidly changes as language develops. The second is the transition from early to middle childhood, at 7 years of age. It is no coincidence that children begin formal schooling at this age—all theories of cognitive development recognize this as a major transition.

Figure 8.13 summarizes these findings. Much genetic influence on *g* involves continuity. That is, genetic factors that affect infancy also affect early childhood and middle childhood. However, some new genetic influence comes into play at the transition from infancy to early childhood. These new genetic factors continue to affect *g* throughout early childhood and into middle childhood. Similarly, new genetic influence also emerges at the transition from early to middle childhood. Still, a surprising amount of genetic influence on general cognitive ability in childhood overlaps with genetic influence even into adulthood, as illustrated in Figure 8.14.

As discussed above, shared environmental influences also affect *g* in childhood. Unlike genetic effects, which contribute to change as well as to continuity, longitudinal analysis suggests that shared environmental effects contribute only to continuity. That is, the same shared environmental factors affect *g* in infancy and in early and middle childhood (Figure 8.13). Socioeconomic factors, which remain relatively constant, might account for this shared environmental continuity.

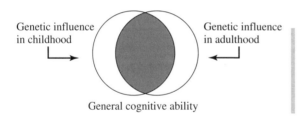

Figure 8.14 Although genetic influences on *g* in childhood are largely the same as those that affect *g* in adulthood, there is some evidence for genetic change.

SUMMING █**UP**█

Heritability of general cognitive ability increases during the life span. The effects of shared environment decrease during childhood to negligible levels after adolescence. Longitudinal genetic analyses of continuity and change indicate that much genetic influence on general cognitive ability contributes to continuity. However, some genetic influence affects change from age to age, especially during the transition from infancy to early childhood.

Identifying Genes

General cognitive ability is a reasonable candidate for molecular genetic research because it is one of the most heritable dimensions of behavior. As for most behaviors, many genes are likely to contribute to genetic influence, which means that no single gene is likely to account for a substantial proportion of the total genetic variance. The important implication is that molecular genetic strategies that can detect genes of small effect size are needed.

As mentioned in Chapter 6, knock-out genes have been shown to affect learning in mice. In our species, more than 100 single-gene disorders include mental retardation among their symptoms (Wahlström, 1990). The major single-gene effects were mentioned in Chapters 6 and 7. The classic example of a single-gene cause of severe mental retardation is PKU. More recently, researchers have identified a gene causing the fragile X type of mental retardation. A gene (apolipoprotein E) on chromosome 19 contributes substantially to risk for the dementia of late-onset Alzheimer's disease.

What about the normal range of general cognitive ability? Differences in the number of fragile X repeats in the normal range do not relate to differences in IQ (Daniels et al., 1994). It is only when the number of repeats expands to more than 200 that mental retardation occurs, as described in Chapter 7. However, it has been reported that apo-E, associated with dementia, is also associated with greater cognitive decline in an unselected population of elderly men (Feskens et al., 1994). In addition, some evidence suggests that carriers for PKU show slightly lowered IQ scores (Bessman, Williamson, & Koch, 1978; Propping, 1987).

Associations between IQ and genetic markers such as blood groups have long been sought, without notable success (e.g., Ashton, 1986; Gibson et al., 1973; Mascie-Taylor et al., 1985). The new DNA markers have begun to be used in research on cognitive abilities. In one study, investigators employed an allelic association strategy that uses DNA markers in or near genes likely to be relevant to neurological functioning, such as genes for neuroreceptors. Allelic frequencies of these DNA markers can be compared for groups high and low in

IQ. In the first report of this type, allelic association results were presented for 100 DNA markers (Plomin et al., 1995). Although several significant associations were found in an original sample, only one association was replicated cleanly in an independent sample. This finding might well be a chance result, because 100 markers were investigated. Another association involves the major histocompatibility region of chromosome 6, which was also found in a study of general cognitive ability in an elderly sample (Jacomb et al., in press) and is the region where a linkage with reading disability was found (Cardon et al., 1994).

Regardless of the ultimate success of this first attempt to identify QTLs associated with cognitive abilities, it indicates the potential for harnessing the power of molecular genetics to identify QTLs responsible for genetic influence on cognitive abilities. Finding genes associated with general cognitive ability will be of limited usefulness for predicting IQs in the general population because of the sheer number of genes that are likely to be involved and their small effect size. Rather, the major implication of identifying QTLs will be for basic research. Even a small handhold on the genetic contribution to individual differences will help in the climb toward understanding how genes interact with neural, psychological, and environmental processes in cognitive development.

Summary

The evidence for a strong genetic contribution to general cognitive ability (g) is clearer than for any other area of psychology. Although g has been central in the nature-nurture debate, few scientists any longer seriously dispute the conclusion that general cognitive ability shows significant genetic influence (Snyderman & Rothman, 1988). The magnitude of genetic influence is still not universally appreciated, however. Taken together, this extensive body of research suggests that about half of the total variance of measures of g can be accounted for by genetic factors. Estimates of heritability are affected by assortative mating (which is substantial for g) and by nonadditive genetic variance (dominance and epistasis).

The heritability of g increases during the life span, reaching levels in adulthood comparable to the heritability of height. The influence of shared environment diminishes sharply after adolescence. Longitudinal genetic analyses of g suggest that genetic factors primarily contribute to continuity, although some evidence for genetic change has been found, for example, in the transition from early to middle childhood.

A new direction for research is to begin to identify genes responsible for the heritability of normal dimensions, not just disorders. Several genes that are associated with or linked to cognitive disabilities have been identified. Similar research is underway to identify QTLs for general cognitive ability.

CHAPTER NINE

Specific Cognitive Abilities

T here is much more to cognitive functioning than general cognitive ability. As discussed in Chapter 8, cognitive abilities are usually considered in a hierarchical model (Figure 8.1). General cognitive ability is at the top of the hierarchy, representing what all tests of cognitive ability have in common. Below general cognitive ability in the hierarchy are broad factors of specific cognitive abilities, such as verbal ability, spatial ability, memory, and speed of processing. These broad factors are indexed by several tests. Verbal ability might be a composite of tests of vocabulary and word fluency (for example, listing words that begin with g and end with t). This and the other specific cognitive abilities correlate moderately with general cognitive ability, but they are also substantially different. At the bottom of the hierarchy are elementary abilities that are thought to be involved in processing information from input to storage and then from retrieval to output.

More is known about the genetics of the broad factors of specific cognitive abilities than about the elementary processes. This chapter presents genetic research on specific cognitive abilities, elementary processes, and their relationship to general cognitive ability. It also considers the genetics of a real-world aspect of cognitive abilities, school achievement.

Broad Factors of Specific Cognitive Abilities

The largest family study of specific cognitive abilities, called the Hawaii Family Study of Cognition, included more than a thousand families (DeFries et al., 1979). Like other work in this area, this study included an analysis of 15 tests of cognitive ability by a technique called factor analysis. Four group factors were derived: verbal (including vocabulary and fluency), spatial (visualizing and rotating objects in two- and three-dimensional space), perceptual speed (simple arithmetic and number comparisons), and visual memory (short-term and longer-term recognition of line drawings). Figure 9.1 summarizes parent-offspring resemblance for the four factors and the 15 cognitive tests for two ethnic groups.

The most obvious fact is that familial resemblance differs for the four factors and for tests within each factor. The data were corrected for unreliability of the tests so that the differences in familial resemblance were not caused by reliability differences among the tests. For Americans of both European ancestry and Japanese ancestry, the verbal and spatial factors show more familial resemblance than do the perceptual speed and memory factors. Other family studies also generally indicate that the greatest familial similarity occurs for verbal ability (DeFries, Vandenberg, & McClearn, 1976). It is not known why one group consistently shows greater parent-offspring resemblance than the other. This study is a good reminder of the principle that the results of genetic research can differ in different populations.

Figure 9.1 also shows another important point: Tests within each factor show dramatic differences in familial resemblance. For example, one spatial

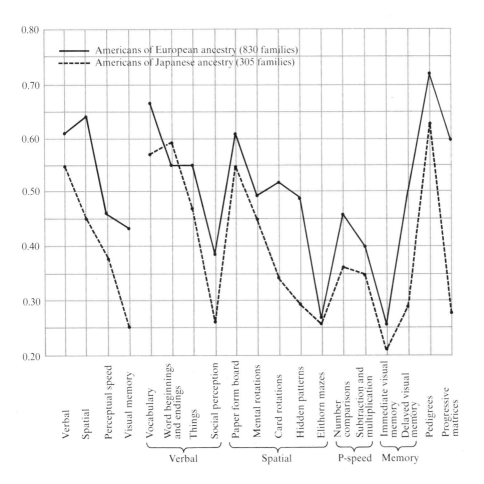

Figure 9.1 Family study of specific cognitive abilities. Regression of midchild on midparent for four group factors and 15 cognitive tests in two ethnic groups. (Data from DeFries et al., 1979.)

TABLE 9.1

Average Twin Correlations for Tests of Specific Cognitive Abilities

Ability	Number of Studies	Twin Correlations Identical Twins	Fraternal Twins
Verbal comprehension	27	.78	.59
Verbal fluency	12	.67	.52
Reasoning	16	.74	.50
Spatial visualization	31	.64	.41
Perceptual speed	15	.70	.47
Memory	16	.52	.36

SOURCE: *Nichols (1978).*

test, *Paper Form Board*, shows high familiality. The test involves showing how to cut a figure to yield a certain pattern—for example, to cut a circle to yield a triangle and three crescents. Another spatial test, *Elithorn Mazes*, shows the lowest familial resemblance. This test involves drawing one line that connects as many dots as possible in a maze of dots. Although these tests correlate with each other and contribute to a broad factor of spatial ability, much remains to be learned about the genetics of the processes involved in each test.

The results of dozens of twin studies of specific cognitive abilities are summarized in Table 9.1 (Nichols, 1978). When we double the difference between the correlations for identical and fraternal twins to estimate heritability (see Chapter 5), these results suggest that specific cognitive abilities show slightly less genetic influence than general cognitive ability. Memory and verbal fluency show lower heritability, about 30 percent; the other abilities yielded heritabilities of 40 to 50 percent. Although the largest twin studies do not consistently find greater heritability for particular cognitive abilities (Brunn, Markkanen, & Partanen, 1966; Schoenfeldt, 1968), it has been suggested that verbal and spatial abilities in general show greater heritability than do perceptual speed and especially memory abilities (Plomin, 1988). Earlier twin studies of specific cognitive abilities have been reviewed in detail elsewhere (DeFries et al., 1976). A recent study of 160 pairs of twins aged 15 to 19 found similar results for tests of verbal and spatial ability. This study is notable because the sample population was Croatian (Bratko, in press), thus broadening the population base of observations on this topic.

Two recent studies of identical and fraternal twins reared apart provide additional support for genetic influence on specific cognitive abilities. One is a U.S. study of 72 reared-apart twin pairs of a wide age range in adulthood

(McGue & Bouchard, 1989), and the other is a Swedish study of older twins (average age of 65 years), including 133 reared-apart twins and 142 control twin pairs reared together (Pedersen et al., 1992). Both studies show significant heritability estimates for all four specific cognitive abilities. As shown in Table 9.2, the heritability estimates are generally higher than implied by the twin results summarized in Table 9.1. This discrepancy may be due to the trend, discussed in Chapter 8, for heritability for cognitive abilities to increase during the life span. In both studies, the lowest heritability is found for memory.

As described in Chapter 8, twin studies of general cognitive ability appear to indicate influence of shared environment because twin resemblance cannot be explained entirely by shared heredity. However, it was noted that both identical and fraternal twins experience more similar environments than do nontwin siblings. For this reason, twin studies inflate estimates of shared environment in studies of general cognitive ability. Adoption designs generally suggest less shared environmental influence.

Influences on specific cognitive abilities appear to be similar to those for general cognitive ability. However, the twin correlations in Table 9.1 imply substantial influence of shared environment, whereas results from the twins-reared-apart adoption studies described in Table 9.2 indicate that shared environment has little influence. Studies of adoptive relatives can provide a direct test of shared environment, but only two adoption studies of specific cognitive abilities have been reported.

One adoption study found little resemblance for adoptive parents and their adopted children or for adoptive siblings on subtests of an intelligence test, except for vocabulary (Scarr & Weinberg, 1978b). Thus, this study supports the results of the two twins-reared-apart adoption studies in suggesting that shared environment has little influence on specific cognitive abilities. Like the twin and twins-reared-apart studies, this adoption study found evidence for genetic influence, in that nonadoptive relatives showed greater resemblance than did adoptive relatives.

TABLE 9.2

Heritability Estimates for Specific Cognitive Abilities in Two Studies of Twins Reared Apart

Ability	Heritability Estimate (%)	
	McGue & Bouchard (1989)	Pedersen et al. (1992)
Verbal	57	58
Spatial	71	46
Speed	53	58
Memory	43	38

Specific cognitive abilities are central to an ongoing longitudinal adoption study called the Colorado Adoption Project (DeFries, Plomin, & Fulker, 1994). Figure 9.2 summarizes parent-offspring results for verbal, spatial, perceptual speed, and recognition memory abilities from early childhood through adolescence. Mother-child and father-child correlations were averaged for both adoptive and control (nonadoptive) families. For each ability, biological mother–adopted child and control parent–control child correlations tend to increase as a function of age. In contrast, adoptive parent–adopted child correlations do not differ substantially from zero at any age. These results indicate increasing heritability and no shared environment.

Developmental genetic analyses of Colorado Adoption Project data from adoptive and nonadoptive siblings indicate that genetically distinct specific cognitive abilities can be found as early as three years of age and show increasing genetic differentiation from three to seven years of age (Cardon, 1994a). Like the findings for general cognitive ability (Chapter 8), new genetic effects are found at seven years, an observation hinting at a genetic transformation of cognitive abilities in the early school years (Cardon & Fulker, 1993). Shared environment shows little effect.

SUMMING UP

Family studies of specific cognitive abilities, most notably the Hawaii Family Study of Cognition, show greater familial resemblance for verbal and spatial abilities than for perceptual speed and memory. Tests within each ability vary in their degree of familial resemblance. Twin studies indicate that most of this familial resemblance is genetic in origin. Although twin studies suggest shared environmental influence, twins are more similar environmentally than are nontwin siblings. In addition, data from twins reared apart show genetic influence but not shared environmental influence. Developmental analyses of adoption data indicate that heritability increases during childhood and that genetically distinct specific cognitive abilities can be found as early as three years of age.

What about creativity? A review of ten twin studies of creativity yielded average twin correlations of .61 for identical twins and .50 for fraternal twins, results indicating only modest genetic influence and substantial influence of shared environment (Nichols, 1978). Some research implies that this modest genetic influence is entirely due to the overlap between tests of creativity and general cognitive ability. That is, when general cognitive ability is controlled,

Figure 9.2 Parent-offspring correlations from infancy to adolescence in the Colorado Adoption Project. BM–AC, biological mother–adopted child; AP–AC, adoptive parent–adopted child; CP–CC, control parent–control child. (J. C. DeFries & R. Plomin, unpublished data.)

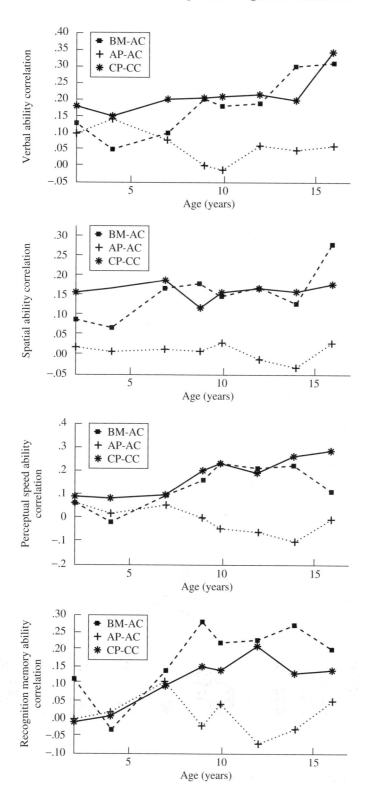

identical and fraternal twin correlations for tests of creativity are similar (Canter, 1973).

Information-Processing Measures

An important direction for research on the genetics of specific cognitive abilities is to capitalize on the laboratory tasks developed by experimental cognitive psychologists to assess how information is processed. The few available twin studies using information-processing measures find some evidence for genetic influence. One twin study focused on speed-of-processing measures such as rapid naming of objects and letters (Ho, Baker, & Decker, 1988). These measures are similar to those used to assess the specific cognitive ability factor of perceptual speed. The results of this twin study yield evidence for moderate genetic influence. More traditional reaction-time measures of information processing also show genetic influence in twin studies (Boomsma & Somsen, 1991; Vernon, 1989) and in a study of twins reared apart (McGue & Bouchard, 1989).

A recent study of 287 twin pairs aged 6 to 13 years (Petrill, Thompson, & Detterman, 1995) used a computerized battery of elementary cognitive tasks designed to test a theory that general cognitive ability is a complex system of independent elementary processes (Detterman, 1986). For example, a speed-of-processing factor was assessed by tasks such as decision time in stimulus discrimination. As shown in Figure 9.3, a probe stimulus is presented above an array of six stimuli, one of which matches the probe. The task is simply to touch as quickly as possible the stimulus that matches the probe. Information-processing tasks can subtract movement time from reaction time to obtain a purer measure of the time required to make the decision.

In this study, a measure of decision time based on stimulus discrimination was highly reliable. Despite the simplicity of the task, it correlates –.42 with IQ. That is, shorter decision times are associated with higher IQ scores. Twin

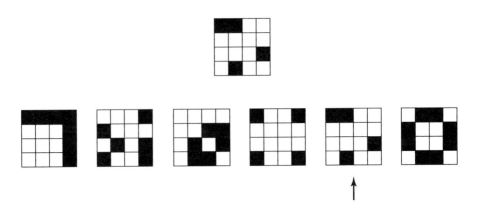

Figure 9.3 Stimulus discrimination display. Arrow indicates correct choice.

correlations for this measure of decision time were .61 for identical twins and .39 for fraternal twins, yielding a heritability of about 45 percent and about 15 percent influence of shared environment. The battery included other measures such as reaction time, learning, and memory, most of which showed more modest heritability, ranging down to zero heritability for reaction time. Estimates of shared environment also varied widely for the various measures.

Multivariate genetic analysis of information-processing measures and their relationship to general and specific cognitive abilities is a special focus of research in this area, as discussed in the following section.

Multivariate Genetic Analysis: Levels of Processing

Genetic studies can go beyond the analysis of the variance of a single variable to consider sources of covariance between traits. This technique is multivariate genetic analysis, described in Appendix B. Multivariate genetic analyses of specific cognitive abilities and their relationship to general cognitive ability have provided some important insights into the organization of cognitive abilities. Although it is generally agreed that levels of abilities are related hierarchically, this conclusion is only a phenotypic description of the relationship among levels.

In terms of genetics, three very different models have been proposed, shown in simplified form in Figure 9.4. A bottom-up model assumes that different genes affect each basic element of information processing. These genetic influences on elementary processes feed into specific cognitive abilities, which in turn converge on general cognitive ability. The implication is that the influence of any gene found to be associated with higher levels of processing comes from that gene's association with a particular basic element of processing. In other words, when we control for genetic effects on basic elements, we find no additional genetic effects for higher levels of processing.

As reasonable as such a reductionistic model seems, it is possible that higher levels of processing involve new genetic effects not found at lower levels. The extreme version of this antireductionistic model is a top-down model, which assumes that genes that affect cognitive abilities primarily affect general cognitive ability, perhaps as a result of some general mechanism such as neural speed or fidelity. These genetic effects on general cognitive ability filter down to specific cognitive abilities and elementary processes. In the extreme, the top-down model implies that the influence of any gene associated with lower levels of processing comes from the association of that gene with general cognitive ability. That is, when we control for genetic effects on general cognitive ability, we find no additional genetic effects for lower levels of processing.

A compromise genetic model could be called a levels-of-processing model. At each level of processing there are unique genetic effects; but there are also genetic effects in common across levels of processing. In other words, as posited by the bottom-up model, there are genes specifically associated with each

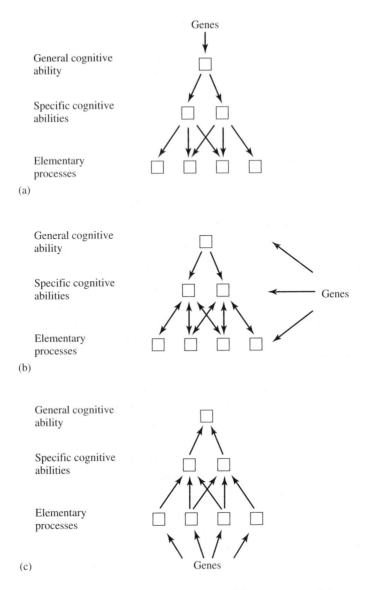

Figure 9.4 Genetic models of cognitive abilities: (a) top-down model; (b) levels-of-processing model; (c) bottom-up model.

elementary process. These genes are associated with higher levels of processing when the elementary processes are controlled. In addition, as implied by the top-down model, there are genes associated with general cognitive ability that are not associated with lower levels of processing when general cognitive ability is controlled. In this sense, the levels-of-processing model suggests that both the bottom-up and top-down models are correct. In addition, each level of processing has unique as well as common genetic effects. For example, some genetic effects will be found at the middle level of specific cognitive abilities that are not found at either the level of elementary processes or at the level of general cognitive ability.

Multivariate genetic analyses have focused on the relationship between general cognitive ability and specific cognitive abilities. The results lean toward the top-down model but also provide some support to the levels-of-processing model (Cardon & Fulker, 1993; Casto, DeFries, & Fulker, 1995; Luo, Petrill, & Thompson, 1994; Pedersen, Plomin, & McClearn, 1994; Tambs et al., 1986). To a considerable degree, genetic effects on cognitive abilities are general. That is, genes that affect one cognitive ability also affect other cognitive abilities, as assumed by the top-down model. However, there are some genetic effects unique to each cognitive ability, as hypothesized by the levels-of-processing model. The top-down model helps to explain an intriguing finding: Heritabilities of tests of specific cognitive abilities are strongly associated with the tests' correlations with general cognitive ability (Jensen, 1987). That is, the higher the heritability of a test, the more that test correlates with general cognitive ability. For example, in the Swedish study of twins reared apart and twins reared together, tests of cognitive abilities differed in their heritability, and they differed in the extent to which they correlate with general cognitive ability. The correlation between the tests' heritabilities and their correlations with general cognitive ability was .77 after controlling for differential reliabilities of the tests (Pedersen et al., 1992). The top-down model would predict this result in terms of the pervasive genetic effect of general cognitive ability.

Much less is known about elementary processes and their genetic relationship with higher levels of processing. Multivariate genetic analyses of elementary processes and general cognitive ability again lean toward a top-down model (Baker, Vernon, & Ho, 1991; Ho, Baker, & Decker, 1988). However, the first genetic analysis that incorporated elementary processes, specific cognitive abilities, and general cognitive ability supports the levels-of-processing model (Petrill et al., in press-a). That is, although there are genetic effects in common across levels, unique genetic effects were found at each level.

SUMMING UP

Information-processing measures also show genetic influence in the few available twin studies. Although information-processing research assumes a bottom-up model in which different genes affect basic elements of information processing, multivariate genetic analyses provide stronger evidence for a top-down model in which genes primarily affect general cognitive ability. A compromise model is a levels-of-processing model with unique genetic effects at each level of processing (bottom-up) but also genetic effects in common across levels of processing (top-down).

The levels-of-processing model thus predicts that when genes that are associated with cognitive abilities are found, most of these genes will be associated with abilities throughout the hierarchy. Some genes, however, will be specific to certain abilities but not others. As mentioned later, the first molecular genetic analyses on specific cognitive abilities provide some support for this prediction.

School Achievement

At first glance, tests of school achievement seem quite different from tests of specific cognitive abilities. School achievement tests focus on performance in specific domains such as grammar, American history, and geometry. Moreover, the word *achievement* itself implies that such tests are due to dint of effort, an environmental influence, in contrast to *ability*, for which genetic influence seems more reasonable. Nonetheless, research is clear in showing that school achievement test performances across diverse topics from grammar to geometry correlate substantially with general cognitive ability. They also show genetic influence.

In elementary school, school achievement tests show strong influence of shared environment (about 60 percent) and modest genetic influence (about 30 percent) in a study of 278 pairs of twins aged 6 to 12 years (Table 9.3; Thompson, Detterman, & Plomin, 1991). Other twin and adoption research focusing on tests of reading ability and spelling in elementary school also yield evidence for modest heritability (Brooks, Fulker, & DeFries, 1990; Stevenson et al., 1987). Reading *disability* was discussed in Chapter 7, but the present discussion considers the normal range of individual differences in reading ability and other school skills.

During the school years, the magnitude of genetic influence appears to increase, and shared environment appears to decrease in importance, as it does

TABLE 9.3

Twin Correlations for School Achievement Tests in Elementary School

	Twin Correlation	
Test Subject	Identical Twins	Fraternal Twins
Reading	.94	.79
Language	.87	.71
Math	.91	.81

SOURCE: *Thompson et al. (1991).*

TABLE 9.4

Twin Correlations for Report Card Grades for 13-Year-Olds

	Twin Correlation	
Subject Graded	**Identical Twins**	**Fraternal Twins**
History	.80	.50
Reading	.72	.57
Writing	.76	.50
Arithmetic	.81	.48

SOURCE: *Husén (1959).*

for cognitive abilities. For example, in a study of a thousand pairs of older (13-year-old) twins in Sweden, heritabilities for report card grades ranged from 30 to over 60 percent, as indicated by the twin correlations shown in Table 9.4 (Husén, 1959). Shared environmental influence accounts for about 25 percent of the variance. As mentioned earlier here and in Chapter 8, twin studies appear to inflate estimates of shared environment for cognitive abilities.

A study of high school–age twins in the United States obtained data from the National Merit Scholarship Qualifying Test for 1300 identical and 864 fraternal twin pairs (Loehlin & Nichols, 1976). The twin correlations shown in Table 9.5 yield heritabilities of about 40 percent. Shared environment is estimated to be about 30 percent. The consistency of results across tests is not so surprising, since the tests correlate highly, an average of about .60. Multivariate genetic analyses of these data indicate that genetic correlations among the tests

TABLE 9.5

Twin Correlations for School Achievement Test in High School

	Twin Correlation	
Test Subject	**Identical Twins**	**Fraternal Twins**
Social studies	.69	.52
Natural sciences	.64	.45
English usage	.72	.52
Mathematics	.71	.51

SOURCE: *Loehlin & Nichols (1976).*

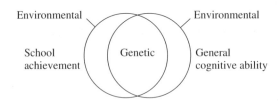

Figure 9.5 Genetic influences on school achievement test scores also influence general cognitive ability, whereas environmental influences are largely independent. The figure does not convey the fact that the heritabilities of tests of school achievement are less than the heritability of general cognitive ability.

are substantial, results implying that a general genetic factor affects performance on the various tests of school achievement (Martin, Jardine, & Eaves, 1984; Plomin & DeFries, 1979).

Could this general genetic factor that affects scores on diverse tests of school achievement be general cognitive ability? Multivariate genetic analyses between tests of school achievement and general cognitive ability suggest that this is the case. That is, genetic effects on school achievement test scores overlap substantially with genetic effects on general cognitive ability, although some genetic influence is specific to achievement (Thompson et al., 1991; Wadsworth, 1994) (see Figure 9.5). An interesting implication is that achievement test scores within the normal range that are independent of ability may be largely environmental in origin.

SUMMING UP

Although school achievement tests are sometimes assumed to assess effort rather than ability, twin studies consistently show genetic influence for such test scores. Developmental analyses indicate that heritability increases and the effects of shared environment diminish during the school years. Multivariate genetic analyses of diverse tests of school achievement indicate that a general genetic factor underlies these measures. Multivariate genetic analyses between tests of school achievement and general cognitive ability suggest that genetic influence on school achievement test scores within the normal range is largely due to genetic influence on general cognitive ability. Differences between the two types of tests are environmental in origin.

Identifying Genes

As discussed in relation to general cognitive ability (Chapter 8), research has begun to identify specific genes responsible for the substantial heritability of cognitive abilities. Research with nonhuman animals has focused on specific cognitive abilities rather than on general cognitive ability. As mentioned in Chapter 6, a targeted mutation in mice that knocks out a particular gene (α-CaMKII) active in the hippocampus impairs spatial learning (Silva et al., 1992). Some of the most sophisticated work on memory has involved fruit flies. For example, a gene (cAMP-responsive element-binding protein, *CREB*) regulates genes responsible for synthesizing proteins involved in long-term memory (Yin et al., 1995). *CREB* produces at least seven different transcription factors, which control whether other genes will be transcribed. Some of these are "repressors," which prevent the transcription of other genes. Other transcription factors are "activators," which promote transcription. *CREB* was altered to produce large amounts of repressors or large amounts of activators. The altered genes were transferred to fruit flies, with the interesting addition of a promoter that turned on the gene only when the flies were exposed to heat. At room temperature, the two types of flies learned normally. But when the temperature was raised, the high-repressors had difficulty learning, whereas the high-activators learned much more quickly.

In the human species, research is underway to identify associations between DNA markers and specific cognitive abilities. In Chapter 7, a replicated QTL linkage with reading disability was described (Cardon et al., 1994). It is possible that this linkage is not for a QTL for reading disability itself but rather for variation in reading ability. Also, as discussed in Chapter 7, allele 4 of the apolipoprotein E gene is strongly associated with late-onset Alzheimer's disease. Research on dementia is also beginning to consider normal variation in cognitive abilities. For example, as mentioned in Chapter 8, this allele has been reported to be associated with decline in cognitive performance in elderly populations (Feskens et al., 1994). The first study of specific cognitive abilities has shown that the allele is associated with normal variation in an elderly population in tests of memory and of perceptual speed, but not in performance on a test of reading ability (Henderson et al., 1995).

Other attempts to identify genes associated with normal variation in specific cognitive abilities in younger populations are also beginning to appear. For example, a DNA marker for a dopamine receptor gene (*DRD2*) was reported to be associated with a particular type of spatial ability (Berman & Noble, 1995), although this marker is not associated with general cognitive ability (Petrill et al., in press-b).

In Chapter 8, we described a study that examined allelic associations between 100 DNA markers and general cognitive ability (Plomin et al., 1995). Although only one association replicated cleanly in an independent sample, five other associations were significant when the original and replication samples were combined. These are associations with general cognitive ability. What about specific cognitive abilities? The levels-of-processing model discussed earlier predicts that most genes associated with cognitive abilities will be associated with all abilities, but that some genes will be specific to a particular ability. The results supported this prediction (Petrill & Plomin, in press). Four of the six

DNA markers were associated with all specific cognitive abilities, but two markers were associated only with spatial and memory abilities. Although these are preliminary results in need of replication, they provide a glimpse of the future of molecular genetic research on specific cognitive abilities.

The intense molecular genetic research on cognitive disabilities such as dementia, combined with the power of transgenic research in mice and fruit flies, seems likely to ignite an explosion of research that attempts to identify genes responsible for the heritability of specific cognitive abilities.

Summary

Many specific cognitive abilities show genetic influence in twin studies, although the magnitude of the genetic effect is generally lower than that for general cognitive ability. Family and twin studies suggest that the genetic contribution may be stronger for some cognitive abilities such as verbal and spatial than for other abilities, especially memory. Recent studies of twins reared apart confirm these findings. Developmental genetic analyses indicate that genetically distinct specific cognitive abilities can be found as early as three years of age and show increasing genetic differentiation from early to middle childhood.

Information-processing measures also show genetic influence in twin studies. Although information-processing research assumes a bottom-up model in which different genes affect basic elements of information processing, multivariate genetic analyses provide stronger evidence for a top-down model in which genes primarily affect general cognitive ability. A compromise model is a levels-of-processing model in which each level of processing has unique as well as common genetic effects.

School achievement tests, and even report card grades, show genetic influence, which appears to increase in magnitude during the school years. Multivariate genetic research indicates that genetic influence on variations within the normal range on school achievement tests is largely due to genetic influence on general cognitive ability.

Genes for learning and memory have been identified in mice and fruit flies, and research is underway to identify associations between DNA markers and specific cognitive abilities in the human species. The levels-of-processing model predicts that most genes associated with cognitive abilities will be associated with general cognitive ability, but some genes may be associated with just one specific cognitive ability.

CHAPTER TEN

Psychopathology

Psychopathology has been the most active area of behavioral genetic research during the past decade, largely because of the social importance of mental illness. One out of two persons in the United States has a serious psychological episode during his or her lifetime, and one out of three persons suffered from the disorder within the last year (Kessler et al., 1994). The costs in terms of suffering to patients and their friends and relatives, as well as the economic costs, make psychopathology one of the most pressing problems today.

This chapter provides an overview of what is known about the genetics of several major categories of adult psychopathology: schizophrenia, mood disorders, and anxiety disorders. Other disorders such as posttraumatic stress disorder, somatoform disorders, and eating disorders are also briefly reviewed, as are disorders usually first diagnosed in childhood: autism, attention-deficit hyperactivity, and tic disorders. Other major categories in the *Diagnostic and Statistical Manual of Mental Disorders-IV* (DSM-IV) include cognitive disorders such as dementia (Chapter 7), personality disorders (Chapter 11), and drug-related disorders (Chapter 12). The DSM-IV includes several other disorders for which no genetic research is as yet available. Much has been written about the genetics of psychopathology, including a recent text (McGuffin et al., 1994) and several edited books (e.g., Gershon & Cloninger, 1994; Hall, 1996).

Schizophrenia

Schizophrenia involves long-term thought disorders (especially delusions), hallucinations (especially hearing voices), and disorganized speech (odd associations and rapid changes of subject). It usually strikes in late adolescence or early adulthood. Early onset in adolescence tends to be gradual but has a worse prognosis.

More genetic research has focused on schizophrenia than on other areas of psychopathology because it is the most severe form of psychopathology and because it is so common, with a lifetime risk of about 1 percent of the population. In the United States alone, about a million people suffer from schizophrenia. Unlike patients of two decades ago, most of these individuals are no longer institutionalized, because drugs control some of their worst symptoms. Nonetheless, schizophrenics still occupy half the beds in mental hospitals, and those released make up about 10 percent of the homeless population (Fischer & Breakey, 1991). It has been estimated that the cost to our society of schizophrenia alone is greater than that of cancer (National Foundation for Brain Research, 1992).

Family Studies

The basic genetic results were described in Chapter 3 (see Figure 3.6) to illustrate genetic influence on complex disorders (see Gottesman, 1991, for details). Forty family studies consistently show that schizophrenia is familial. A handful of studies that do not show familial resemblance are small studies that lack the power to detect resemblance (Kendler, 1988). In contrast to the base rate of 1 percent lifetime risk in the population, the risk for relatives increases with genetic relatedness to the schizophrenic proband: 4 percent for second-degree relatives and 9 percent for first-degree relatives.

The average risk of 9 percent for first-degree relatives differs for parents, siblings, and offspring of schizophrenics. In 14 family studies of over 8000 schizophrenics, the median risk was 6 percent for parents, 9 percent for siblings, and 13 percent for offspring. The low risk for parents of schizophrenics (6 percent) is probably due to the fact that schizophrenics are less likely to marry and those who do marry have relatively few children. For this reason, parents of schizophrenics are less likely than expected to be schizophrenic. When schizophrenics do become parents, the rate of schizophrenia in their offspring is high (13 percent). The risk is the same regardless of whether the mother or the father is schizophrenic. When both parents are schizophrenic, the risk for their offspring shoots up to 46 percent. Siblings provide the least biased risk estimate, and their risk (9 percent) is in between the estimates for parents and for offspring. Although the risk of 9 percent is high, nine times the population risk of 1 percent, it should be remembered that the great majority of schizophrenics do not have a schizophrenic first-degree relative.

The family design provides the basis for *genetic high-risk studies* of the development of children whose mothers were schizophrenic. In one of the first such studies, begun in the early 1960s in Denmark, 200 such offspring are now between 40 and 50 years old (Parnas et al., 1993). In the high-risk group, 16 percent have been diagnosed as schizophrenic (whereas 2 percent in the low-risk group are schizophrenic), and the children who eventually became schizophrenic had mothers whose schizophrenia was more severe. These children experienced a less stable home life and more institutionalization, reminding us that family studies do not disentangle nature and nurture the way an adoption study does. The schizophrenic children were also more likely to have birth complications, particularly prenatal viral infection (Cannon et al., 1993). They

also showed attention problems in childhood, especially problems in "tuning
out" incidental stimuli like the ticking of a clock (Hollister et al., 1994). Similar
results were found in childhood in one of the best U.S. genetic high-risk studies,
which also found more personality disorders in the offspring of schizophrenic
parents when the offspring were young adults (Erlenmeyer-Kimling et al.,
1995). Fifteen long-term genetic high-risk studies have cooperated in the Risk
Research Consortium, which includes 1200 children with at least one schizophrenic parent and 1400 normal control subjects (Watt et al., 1984). Much
more will be learned from these studies as their subjects complete the age of
risk for schizophrenia.

Twin Studies

Twin studies show that genetics contributes importantly to familial resemblance
for schizophrenia. Twin concordances are 48 percent for identical twins and
17 percent for fraternal twins, results averaged over the four most recent twin
studies (Gottesman, 1991). Translated into liability-threshold model correlations, these concordances suggest a liability heritability of about 60 percent, as
explained in Chapter 5.

A dramatic case study involved identical quadruplets, called the Genain
quadruplets, all of whom were schizophrenic, although they varied considerably
in severity of the disorder (DiLisi et al., 1984). In another study of 14 pairs of
identical twins reared apart before two years of age in which at least one member of each pair became schizophrenic, 9 pairs (64 percent) were concordant
(Gottesman, 1991).

Despite the strong and consistent evidence for genetic influence provided by
the twin studies, it should be remembered that the average concordance for
identical twins is only about 50 percent, which means that half of the time, these

CLOSE UP

Irving I. Gottesman earned his doctorate in 1960 in child and adult clinical psychology from the University of Minnesota. Mentored by the geneticist Sheldon C. Reed of the Dight Institute of Human Genetics, he completed a dissertation on a twin study of personality and of intelligence. Since 1985, he has been at the University of Virginia, where he is the Sherrell J. Aston Professor of Psychology and professor of pediatrics (medical genetics). In the early 1960s, Gottesman joined forces with James Shields as a postdoctoral fellow in psychiatric genetics in Eliot Slater's MRC Unit at the Institute of Psychiatry in London. This collaboration led to their classic 1972 monograph on schizophrenia and genetics, a twin study based on 16 years of consecutive admissions to the inpatient and outpatient beds of the Maudsley and Royal Bethlem hospitals. With Danish colleagues, he has reported high risks for schizophrenia in the offspring of normal co-twins of probands, and he has found appreciable heritability for criminal offending in a birth cohort of twins. More recently, Gottesman has joined the hunt for the QTLs involved in schizophrenia by using genome-wide scans with Hans Moises in Germany and an international collaborative team.

genetically identical pairs of individuals are *dis*cordant for schizophrenia, an outcome that provides strong evidence for the importance of nongenetic factors.

Because differences within pairs of identical twins cannot be genetic in origin, the *co-twin control method* can be used to study nongenetic reasons why one identical twin is schizophrenic and the other is not. One early study of discordant identical twins found few life history differences except that the schizophrenic co-twins were more likely to have had birth complications and some neurological abnormalities (Mosher, Pollin, & Stabenau, 1971). A recent study also found changes in brain structures and more frequent birth complications for the schizophrenic co-twin in discordant twin pairs (Torrey et al., 1994).

An interesting finding has emerged from another use of discordant twins: studying their offspring. The simplest conclusion is that discordant identical twins are direct proof of nongenetic influences, because the twins are identical genetically yet discordant for schizophrenia. That is, it is possible that the unaffected twins, for environmental reasons, managed to escape the genetic risk to which their twin partner succumbed. If this were the case, offspring of identical twins will inherit the same genetic risk factors regardless of whether their parent was the schizophrenic or nonschizophrenic twin. One study supports this hypothesis (Gottesman & Bertelsen, 1989). The incidence of schizophrenia and schizophrenic-like disorders in the offspring of the nonschizophrenic twin was just as great as in the offspring of the schizophrenic twin. In other words, the identical twin who does not succumb to schizophrenia nonetheless transmits the illness to offspring to the same extent as does the schizophrenic twin.

For the offspring of discordant fraternal twins, the offspring of the twin who was schizophrenic were at much greater risk than were offspring of the nonschizophrenic twin. Because members of discordant fraternal twin pairs, unlike identical twins, differ genetically as well as environmentally, the offspring of the schizophrenic twin may be at increased genetic risk for schizophrenia. However, the sample sizes are small (another small study by Kringlen & Cramer, 1989, found different results), and the study has been criticized on other grounds (Torrey, 1990). Nonetheless, these data provide food for thought about the complex interactions between nature and nurture.

Adoption Studies

Results of adoption studies agree with those of family and twin studies in pointing to genetic influence in schizophrenia. As described in Chapter 5, the first adoption study of schizophrenia by Leonard Heston in 1966 is a classic study. The results (see Box 5.1) showed that the risk of schizophrenia in adopted-away offspring of schizophrenic biological mothers was 11 percent (5 of 47), much greater than the 0 percent risk for 50 adoptees whose birth parents had no known mental illness. The risk of 11 percent is similar to the risk for offspring reared by their schizophrenic biological parents. This finding not only indicates that family resemblance for schizophrenia is largely genetic in origin but it also implies that growing up in a family with schizophrenics does not increase the risk for schizophrenia beyond the risk due to heredity.

Box 5.1 also mentioned that Heston's results have been confirmed and extended by other adoption studies. Two Danish studies began in the 1960s with 5500 children adopted between 1924 and 1947 and 10,000 of their 11,000 biological parents. One of the studies (Rosenthal et al., 1968, 1971) used the adoptees' study method. This method is the same as that used in Heston's study, but important experimental controls were added. Because biological parents typically relinquish children for adoption when the parents are in their teenage years, but schizophrenia does not usually occur until later, the adoption agencies and the adoptive parents were not aware of the diagnosis in most cases. In addition, schizophrenic fathers as well as mothers were studied to assess whether Heston's results, which only involved mothers, were influenced by prenatal maternal factors.

Biological parents who had been admitted to a psychiatric hospital were identified. Biological mothers or fathers who were diagnosed as schizophrenic and whose children had been placed in adoptive homes were selected. This

procedure yielded 44 birth parents (32 mothers and 12 fathers) who were diagnosed as chronic schizophrenics. Their 44 adopted-away children were matched
to 67 control adoptees whose birth parents had no psychiatric history, as indicated by the records of psychiatric hospitals. The adoptees, with an average age
of 33 years, were interviewed for three to five hours by an interviewer blind to
the status of their birth parents.

Three (7 percent) of the 44 proband adoptees were chronic schizophrenics, whereas none of the 67 control adoptees were (Figure 10.1). Moreover, 27
percent of the probands showed schizophrenic-like symptoms, whereas 18 percent of the controls had similar symptoms. Results were similar for 69 proband
adoptees whose parents were selected by using broader criteria for schizophrenia. Results were also similar, regardless of whether the mother or the father
was schizophrenic. The unusually high rates of psychopathology in the Danish
control adoptees may have occurred because the study relied on hospital
records to assess psychiatric status of the birth parents. For this reason, the
study may have overlooked psychiatric problems of control parents that had
not come to the attention of psychiatric hospitals. To follow up this possibility,
the researchers interviewed the birth parents of the control adoptees and
found that one-third fell in the schizophrenic spectrum. Thus, the researchers
concluded that "our controls are a poor control group and . . . our technique
of selection has minimized the differences between the control and index
groups" (Wender et al., 1974, p. 127). This bias is conservative in terms of
demonstrating genetic influence.

An ongoing adoptees' study in Finland confirms these results (Tienari et al.,
1994). About 10 percent of adoptees who had a schizophrenic biological parent
showed some form of psychosis, whereas 1 percent of control adoptees had similar
disorders. This study also suggested genotype-environment interaction in that
adoptees whose biological parents were schizophrenic were more likely to have
schizophrenia-related disorders when the adoptive families functioned poorly.

The second Danish study (Kety et al., 1994) used the adoptees' family
method, focusing on 47 of the 5500 adoptees diagnosed as chronic schizophrenic.

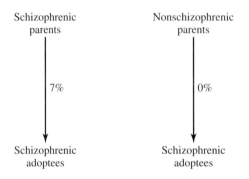

Figure 10.1 Danish adoption
study of schizophrenia:
adoptees' study method.

A matched control group of 47 nonschizophrenic adoptees was also selected. The biological and adoptive parents and siblings of the index and control adoptees were interviewed. For the schizophrenic adoptees, the rate of chronic schizophrenia was 5 percent (14 of 279) for their first-degree biological relatives and 0 percent (1 of 234) for the biological relatives of the control adoptees. The adoptees' family method also provides a direct test of the influence of the environmental effect of having a schizophrenic relative. If familial resemblance for schizophrenia were caused by family environment brought about by schizophrenic parents, schizophrenic adoptees should be more likely to come from adoptive families with schizophrenia, relative to the control adoptees. To the contrary, 0 percent (0 of 111) of the schizophrenic adoptees had adoptive parents or siblings who were schizophrenic—like the 0 percent incidence (0 of 117) for the control adoptees (see Figure 10.2).

This study also included many biological half siblings of the adoptees (Kety, 1987). Such a situation arises when biological parents relinquish a child for adoption and then later have another child with a different partner. The comparison of biological half siblings who have the same father (paternal half siblings) with those who have the same mother (maternal half siblings) is particularly useful for examining the possibility that the results of adoption studies may be affected by prenatal factors, rather than by heredity. Data on paternal half siblings cannot be influenced by prenatal factors because they were born to different mothers. For half siblings of schizophrenic adoptees, 16 percent (16 of 101) were schizophrenic; for half siblings of control adoptees, only 3 percent (3 of 104) were schizophrenic. The results were the same for maternal and paternal half siblings, suggesting that prenatal factors are not likely to be of major importance in the origin of schizophrenia.

In summary, the adoption studies clearly point to genetic influence. Moreover, adoptive relatives of schizophrenic probands do not show increased risk for schizophrenia. These results imply that familial resemblance for schizophrenia is due to heredity rather than to shared family environment.

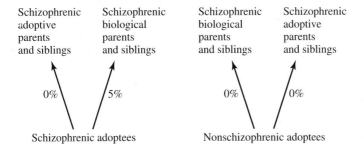

Figure 10.2 Danish adoption study of schizophrenia: adoptees' family method.

Schizophrenia or Schizophrenias?

Is schizophrenia one disorder or is it a heterogeneous collection of disorders? Multivariate genetic analysis can address this fundamental issue of heterogeneity. The classical subtypes of schizophrenia—such as catatonic schizophrenia (disturbance in motor behavior) and paranoid schizophrenia (persecution delusions)—are not supported by genetic research. That is, although schizophrenia runs in families, the particular subtype does not. This result is seen most dramatically in a follow-up of the Genain quadruplets (DeLisi et al., 1984). Although they were all diagnosed as schizophrenic, their symptoms varied considerably.

There is evidence that more severe schizophrenia is more heritable than milder forms (Gottesman, 1991). An alternative to the classical subtypes is a distinction largely based on severity (Crow, 1985). Type I schizophrenia, which has a better prognosis, involves active symptoms such as hallucinations and good response to drugs. Type II schizophrenia, which is more severe, has a poorer prognosis and passive symptoms such as withdrawal and lack of emotion. Type II schizophrenia appears to be more heritable than Type I (Dworkin & Lenzenweger, 1984). Another approach to the problem of heterogeneity divides schizophrenia on the basis of family history (Murray, Lewis, & Reveley, 1985), although there are problems with this approach (Eaves, Kendler, & Schulz, 1986) and there is clearly no simple dichotomy (Jones & Murray, 1991). These typologies seem more likely to represent a continuum from less to more severe forms of the same disorder, rather than genetically distinct disorders (McGuffin, Farmer, & Gottesman, 1987).

A related strategy is the search for behavioral markers of genetic liability, called *endophenotypes* (Gottesman & Shields, 1972). One example of current interest in genetic research is called smooth-pursuit eye tracking. This term refers to the ability to follow a moving object smoothly with one's eyes without moving the head (Levy et al., 1993). Some studies have shown that schizophrenics whose eye tracking is jerky tend to have more negative symptoms, and their relatives with poor eye tracking are more likely to show schizophrenic-like behaviors (Clementz, McDowell, & Zisook, 1994). However, some research does not support this hypothesis (Torrey et al., 1994). The hope is that such endophenotypes can clarify the inheritance of schizophrenia and assist attempts to find specific genes responsible for schizophrenia (Iacono & Grove, 1993).

Although some researchers assume that schizophrenia is heterogeneous and needs to be split into subtypes, others argue in favor of the opposite approach, lumping schizophrenia-like disorders in a broader spectrum of schizoid disorders (Farmer, McGuffin, & Gottesman, 1987; McGue & Gottesman, 1989). It is possible that schizophrenia represents the extreme of a quantitative dimension that extends into normality.

Ultimately, these crucial issues about splitting and lumping will be resolved by molecular genetics. When genes that are associated with schizophrenia are found, the question is whether they will relate to a particular type of schizophrenia, as assumed by the "splitters." Or, at the other extreme, will these genes for schizophrenia relate to a continuum of thought disorders that extends into normal behavior, such as social withdrawal, attention problems, and magical thinking?

Identifying Genes

Before the new DNA markers were available, attempts were made to associate classical genetic markers such as blood groups with schizophrenia. A weak association was found with the major gene involved in immune response (HLA), a gene associated with many diseases. In seven of nine studies, this gene was associated with schizophrenia, especially schizophrenia marked by paranoid delusions (McGuffin & Sturt, 1986); the association was not found in a more recent study, however (Alexander et al., 1990). If the association exists, it is a weak association, accounting for about 1 percent of the genetic liability to schizophrenia. Recent association studies focus on candidate genes, such as those involved in the dopamine system, but have not yet found replicated associations.

Although schizophrenia was one of the first behavioral domains put under the spotlight of molecular genetic analysis, it has been slow to reveal evidence for specific genes. In the 1980s' euphoria of using the new DNA markers to find genes for complex traits, some claims were made for linkage, but they could not be replicated. The first was a claim for linkage with an autosomal dominant gene on chromosome 5 for Icelandic and British families (Sherrington et al., 1988). However, combined data from five other studies in other countries failed to confirm the linkage (McGuffin et al., 1990).

Large collaborative efforts have been launched in Europe and in North America to conduct systematic linkage studies across the genome. If there is a major gene responsible for schizophrenia, these studies will detect it within the next few years. The first payoff of these huge efforts may be a locus on the short arm of chromosome 6 (6p24–22). A linkage study of 265 pedigrees reported linkage in this region for about a quarter of families with multiple schizophrenics (Straub et al., 1995). Evidence suggesting linkage in this region was also found in two other pedigree studies (Antonarakis et al., 1995; Moises et al., 1995) and in a study of affected sibling pairs (Schwab et al., 1995). However, the studies differed in whether they found strongest evidence for linkage for narrow or broad definitions of schizophrenia and for dominant or recessive hypotheses. Moreover, despite positive evidence from these four studies, two other studies did not find evidence for linkage in this region of chromosome 6. More research is needed to confirm this linkage, but the results so far warrant cautious optimism (Peltonen, 1995).

S U M M I N G UP

For schizophrenia, lifetime risk is about 1 percent in the general population, 10 percent in first-degree relatives whether reared together or adopted apart, 17 percent for fraternal twins, and 48 percent for identical twins. This pattern of results indicates substantial genetic influence as well as nonshared family environmental influence. Genetic high-risk studies and co-twin control studies suggest that, within genetically high-risk groups, birth complications and attention problems in childhood predict schizophrenia, which usually strikes in early adulthood. Genetic influence has been found for both the adoptees' study method, like that used in the first adoption study by Heston, and the adoptees' family method. More severe schizophrenia (Type II) may be more heritable than less severe forms (Type I). Schizophrenia may be linked to chromosome 6.

Mood Disorders

Mood disorders involve severe swings in mood, not just the "blues" that all people feel on occasion. For example, the lifetime risk for suicide for people diagnosed as having mood disorders has been estimated as 19 percent (Goodwin & Jamison, 1990). There are two major categories of mood disorders: depressive disorders and bipolar disorders.

Major depressive disorder usually has a slow onset over weeks or even months. Each episode typically lasts several months and ends gradually. Characteristic features include depressed mood, loss of interest in usual activities, disturbance of appetite and sleep, loss of energy, and thoughts of death or suicide. Major depressive disorder affects an astounding number of people. In a recent U.S. survey, the lifetime risk is about 17 percent, with a risk two times greater for women than for men after adolescence (Blazer et al., 1994). Moreover, the problem is getting worse: Each successive generation born since World War II has higher rates of depression (Burke et al., 1991). These temporal trends indicate environmental influence.

Bipolar disorder, as its name implies, alternates between the depressive pole and the other pole of mood, called mania. Mania involves euphoria, inflated self-esteem, sleeplessness, talkativeness, racing thoughts, distractibility, hyperactivity, and reckless behavior. Mania typically begins and ends suddenly and lasts from several days to several months. Mania is sometimes difficult to diagnose and, for this reason, DSM-IV has distinguished bipolar I disorder with a clear manic episode and bipolar II disorder with a less clearly defined manic episode. Bipolar disorder is much less common than major depression, with an incidence of about 1 percent of the adult population and no gender difference (Kessler et al., 1994).

Family Studies

For 70 years, family studies have shown increased risk for first-degree relatives of individuals with mood disorders (Slater & Cowie, 1971). Since the 1960s, researchers have considered major depression separately from bipolar depression. A review of a dozen family studies of bipolar depression yielded an average risk of about 8 percent in first-degree relatives; the base rate is 1 percent (McGuffin & Katz, 1986). In seven studies of major depression, the family risk was 9 percent, whereas the base rate was about 3 percent (Figure 10.3). The risks in these studies are low relative to the frequency of the disorder mentioned earlier, because these studies focused on severe depression, often requiring hospitalization. The range of family risks estimated from these studies is great, probably reflecting differences in diagnostic criteria. The affected offspring of severely depressed parents, especially those with early onset of the disorder, appear to show depression sometimes even in childhood (Weissman et al., 1992).

It has been hypothesized that the distinction between major depression (often called *unipolar* depression) and bipolar depression is primarily a matter of severity, that is, bipolar depression may be a more severe form of depression (McGuffin & Katz, 1986). The basic multivariate finding from family studies is that relatives of unipolar probands are not at increased risk for bipolar depression (less than 1 percent in the above review), but relatives of bipolar probands are at increased risk (11 percent) for unipolar depression. If we postulate that

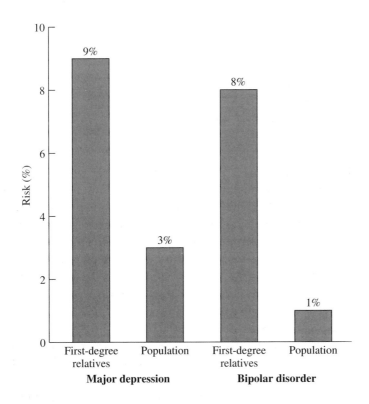

Figure 10.3 Family studies of mood disorders.

CLOSE **UP**

Peter McGuffin is professor and head of the Division of Psychological Medicine at the University of Wales College of Medicine, Cardiff. He graduated from Leeds University Medical School in 1972 and underwent a period of postgraduate training in internal medicine before specializing in psychiatry at the Bethlem Royal and Maudsley hospitals, London. In 1979 he was awarded a Medical Research Council Fellowship to train in genetics at the Institute of Psychiatry in London and at Washington University Medical School, St. Louis, Missouri. During this time, he completed the work for his doctoral dissertation, which constituted one of the first genetic linkage studies on schizophrenia. He went on to carry out family and twin studies of depression and other psychiatric disorders, attempting to integrate the investigation of genetic and environmental influences. Although he continues with this general theme, his current work focuses more specifically on molecular genetic techniques and their applications in the study of both normal and abnormal behaviors.

bipolar depression is a more severe form of depression, this model would explain why familial risk is greater for bipolar depression, why bipolar probands have an excess of unipolar relatives, and why unipolar probands do not have many relatives with bipolar depression. It is possible that the continuum extends down to normal sadness (Vrendenberg, Flett, & Krames, 1993).

Are there subtypes of major depressive disorder? There is a long history of trying to subdivide depression into reactive (triggered by an event) and endogenous (coming from within) subtypes. Family studies provide little support for this distinction. Although one would expect that endogenous depression should show greater familiality than reactive depression, no difference in family history is found for the two types of depression (Rush & Weissenburger, 1994).

One distinction is that major depression is twice as likely in offspring when the parents' first depressive episode occurred before they were 20 years old (Weissman et al., 1988). Interestingly, this increased familial resemblance is not found with respect to onset in childhood (Harrington, Rutter, & Fombonne, in press). In contrast, when the first depressive episode occurred after 40 years of age, relatives are at no increased risk for depression (Price, Kidd, & Weissman, 1987). Another promising direction for subdividing depression is in terms of response to drugs, because the therapeutic response to specific antidepressants tends to run in families (Tsuang & Faraone, 1990).

Twin Studies

Twin studies yield evidence for substantial genetic influence for mood disorders. For unipolar depression, a summary of earlier twin studies, typically involving severe hospitalized cases, yielded average twin concordances of 40 and 11 percent for identical and fraternal twins, respectively (Allen, 1976). For bipolar depression, average twin concordances were 72 and 40 percent (see Figure 10.4). A subsequent twin study of both unipolar and bipolar depression yielded identical and fraternal twin concordances of 43 and 18 percent for unipolar depression and 62 and 8 percent for bipolar depression (Bertelsen, Harvald, & Hauge, 1977). Two other twin studies of hospitalized unipolar depressives found similar results, with average identical and fraternal twin concordances of 53 and 24 percent (McGuffin, Katz, & Rutherford, 1991; Torgersen, 1986). Twelve pairs of identical twins reared apart in which at least one member of each pair had suffered from major depression have been reported. Eight of the 12 pairs (67 percent) were concordant for major depression (Bertelsen, 1985).

Less severe unipolar depression may show less genetic influence. For example, in a population-based sample of female twins, 30 percent of the individuals had been diagnosed as having had a major depressive episode (Kendler et al., 1992a). Identical and fraternal twin concordances for this less severe diagnosis of depression were 49 and 42 percent, respectively. Although these concordances suggest little genetic influence, a liability-threshold analysis estimated moderate

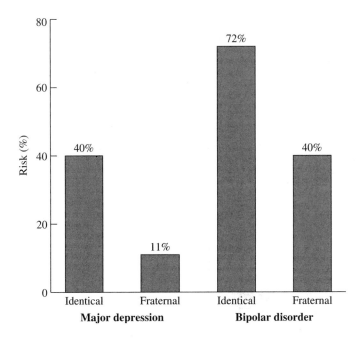

Figure 10.4 Approximate twin results for mood disorders.

genetic influence. A reanalysis of these data focusing on more severe cases (a life-time prevalence of 4 percent rather than 30 percent) found greater evidence for genetic influence (McGuffin et al., 1993). Other evidence is consistent with the hypothesis that more severe depression shows greater genetic influence (McGuffin et al., 1994, 1996). Twin studies of the normal range of depressive symptoms in unselected samples show only modest genetic influence, as discussed in Chapter 12.

Like the research on schizophrenia, a study of offspring of identical twins discordant for bipolar depression has been reported (Bertelsen, 1985). And as in the results for schizophrenia, the same 10 percent risk for mood disorder was found in the offspring of the unaffected twin and in the offspring of the affected twin. This outcome implies that the identical twin who does not succumb to bipolar depression nonetheless transmits a liability for the illness to offspring to the same extent as does the ill twin.

Adoption Studies

Results of adoption research on mood disorders are mixed. One study began with 71 adoptees with a broad range of mood disorders (Wender et al., 1986). Mood disorders were found in 8 percent of the 387 biological relatives of the probands, a risk only slightly greater than the risk of 5 percent for the 344 bio-logical relatives of control adoptees. The biological relatives of the probands showed somewhat greater rates of alcoholism (5 percent versus 2 percent) and of attempted or actual suicide (7 percent versus 1 percent). Two other adoption studies relying on medical records of depression found little evidence for genetic influence (Cadoret et al., 1985a; Von Knorring et al., 1983).

An adoption study that focused on adoptees with bipolar depression found stronger evidence for genetic influence (Mendlewicz & Rainer, 1977). The rate of bipolar disorders in the biological parents of the bipolar adoptees was 7 percent, as compared to 0 percent for the parents of control adoptees. As in the family studies, biological parents of these bipolar adoptees also showed ele-vated rates of unipolar depression (21 percent) relative to the rate for biologi-cal parents of control adoptees (2 percent), a result suggesting that the two dis-orders are not distinct genetically. Adoptive parents of the bipolar and control adoptees differed little in their rates of mood disorders.

In summary, the genetic results for mood disorders are less clear-cut than for schizophrenia, perhaps because of greater problems in diagnosis. (For details, see Tsuang & Faraone, 1990.) Nonetheless, the evidence for genetic influence for bipolar depression is reasonably consistent. More severe major depression also seems to show moderate genetic influence.

SUMMING UP

Family, twin, and adoption data provide a consistent case for genetic influence on bipolar disorder. The case for genetic influence is less clear for major depression, although more severe depression (for example, depression that requires hospitalization) and earlier onset depression (but not childhood depression) show greater genetic influence. Bipolar disorder appears to be a more severe form of depression. As shown in research on schizophrenia, offspring of identical twins discordant for bipolar depression also show the same 10 percent risk for mood disorder, whether they are the offspring of the unaffected twin or the offspring of the affected twin.

Identifying Genes

For decades, the greater risk of major depression for females led to the hypothesis that a dominant gene on the X chromosome might be involved. As explained in Chapter 3, females can inherit the gene on either of their two X chromosomes, whereas males can inherit the gene only on the X chromosome they receive from their mother. Although initially linkage was reported between depression and color blindness, which is caused by genes on the X chromosome (Chapter 3), studies of DNA markers on the X chromosome failed to confirm linkage (Baron et al., 1993). Father-to-son inheritance is common for both major depression and bipolar depression, which argues against X-linkage inheritance. Moreover, as mentioned earlier, bipolar depression shows little sex difference. For these reasons, X linkage seems unlikely (Hebebrand, 1992).

In 1987, linkage was reported for bipolar depression on chromosome 11 in a genetically isolated community of Old Order Amish in Pennsylvania (Egeland et al., 1987). Unfortunately, this highly publicized finding was not replicated in other studies. The original report was withdrawn when follow-up research on the original pedigree with additional data showed that the evidence for linkage disappeared (Kelsoe et al., 1989).

These false starts led to greater caution in the search for genes for mood disorders. As yet, no replicated linkages have been reported for major depressive disorder or bipolar depression, although large-scale systematic linkage studies of bipolar depression are underway in the United States and Europe.

Anxiety Disorders

A wide range of disorders involve anxiety (panic disorder, generalized anxiety disorder, phobias) or attempts to ward off anxiety (obsessive-compulsive disorder). In panic disorder, recurrent panic attacks come on suddenly and unexpectedly, usually lasting for several minutes. Panic attacks often lead to a fear of being in a situation that might bring on more panic attacks (agoraphobia, which literally means "fear of the marketplace"). Generalized anxiety refers to a more chronic state of diffuse anxiety marked by excessive and uncontrollable worrying. In phobia, the fear is attached to a specific stimulus, such as fear of heights (acrophobia) and enclosed places (claustrophobia), or fear of social situations (social phobia). In obsessive-compulsive disorder, anxiety occurs when the person does not perform some compulsive act driven by an obsession—for example, repeated hand-washing to ward off an obsession with hygiene.

Anxiety disorders are not as crippling as schizophrenia or depressive disorders, but they are the most common form of mental illness (Kessler et al., 1994) and can lead to other disorders, notably depression and alcoholism. The lifetime risk for panic disorder is about 3 percent, although about 7 percent of the population experiences at least one panic attack (Kessler et al., 1994). The other anxiety disorders are also very common: generalized anxiety disorder, 5 percent; specific phobias, 11 percent; social phobia, 13 percent; and obsessive-compulsive disorder, 3 percent. Panic disorder, generalized anxiety disorder, and specific phobias are about twice as common in women as in men.

There has been much less genetic research on anxiety disorders than on schizophrenia and mood disorders. For anxiety disorders in general, family studies find familial resemblance (Marks, 1986). Two twin studies suggested genetic influence (Slater & Shields, 1969; Torgersen, 1983), but another did not (Allgulander, Nowak, & Rice, 1991).

For specific anxiety disorders, the evidence for genetic influence is also mixed. A family study of probands hospitalized for panic disorder found a risk of 25 percent in first-degree relatives and 2 percent for controls (Crowe et al., 1983). In one twin study of panic disorder, the concordance rates for identical and fraternal twins were 31 and 10 percent, respectively (Torgersen, 1983), although another twin study of a nonclinical sample found no evidence for genetic influence (Andrews et al., 1990). Family studies indicate that panic disorder and agoraphobia are related but that panic disorder and generalized anxiety disorder are not (Noyes et al., 1986).

Generalized anxiety disorder is also familial. First-degree relatives have a 20 percent risk of generalized anxiety disorder (much higher than the 5 percent risk in the general population), with no increased risk for panic disorder (Noyes et al., 1987). Two twin studies found no evidence for genetic influence (Andrews et al., 1990; Torgersen, 1983); another suggested modest genetic influence (Kendler et al., 1992b).

For specific phobias excluding agoraphobia, a recent family study found risks of 31 percent for first-degree relatives and 11 percent for controls (Fyer et al., 1990). Social phobias also show familial resemblance (Reich & Yates, 1988). A study of social phobia in female twins yielded evidence for modest genetic influence, with concordances of 24 percent for identical twins and 15 percent for fraternal twins (Kendler et al., 1992c). There have been no substantial twin studies of diagnosed phobias, although twin studies of minor phobias in unselected samples indicated genetic influence (Rose & Ditto, 1983; Torgersen, 1979).

For obsessive-compulsive disorder (OCD), family studies provide a wide range of results because of differences in diagnostic criteria and the failure to include control groups (Carter, Pauls, & Leckman, 1995). No adoption data are available on OCD. Twin studies of OCD suggested some genetic influence, but the sample sizes were too small to permit any confidence in the results to date (Carey & Gottesman, 1981). A study of a nonclinical population of twins selected for OCD symptoms found no evidence for genetic influence (Andrews et al., 1990). By using a dimensional rather than a categorical approach, researchers found modest heritability for OCD symptoms in an unselected sample of twins (Clifford, Murray, & Fulker, 1984). In summary, the picture of genetic influence on OCD is not yet clear.

DSM-IV also includes posttraumatic stress disorder (PTSD) as an anxiety disorder, even though its diagnosis depends on a traumatic event that threatens death or serious injury, such as war, assault, or natural disaster. PTSD symptoms include reexperiencing the trauma (intrusive memories and nightmares) and denying the trauma (emotional numbing). One survey estimated that the lifetime risk for one episode is about 1 percent (Davidson et al., 1991). The risk is much higher, of course, in those who have experienced trauma. For example, after a plane crash, as many as a half of the survivors develop PTSD (Smith et al., 1990). About 10 percent of U.S. veterans of the Vietnam War still suffer from PTSD many years later (Weiss et al., 1992). Individual differences in response to trauma appear in part to be due to family factors. PTSD is more likely to afflict people whose families have more psychopathology in general (McFarlane, 1989) and more anxiety disorders in particular (Davidson, Smith, & Kudler, 1989). A study of twins who are veterans of the Vietnam War found genetic influence on PTSD (True et al., 1993).

Molecular genetic research on anxiety disorders is scarce, perhaps because the evidence for genetic influence is not clear. However, linkage studies of panic disorder are underway (Crowe, 1994; Vieland et al., 1994).

Other Disorders

DSM-IV includes many other categories of disorders, mentioned earlier, for which next to nothing is known about their genetics. Interesting results are emerging, however, from the early stages of genetic research on two of these categories of disorders, somatoform disorders and eating disorders.

In somatoform disorders, psychological conflicts lead to physical (somatic) symptoms such as stomach pains. Somatoform disorders include somatization disorder, hypochondriasis, and conversion disorder. Somatization disorder involves multiple symptoms with no apparent physical cause. Hypochondriacs worry that a specific disease is about to appear. Conversion disorder, which was formerly called hysteria, involves a specific disability such as paralysis with no physical cause. Somatoform disorders show some genetic influence in family, twin, and adoption studies (Guze, 1993). Somatization disorder, which is much more common in women than in men, shows strong familial resemblance for women, but for men it is related to increased family risk for antisocial personality (Guze et al., 1986; Lilienfeld, 1992). An adoption study suggests that this

CLOSE UP

Kenneth S. Kendler, M.D., is Rachel Brown Banks Distinguished Professor of Psychiatry, professor of human genetics, and director of the Psychiatry Genetics Research Program (PGRP) in the Department of Psychiatry in the Medical College of Virginia/Virginia Commonwealth University (MCV/VCU) in Richmond. The PGRP conducts research aimed at clarifying the genetic and environmental risk factors for mental illness. Kendler received a B.A. in 1972 from the University of California at Santa Cruz and an M.D. from Stanford University School of Medicine in 1977. After completing a residency in psychiatry at Yale University School of Medicine in 1980, he served as assistant professor of psychiatry at the Mt. Sinai School of Medicine and as research associate at the Bronx Veterans Administration Medical Center. He came to MCV/VCU in 1985. Kendler's research employs methods from genetic epidemiology, psychiatric genetics, and molecular genetics in the design and implementation of large epidemiological family, linkage, twin, and combined twin-family studies aimed at increasing understanding of the complex and interrelated roles of genetic and environmental risk factors in the etiology of psychiatric and substance abuse disorders.

link between somatization disorder in women and antisocial behavior in men may be genetic in origin (Bohman et al., 1984; Cloninger et al., 1984). Biological fathers of adopted women with somatization disorder showed increased rates of antisocial behavior and alcoholism.

Eating disorders include anorexia nervosa (extreme dieting and avoidance of food) and bulimia nervosa (binge eating followed by vomiting), which occur mostly in adolescent girls and young women. Eating disorders appear to run in families (Spelt & Meyer, 1995). Although some family studies report that families with eating disorders are also at increased risk for mood disorders, the two types of disorders appear to be separate (Rutter et al., 1990). The first twin study of eating disorders found identical and fraternal twin concordances of 59 and 8 percent, respectively, for anorexia, implying strong genetic influence (Treasure & Holland, 1991). In contrast, bulimia showed no genetic influence, with concordances of 36 percent for identical twins and 38 percent for fraternal twins, results suggesting shared family environmental influence. In a study of unselected twins, symptoms of eating disorders showed moderate heritability, again with the exception of bulimic symptoms, which showed shared family environmental influence (Rutherford et al., 1993). However, more research is needed to clarify this issue, because two other twin studies found some genetic influence for bulimia (Fichter & Noegel, 1990; Kendler et al., 1991).

The evidence for substantial heritability of anorexia has motivated several molecular genetic studies, although no consistent linkages or associations have as yet emerged.

SUMMING UP

Genetic research on anxiety disorders has just begun, and results are mixed. Some evidence for genetic influence has been reported for panic disorder, generalized anxiety disorder, obsessive-compulsive disorder, and posttraumatic stress disorder. For many other DSM-IV categories of disorders, no genetic research has as yet been reported, although some evidence for genetic influence has been reported for somatoform disorders and eating disorders, especially anorexia.

Disorders of Childhood

Schizophrenia, mood disorders, and anxiety disorders are typically diagnosed in adulthood. Other disorders emerge in childhood. Mental retardation, learning disorders, and communication disorders were discussed in Chapter 7. Other DSM-IV diagnostic categories that first appear in childhood include pervasive developmental disorders (e.g., autistic disorder), attention-deficit and disruptive behavior disorders (e.g., attention-deficit hyperactivity disorder, conduct disorder), tic disorders (e.g., Tourette's disorder), and elimination disorders (e.g., enuresis). It has been estimated that as many as one out of four children has a diagnosable disorder (Cohen et al., 1993), and one in five has a moderate or severe disorder (Brandenberg, Friedman, & Silver, 1990). Only recently has genetic research begun to focus on disorders of childhood (Rutter et al., 1990).

Autism

Autism was once thought to be a childhood version of schizophrenia but it is now known to be a distinct disorder marked by abnormalities in social relationships, communication deficits, and stereotyped behavior. As traditionally diagnosed, it is relatively uncommon, occurring in 3 to 6 individuals out of every 10,000; and it occurs several times more often in boys than in girls. Most autistic children are delayed in the development of language, and the majority show cognitive impairment. At first, autism was thought to be environmentally caused, either by cold, rejecting parents or by brain damage. Genetics did not seem to be important, because there were no reported cases of an autistic child having an autistic parent and because the risk to siblings was only about 3 to 6 percent (Bolton et al., 1994; Smalley, Asarnow, & Spence, 1988). However, this rate of .03 to .06 is 100 times greater than the population rate of .0003, implying strong familial resemblance. The reason why autistic children do not have autistic parents is that very few autistic individuals marry and have children.

In 1977, the first systematic twin study of autism began to change the view that autism was environmental in origin (Folstein & Rutter, 1977). Four of 11 pairs of identical twins were concordant for autism, whereas none of 10 pairs of fraternal twins were. These pairwise concordance rates of 35 and 0 percent rose to 82 and 10 percent when the diagnosis was broadened to include cognitive disabilities. Co-twins of autistic children are more likely to have speech and language disorders as well as social difficulties. In a follow-up of the twin sample into adult life, problems with social relationships were prominent (Le Couteur et al., 1996). These findings were replicated in other twin studies (Bailey et al., 1995; Steffenburg et al., 1989), and a conservative estimate of the concordance in monozygotic pairs is 60 percent—a thousandfold increase in risk over the general population base rate.

The findings of twin and family studies of autism are broadly consistent and point to several important conclusions that go beyond demonstration of a strong genetic component (Bailey, Phillips, & Rutter, 1996). First, although the underlying liability for autism has a heritability exceeding 90 percent, the very marked falloff in rate from identical to fraternal twins, and from first-degree to second-degree relatives, suggests that a small number of interacting genes are probably responsible (Pickles et al., 1995). Second, the slight increase in obstetric complications seems to have arisen as a result of a genetically abnormal fetus, rather than as a reflection of an environmentally mediated risk factor. Third, the findings have indicated the need to reconceptualize both the frequency and the characteristics of autism. It seems now that the autism phenotype occurs much more often in individuals of normal intelligence than used to be appreciated.

On the basis of these twin and family findings, views regarding autism have changed radically. Instead of being seen as an environmentally caused disorder, it is now considered to be one of the most heritable mental disorders (Rutter et al., 1993). An international collaboration to identify the genes responsible for autism has begun (Rutter, 1996).

Attention-Deficit and Disruptive Behavior Disorders

The DSM-IV grouping of attention-deficit and disruptive behavior disorders is interesting because it includes a disorder that appears to be substantially heritable, attention-deficit hyperactivity disorder, and a disorder that shows only modest genetic influence, conduct disorder, when it occurs in the absence of overactivity/inattention. Although all children have trouble learning self-control, most have made considerable progress by the time they enter school. Those who have not learned self-control are often disruptive, impulsive, and aggressive and have problems adjusting to school.

Attention-deficit hyperactivity disorder (ADHD), as defined by DSM-IV, refers to children who are very restless and have a poor attention span, combined with impulsivity. Estimates of the prevalence of ADHD in North America are quite high, about 4 percent of elementary school children, with boys greatly outnumbering girls (Barkley, 1990). European psychiatrists have tended to take a more restricted approach to diagnosis, with an emphasis on hyperactivity that not only is severe and pervasive across situations but also is of early onset and unaccompanied by high anxiety (Taylor, 1995). There is continuing uncertainty about the merits and demerits of these narrower and broader approaches to diagnosis. However conceptualized, ADHD usually continues into adolescence and, in about a third of cases, continues through into adulthood (Klein & Mannuzza, 1991).

Twin studies have been quite consistent in showing a strong genetic effect on hyperactivity regardless of whether it is measured by questionnaire (Goodman & Stevenson, 1989; Silberg et al., 1996) or by standardized and detailed interviewing (Eaves et al., 1996), and regardless of whether it is treated as a continuously distributed dimension (Thapar, Hervas, & McGuffin, 1995) or as a clinical diagnosis (Gillis et al., 1992). Putting the findings together, a heritability of about 70 percent seems to be a reasonable estimate (Eaves et al., submitted). This value implies that the genetic component is stronger for this disorder than for most other types of psychopathology, other than autism. Adoption studies also lend some support to the hypothesis of genetic influence for ADHD (e.g., Cantwell, 1975), although adoption studies to date have been quite limited methodologically (McMahon, 1980).

In contrast to ADHD, twin studies of juvenile delinquency yield concordance rates of 87 percent for identical twins and 72 percent for fraternal twins, rates that suggest only modest genetic influence (McGuffin & Gottesman, 1985). What is striking about these results is that they imply the strongest influence of shared family environment found for any behavioral disorder. Of course, juvenile delinquency is very common; in inner city areas, at least a quarter of boys acquire a court record. Most of these do not have any form of psychiatric disorder and many grow up to be ordinary, well-functioning adults. Nevertheless, some show a broader pattern of social malfunction that may amount to a diagnosis of conduct disorder. DSM-IV criteria for conduct disorder include aggression, destruction of property, deceitfulness or theft, and other serious violations of rules such as running away from home. Some 5 to 10 percent of children and adolescents meet these diagnostic criteria, with boys again greatly outnumbering girls (Cohen et al., 1993; Rutter et al., in press-b).

Despite this broader pattern of social malfunction, the limited available evidence suggests that genetic influences are only modest. However, the overall heritability figure conceals more than it reveals, because antisocial behavior is so heterogeneous. To begin with, it seems that antisocial behavior that continues into adult life shows a substantially stronger genetic component than that which is confined to the childhood years (DiLalla & Gottesman, 1989; Lyons et al., 1995; Rutter et al., in press-b). Also, however, there is a considerable overlap between hyperactivity and conduct disorder symptoms, as shown both by cross-sectional data and by longitudinal studies (Rutter et al., in press-b). It seems that early-onset hyperactivity is associated with a greatly increased risk for the later onset of conduct problems and that genetic factors play a key role in this co-occurrence of two supposedly separate patterns of symptomatology (Silberg et al., in press). Family studies, too, indicate familial resemblance for ADHD (e.g., Biederman et al., 1986) and an increased risk for conduct disorders in family members of ADHD probands (e.g., Faraone et al., 1991), although such comorbidity is diminished for "pure" ADHD probands who do not show symptoms of conduct disorder (Biederman, Munir, & Knee, 1987).

Genetic findings have been informative, therefore, in showing how the heterogeneous broad category of antisocial behavior can be subdivided. A recent twin study of adolescent males uses a new technique called *latent class analysis*, which attempts to account for patterning among symptoms by hypothesizing underlying (latent) classes (Eaves et al., 1993). One class involves symptoms from both ADHD and conduct disorder, for which a strong genetic influence was found (Silberg et al., 1996). By sharp contrast, there was almost no significant genetic influence for a "pure" class of conduct disorder without hyperactivity, for which there was a strong shared environmental influence (Silberg et al., 1996).

This demonstrated heterogeneity in antisocial behavior probably accounts for some of the inconsistencies in the published research findings. For example, some twin studies of delinquent acts in normal samples of adolescents have shown genetic influence (Rowe, 1983a, 1986), and a study of adopted children with aggressive conduct disorder found increased rates of psychopathology in their biological mothers (Jary & Stewart, 1985). Family studies show that criminality in the parent is a risk factor for juvenile delinquency (Rutter & Giller, 1983), but the evidence suggests that parent criminality functions as both a genetic and an environmental risk factor. Environmentally mediated risks are probably strongest with respect to juvenile delinquency that has an onset in the adolescent years and does not persist into adult life, and genetic effects are probably greatest with respect to early-onset antisocial behavior that is accompanied by hyperactivity and shows a strong tendency to persist into adulthood as an antisocial personality disorder (Moffitt, 1993; Robins & Price, 1991; Rutter et al., in press-b). (See Chapter 11 for a discussion of personality disorders, including antisocial personality disorder.)

Other Disorders

Enuresis (bedwetting) in children after four years of age is common, about 7 percent for boys and 3 percent for girls. An early family study found substantial familial resemblance (Hallgren, 1957). Strong genetic influence was found in two small twin studies (Bakwin, 1971; Hallgren, 1960).

Tic disorders involve involuntary twitching of certain muscles, especially of the face, that typically begin in childhood. Genetic research has focused on the most severe form, called *Tourette's disorder.* Tourette's disorder is rare (about 0.4 percent), whereas simple tics are much more common. Family studies show little familial resemblance for simple tics. However, relatives of probands with chronic, severe tics characteristic of Tourette's disorder are at increased risk for tics of all kinds (Pauls et al., 1990), for obsessive-compulsive disorder (Pauls et al., 1986), and for ADHD (Pauls, Leckman, & Cohen, 1993). A twin study of Tourette's disorder found concordances of 53 percent for identical twins and 8 percent for fraternal twins (Price et al., 1985). Molecular genetic studies are underway (Peterson, Leckman, & Cohen, 1995),

CLOSE UP

David **Pauls** is an associate professor of human genetics in the Child Study Center at the Yale University School of Medicine and holds a joint appointment in the Department of Psychology at Yale. He received his doctorate in genetics and cell biology in 1972 from the University of Minnesota, Minneapolis, with an emphasis in human population genetics. From 1971 through 1977, Pauls was on the faculty of Fresno Pacific College, in the Division of Natural Sciences. In 1977, he went to the University of Iowa, where he completed a postdoctoral fellowship in psychiatric genetics with Raymond Crowe. In 1979, Pauls moved to Yale and there completed a postdoctoral fellowship in psychiatric epidemiology with Kenneth Kidd and Myrna Weissman. In 1983, he assumed his current positions in the Child Study Center and Department of Psychology. The focus of Pauls's research is the inheritance of childhood psychopathology, with primary emphasis on Gilles de la Tourette's disorder, obsessive-compulsive disorder, and related conditions. Recently, he has become involved in work on specific reading disability and Asperger's syndrome. Pauls's research combines the use of quantitative genetic and molecular genetic techniques.

and a tentative association with a dopamine receptor gene has recently been reported (Grice et al., 1996).

SUMMING UP

Genetic research has begun to be applied to disorders that appear in childhood. Twin studies indicate that autism is highly heritable. The DSM-IV category of attention-deficit and disruptive behavior includes attention-deficit hyperactivity disorder, which is substantially heritable, and conduct disorder, which shows only modest genetic influence and substantial influence of shared family environment. Some evidence for genetic influence has also been reported for enuresis and chronic tics.

Co-Occurrence of Disorders

To what extent are mental disorders distinct genetically? That is, are the genes that affect one disorder completely different from the genes that affect another disorder, or do they overlap? Multivariate genetic research (Appendix B) is well suited to address this fundamental question.

One of the few reasonably clear distinctions is between schizophrenia and bipolar disorder. Relatives of schizophrenics are not at increased risk for bipolar disorder. For example, twin partners of identical twins who are schizophrenic have nearly a 50 percent chance of being schizophrenic, but they are at no increased risk for bipolar disorder. However, the genetic distinction between schizophrenia and other mood disorders is not so clear (Crow, 1994; McGuffin et al., 1994). DSM-IV acknowledges a mixed category called schizoaffective disorder.

For less severe disorders, the overlap may be considerable. For example, having two or more psychological disorders is more common than having just one. When one disorder was reported, 80 percent of the time another disorder was also reported in the National Comorbidity Survey of 8000 adults in the United States (Kessler et al., 1994). These co-occurring (often called comorbid) disorders are concentrated in a relatively small number of people. More than half of all reported disorders occurred in the 14 percent of the population who reported three or more disorders. This group also tended to have the most severe disorders.

Are these really different disorders that co-occur, or does this finding call into question current diagnostic systems? Diagnostic systems are based on phenotypic descriptions of symptoms rather than causes. Genetic research offers the hope of systems of diagnosis that take into account evidence on causation. As explained in Appendix B, multivariate genetic analysis of twin and adoptee data can be used to ask whether genes that affect one trait also affect another trait. For example, to what extent are the genes that affect anxiety the same genes that affect depression? Family and twin studies suggest considerable overlap because probands selected for anxiety are very likely to be depressed as well. When probands who are anxious but not depressed are selected, their relatives are at increased risk for anxiety but not for depression in family studies (e.g., Noyes et al., 1987) and twin studies (e.g., Torgersen, 1990). However, anxiety is not typically diagnosed excluding depression. Multivariate genetic analysis in a large twin study of generalized anxiety disorder and major depression found that the overlap in genetic factors that affect the two disorders is nearly complete (Kendler et al., 1992b).

Although current methods for diagnosing disorders show comorbidity, this overlap between disorders may be misleading. For example, most people who suffer from a depressive disorder are also generally anxious. In other words, depressive and anxious symptoms are in part manifestations of the same underlying liability. On the other hand, many people with specific fears do not show depression at all. When more specific anxiety disorders are considered, a more complicated pattern of genetic specificity and genetic generality is likely to emerge.

Nonetheless, on the basis of current diagnostic schemes, evidence for genetic generality across disorders means that when specific genes that are associated with general anxiety are found, we can predict that the same genes will be associated with depression. Molecular genetic research will provide the most direct test of genetic overlap. When genes for a particular disorder are found, will they be specific to that disorder or will they be associated with other disorders as well? The degree of co-occurrence of disorders lends support to the latter hypothesis. If genetic overlap is found to be extensive, as appears to be the case for generalized anxiety and depression, this finding will lead to major changes in the diagnosis of psychopathology.

Summary

Psychopathology is the most active area of research in behavioral genetics. Family, twin, and adoption studies generally point to genetic influence, especially for severe disorders such as schizophrenia and bipolar mood disorder. Psychopathology is also the most active area of behavioral research in terms of attempts to identify specific genes, although such genes remain elusive.

The lifetime risks for schizophrenia are about 1 percent in the population and 10 percent in first-degree relatives whether reared together or adopted apart, 15 percent for fraternal twins, and 50 percent for identical twins. Twin and adoption studies consistently find substantial genetic influence, although the concordance of 50 percent for identical twins provides powerful evidence for the importance of environmental factors as well. Currently there is interest in the possibility that schizophrenia may be due to a gene on chromosome 6.

Two categories of severe mood disorders are major depressive disorder and bipolar disorder. Family, twin, and adoption data provide a consistent case for genetic influence on bipolar disorder. The case is less strong for genetic influence on major depression, although more severe depression (for example, depression that requires hospitalization) and earlier onset depression (but not childhood depression) appear to show stronger genetic influence.

Less genetic research is available for anxiety disorders than for schizophrenia or mood disorders, and its results are more mixed. There is some evidence for genetic influence on panic disorder, generalized anxiety disorder, and post-traumatic stress disorder. It is not yet possible to reach a clear conclusion concerning genetic influence for obsessive-compulsive disorder. Little is known about the genetics of specific phobias, although two twin studies of normal phobias in unselected samples found some genetic influence.

One twin study and one adoption study found genetic influence for somatoform disorders. A twin study of eating disorders suggested the interesting hypothesis that anorexia shows genetic influence but bulimia does not. For many other categories of psychopathology, no genetic research at all has as yet been reported.

Disorders that appear in childhood have only recently received attention in genetic research. Most striking are the results of genetic research on autism. Two decades ago, autism was thought to be an environmental disorder. Now, twin studies suggest that it is one of the most heritable disorders. The DSM-IV category of attention-deficit and disruptive behavior disorders includes a heritable disorder, attention-deficit hyperactivity disorder (ADHD). This category also includes conduct disorder, which shows modest genetic influence and extremely strong influence of shared family environment. Enuresis and chronic tics appear to show genetic influence.

Many disorders co-occur, especially less severe disorders. Multivariate genetic research indicates that genetic overlap between disorders may be responsible for this comorbidity. If molecular genetic research verifies that genes associated with one disorder are also typically associated with other disorders, this finding will revolutionize how psychopathology is diagnosed.

CHAPTER ELEVEN

Personality and Personality Disorders

If you were asked what someone is like, you would probably describe various personality traits, especially those depicting extremes of behavior. "Jennifer is full of energy, very sociable, and unflappable." "Steve is conscientious, quiet, but quick tempered." Genetic researchers have been drawn to the study of personality because, within psychology, personality has always been the major domain for studying the normal range of individual differences, with the abnormal range being the province of psychopathology. Personality traits are relatively enduring individual differences in behavior that are stable across time and across situations (Pervin, 1990). In the 1970s, there was an academic debate about whether personality exists, a debate reminiscent of the nature-nurture debate. Some psychologists argued that behavior is more a matter of the situation than of the person, but it is now generally accepted that both are important and can interact (Kenrick & Funder, 1988; Rowe, 1987). Cognitive abilities (Chapters 8 and 9) also fit the definition of enduring individual differences, but they are usually considered separately from personality. Another definitional issue concerns temperament, personality traits that emerge early in life and, according to some researchers (e.g., Buss & Plomin, 1984), may be more heritable. However, there are many different definitions of temperament (Goldsmith et al., 1987), and the supposed distinction between temperament and personality will not be emphasized here.

Genetic research on personality is extensive, as described in several books (Cattell, 1982; Eaves, Eysenck, & Martin, 1989; Loehlin, 1992; Loehlin & Nichols, 1976) and hundreds of research papers. We shall provide only an overview of this huge literature, in part because its basic message is quite simple: Genes make a major contribution to individual differences in personality, especially when assessed by a self-report questionnaire. After a brief overview

of these results, we shall describe other findings from genetic research on personality and on personality disorders, and recent reports of specific genes associated with personality.

Genetic research on animal personality has focused on traits such as fearfulness and activity level. Some of this work was described in Chapter 5. Animal research is especially useful in identifying specific genes related to personality, as described in Chapter 6. The present chapter focuses on human personality.

Self-Report Questionnaires

The vast majority of genetic research on personality involves self-report questionnaires administered to adolescents and adults. Such questionnaires include from dozens to hundreds of items like, "I am usually shy when meeting people I don't know well" or "I am easily angered." People's responses to such questions are remarkably stable, even over several decades (Costa & McCrae, 1994).

Twenty years ago, a landmark study involving nearly 800 pairs of adolescent twins and dozens of personality traits reached two major conclusions that have stood the test of time (Loehlin & Nichols, 1976). First, nearly all personality traits show moderate heritability. This conclusion might seem surprising, because you would expect some traits to be more heritable than other traits. Second, although environmental variance is also important, virtually all the environmental variance makes children growing up in the same family different from one another. This category of environmental effects is called *nonshared environment.* The second conclusion is also surprising, because theories of personality from Freud onward assumed that parenting played a critical role in personality development. This important finding is discussed in Chapter 14.

Genetic research on personality has focused on five broad dimensions of personality, called the "big five," that encompass many aspects of personality (Goldberg, 1990). The most well studied of these are extraversion and neuroticism. Extraversion encompasses sociability, impulsiveness, and liveliness. Neuroticism (emotional instability) includes moodiness, anxiousness, and irritability. The other three traits included in the big five are agreeableness (likability, friendliness), conscientiousness (conformity, will to achieve), and culture (openness to experience).

Genetic results for extraversion and neuroticism are summarized in Table 11.1 (Loehlin, 1992). For five large and recent twin studies in five different countries, with a total sample size of 24,000 pairs of twins, results indicate moderate genetic influence in all five studies. Correlations are about .50 for identical twins and about .20 for fraternal twins. Studies of twins reared apart also indicate genetic influence, as do adoption studies of extraversion. For neuroticism, adoption results point to less genetic influence than do the twin studies—indeed, the sibling data indicate no genetic influence at all. Lower heritability in adoption than in twin studies could be due to nonadditive genetic variance, which makes identical twins more than twice as similar as first-degree relatives. It could also be due to a special environmental effect that boosts identical twin similarity (Plomin, Chipuer, & Loehlin, 1990a). Model-fitting analyses across these twin

TABLE 11.1

Twin, Family, and Adoption Results for Extraversion and Neuroticism

Type of Relative	Correlation	
	Extraversion	**Neuroticism**
Identical twins reared together	.51	.46
Fraternal twins reared together	.18	.20
Identical twins reared apart	.38	.38
Fraternal twins reared apart	.05	.23
Nonadoptive parents and offspring	.16	.13
Adoptive parents and offspring	.01	.05
Nonadoptive siblings	.20	.09
Adoptive siblings	−.07	.11

SOURCE: *Loehlin (1992).*

and adoption designs produce heritability estimates of 49 percent for extraversion and 41 percent for neuroticism (Loehlin, 1992). The fact that the heritability estimates are much less than 100 percent implies that environmental factors are important, but, as mentioned earlier, this environmental influence is almost entirely due to nonshared environmental effects.

Heritabilities in the 30 to 50 percent range are typical of personality results (Figure 11.1). Much less genetic research has been done on the other three traits of the big five. Also, agreeableness, conscientiousness, and culture have been measured differently in different studies because, until recently, no standard measures were available. A model-fitting summary of family, twin, and adoption data for scales of personality thought to be related to these three traits yielded heritability estimates of 35 percent for agreeableness, 38 percent for conscientiousness, and 45 percent for culture (Loehlin, 1992). The first genetic study to use a measure specifically designed to assess the big five factors found somewhat different estimates in an analysis of twins reared together and twins reared apart (Bergeman et al., 1993). Heritability estimates were 12 percent for agreeableness, 40 percent for conscientiousness, and 29 percent for culture.

Do these broad big five factors represent the best level of analysis for genetic research? Multivariate genetic research indicates that a narrower focus will add to our understanding because subtraits within each big five factor show some unique genetic variance not shared with other traits in the factor (Loehlin, 1992). For example, extraversion includes diverse traits such as sociability, impulsiveness, and liveliness, as well as activity, dominance, and sensation seeking. Each of these traits has received some attention in genetic research but not nearly as much as the more global traits of extraversion and neuroticism. In addition, there are other theories about the ways in which personality should be sliced. For example, a recent neurobiologically oriented theory organizes personality into four different domains: novelty seeking, harm avoidance, reward dependence, and persistence (Cloninger, 1987a). Several theories of personality development have been proposed, and some support for genetic influence has

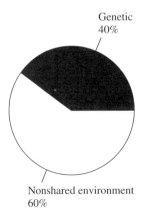

Genetic
40%

Nonshared environment
60%

Figure 11.1 Genetic results for personality traits assessed by self-report questionnaires are remarkably similar, suggesting that 30 to 50 percent of the variance is due to genetic factors. Environmental variance is also important, but hardly any environmental variance is due to shared environmental influence.

been found for the different traits highlighted in these theories (Kohnstamm, Bates, & Rothbart, 1989).

Sensation seeking, which is related to conscientiousness as well as to extraversion (Zuckerman, 1994), is especially interesting because it is the domain of the first association reported between a specific gene and normal personality, as described later. In one large twin study, twin correlations were .60 for identical twins and .21 for fraternal twins for a measure of general sensation seeking (Fulker, Eysenck, & Zuckerman, 1980). This evidence for substantial genetic influence is supported by results from a study of identical twins reared apart, which yielded a correlation of .54 (Tellegen et al., 1988). Sensation seeking itself can be broken down into components, such as disinhibition (seeking sensation through social activities such as parties), thrill seeking (desire to engage in physically risky activities), experience seeking (seeking novel experiences through the mind and senses), and boredom susceptibility (intolerance for repetitive experience). Each of these subscales also shows moderate heritability. Multivariate genetic analyses indicate that genetic factors that affect experience seeking overlap completely with other subscales, whereas genetic influence on boredom susceptibility is independent of the other subscales (Eysenck, 1983).

One of the most surprising findings from genetic research on personality questionnaires is that the many traits that have been studied all show moderate genetic influence. It is surprising that studies have not found any personality traits assessed by self-report questionnaire that consistently show low or no heritability. Can this be true? One way to explore this issue is to use measures of personality other than self-report questionnaires to investigate whether this result is somehow due to self-report measures.

Other Measures of Personality

A recent study of adult twins in Germany and Poland compared twin results for self-report questionnaires and ratings by peers for a measure of the big five personality factors for nearly a thousand pairs of twins (Angleitner, Riemann, & Strelau, 1995). Each twin's personality was rated by two different peers. Correlations between the two peer ratings were .63, a result indicating substantial agreement concerning each twin's personality. The averaged peer ratings correlated .55 with the twins' self-report ratings, a result indicating moderate validity of self-report ratings. Twin correlations for the big five personality traits are shown in Table 11.2. The results for self-report ratings are similar to those in Table 11.1. The novel finding involved peer ratings, which show almost as much genetic influence as the self-report ratings. Again, no evidence for shared environmental influence emerged. Another important finding came from multivariate genetic analyses of these data: The *same* genetic factors are largely involved in self-report ratings and in peer ratings of personality.

Genetic researchers interested in personality in childhood were forced to use measures other than self-report questionnaires. For the past 20 years, this

TABLE 11.2

Twin Study Using Self-Report and Peer Ratings of Personality Traits

Personality Trait	Correlation from Self-Report Ratings		Correlation from Peer Ratings	
	Identical	Fraternal	Identical	Fraternal
Extraversion	.56	.28	.40	.17
Neuroticism	.53	.13	.43	−.03
Agreeableness	.42	.19	.32	.18
Conscientiousness	.54	.18	.43	.18
Culture	.54	.35	.48	.31

SOURCE: *Angleitner et al. (1995).*

research has primarily relied on ratings by parents, but twin studies using parent ratings have yielded odd results. Correlations for identical twins are high and correlations for fraternal twins are very low, sometimes even negative. It is likely that these results are due to contrast effects in which parents of fraternal twins contrast the twins (Plomin et al., 1990a). For example, parents might say that one twin is the active twin and the other is the inactive twin, even though, relative to other children that age, the twins are not really very different from each other (Carey, 1986; Eaves, 1976; Neale & Stevenson, 1989). Furthermore, adoption studies using parent ratings in childhood find little evidence for genetic influence (Loehlin, Willerman, & Horn, 1982; Plomin et al., 1991; Scarr, Webber, & Wittig, 1981; Schmitz, 1994). A combined twin study and stepfamily study of parent ratings of adolescents found significantly greater heritability estimates for twins than for nontwins and confirmed that parent ratings are subject to contrast effects (Saudino et al., in press-a). As mentioned in relation to self-report questionnaires, such findings might also be due to nonadditive genetic variance, which makes identical twins more similar than other first-degree relatives. However, the weight of evidence indicates that genetic results for parent ratings of personality are due to contrast effects.

Other measures of children's personality, such as behavioral ratings by observers, show more reasonable patterns of results in both twin and adoption studies (Braungart et al., 1992a; Cherny et al., 1994; Goldsmith & Campos, 1986; Matheny, 1980; Plomin & Foch, 1980; Plomin, Foch, & Rowe, 1981; Plomin et al., 1993; Saudino, Plomin, & DeFries, 1996; Wilson & Matheny, 1986). For example, genetic influence has been found in observational studies of young twins for a dimension of fearfulness called behavioral inhibition (Matheny, 1989; Robinson et al., 1992), for shyness observed in the home and the laboratory (Cherny et al., 1994), and for activity level that is measured by using actometers, which record movement (Saudino & Eaton, 1991). In contrast, the only twin study of observer ratings of personality in the first few days of life found no evidence for genetic influence (Riese, 1990). Individual differences in smiling in infancy also show no genetic influence (Plomin, 1987).

SUMMING UP

Twin studies using self-report questionnaires of personality typically find heritabilities ranging from 30 to 50 percent, with no evidence for shared environmental influence. Adoption studies find somewhat less genetic influence, perhaps as a result of the presence of nonadditive genetic variance. Extraversion and neuroticism have been studied most and yield heritability estimates of 50 and 40 percent, respectively, across both twin and adoption studies. A recent twin study of peer ratings yielded similar results. Parent ratings of children's personality are affected by contrast effects that exaggerate estimates of genetic influence in twin studies. Observational measures of children's personality also show genetic influence in twin and adoption studies.

More research that uses measures other than self-report questionnaires is needed. In addition to laboratory-based observations and tests, other methods of measurement have yet to be explored, such as time-sampling methods using "beepers," story-telling techniques, and examination of autobiographical material (Goldsmith, 1993). Nonetheless, the results so far are encouraging in that the pervasive evidence for genetic influence on personality gleaned from self-report questionnaires can be confirmed using other measures.

Other Findings

There is a renaissance of genetic research on personality, which will be accelerated by research showing the association between personality and psychopathology and by reports of specific genes associated with personality. One example of new directions for research is increasing interest in measures other than self-report questionnaires, as just described. Three other examples include research on personality in different situations, developmental change and continuity, and the role of personality in the interplay between nature and nurture.

Situations

It is interesting, in relation to the person-situation debate mentioned earlier, that some research suggests that genetics is involved in situational change as well as in stability of personality. For example, in one study, observers rated the adaptability of infant twins in two laboratory settings, unstructured free play and test taking (Matheny & Dolan, 1975). Adaptability differed to some extent across these situations, but identical twins changed in more similar ways than fraternal twins did, an observation implying that genetics contributes to change as well as to continuity across situations for this personality trait. Such results might differ for other personality traits. For example, a recent twin study of shyness found that genetic factors largely contribute to stability across observations in the home and in the laboratory; environmental factors account for shyness differences between these situations (Cherny et al., 1994). A twin study using a questionnaire to assess personality in different situations found that genetic factors contribute to personality changes across situations (Dworkin, 1979). Even patterns of responding across items of personality questionnaires show genetic influence (Eaves & Eysenck, 1976; Hershberger, Plomin, & Pedersen, in press).

Development

Does heritability change during development? Unlike general cognitive ability, which shows increases in heritability throughout the life span (Chapter 8), it is more difficult to draw general conclusions concerning personality development, in part because there are so many personality traits. In general, heritability appears to increase during infancy (Goldsmith, 1983; Loehlin, 1992), starting with zero heritability for personality during the first days of life (Riese, 1990). Of course, what is assessed as personality during the first few days of life is quite different from what is assessed later in development, and the sources of individual differences might also be quite different in neonates. Throughout the rest of the life span, it is clear that twins become less similar as time goes by, but this decreasing similarity occurs for identical twins as much as for fraternal twins for most personality traits, suggesting that heritability does not change (McCartney, Harris, & Bernieri, 1990). However, other evidence suggests that, when heritability changes during development, it tends to increase (Plomin & Nesselroade, 1990).

A second important question about development concerns the genetic contribution to continuity and change from age to age. For cognitive ability, genetic factors largely contribute to stability from age to age rather than to change, although some evidence can be found, especially in childhood, for genetic change (Chapter 8). Although less well studied than cognitive ability, developmental findings for personality appear to be similar (Loehlin, 1992). Twin studies using self-report questionnaires show little evidence for genetic influence on change from age to age during adulthood (e.g., McGue, Bacon & Lykken, 1993b; Pogue-Geile & Rose, 1985). A ten-year follow-up adoption analysis of parent ratings also found no evidence for genetic influence on personality change (Loehlin, Horn, & Willerman, 1990). However, an analysis of parents and their adolescent twins suggests that genetic factors may be

As an undergraduate, **Hill Goldsmith** was a biology major. During his senior year, he developed interests in human genetics and the psychology of individual differences. These interests eventually led to graduate work in behavioral genetics at the University of Minnesota, where his adviser was Irving Gottesman. Goldsmith shared Gottesman's interest in psychiatric genetics, but his research during graduate school focused on normal personality and its developmental course. After receiving his doctorate in 1978, Goldsmith moved to Denver. Working in the Department of Psychiatry at the University of Colorado and in the Department of Psychology at the University of Denver, Goldsmith specialized in temperament during infancy and early childhood. Partly because of this experience, Goldsmith believes strongly that aspiring behavioral geneticists should plan to become expert in both genetics and in a particular area of psychology. Goldsmith has held faculty positions at the University of Texas, Austin, and the University of Oregon. In 1992, he moved to the Department of Psychology at the University of Wisconsin-Madison, where he is currently chair. Goldsmith's recent research elucidates the role of emotion in early personality, social, and moral development. Most of his studies involve young twins and their families.

involved in long-term change from adolescence to adulthood (Eaves, Eysenck, & Martin, 1989). Greater genetic change also appears likely in childhood (Loehlin, 1992). For example, one twin study of observer ratings of reactivity reported genetic influence on change from four to seven years (Goldsmith & Gottesman, 1981), and twin analyses of profiles of age-to-age change in early childhood also found evidence of genetic influence (Wilson & Matheny, 1986). However, longitudinal twin studies of shyness found that genetic factors predominantly contribute to continuity from age to age (Cherny et al., 1994; Saudino et al., 1996).

Nature-Nurture Interplay

Another new direction for genetic research on personality involves its role in explaining a fascinating finding: Environmental measures widely used in psychological research show genetic influence. As discussed in Chapter 14, genetic research consistently shows that family environment, peer groups, social support, and life events often show as much genetic influence as measures of personality. The finding is not as paradoxical as it might seem at first. Measures of psychological environments in part assess genetically influenced characteristics of the individual. Personality is a good candidate to explain this genetic influence, because personality can affect how people select, modify, construct, or perceive their environments. Multivariate genetic analysis has shown that a significant amount of genetic influence on adults' perceptions of their current family's environment can be explained by genetic influence on extraversion and neuroticism (Chipuer et al., 1992). Genetic influence on perceptions of life events can be entirely accounted for by the big five personality factors (Saudino et al., in press-b). These findings are not limited to self-report questionnaires. For example, genetic influence found on an observational measure of home environments can be explained entirely by genetic influence on a tester-rated measure of attention called task orientation (Saudino & Plomin, in press).

Personality and Social Psychology

Social psychology focuses on the behavior of groups, whereas individual differences are in the spotlight for personality research. For this reason, there is not nearly as much genetic research relevant to social psychology as there is for personality. However, some areas of social psychology border on personality, and genetic research has begun to sprout at these borders. Three examples are relationships, self-esteem, and attitudes.

Relationships

Genetic research has been done on parent-offspring relationships, romantic relationships, and sexual orientation.

Parent-offspring relationships Relationships between parents and offspring vary widely in their warmth (such as affection and support) and control (such as monitoring and organization). To what extent do genetic influences on parents and on offspring contribute to relationships? If identical twins are more similar in the qualities of their relationships than fraternal twins, this difference indicates genetic influence on relationships. For example, the first research of this sort involved adolescent twins' perceptions of their relationships with their parents. In two studies with different samples and different measures, genetic influence was found for twins' perceptions of their mothers' and fathers' warmth toward them (Rowe, 1981, 1983b). In contrast, adolescents' perceptions of their parents' control did not show genetic influence. One possible explanation is that parental warmth reflects genetically influenced characteristics of their children, but parental control does not (Lytton, 1991). Dozens of subsequent twin and adoption studies have found similar results that point to substantial genetic influences in most aspects of relationships, not just between parents and offspring, but also between siblings and friends (Plomin, 1994a).

A major area of developmental research on parent-offspring relationships involves attachment between infant and caregiver, as assessed in the so-called Strange Situation (Ainsworth et al., 1978). A sibling study found a concordance of 57 percent for attachment classification (Ward, Vaughn, & Robb, 1988). A twin analysis of attachment reported that genetic factors affect some aspects of attachment but not others (Ricciuti, 1993).

Another component of relationships is empathy. One twin study of infants used videotape observations of the empathic responding of infant twins following simulations of distress in the home and the laboratory (Zahn-Waxler, Robinson, & Emde, 1992). Evidence was found for genetic influence for some aspects of the infants' empathic responses.

Romantic relationships Like parent-offspring relationships, romantic relationships differ widely in various aspects such as closeness and passion. The first genetic study of styles of romantic love is interesting because it showed *no* genetic influence (Waller & Shaver, 1994). The average twin correlations for six scales (for example, companionship and passion) were .26 for identical twins and .25 for fraternal twins, results implying some shared environmental influence but no genetic influence. In other words, genetics plays no role in the type of romantic relationships we choose. Perhaps love *is* blind, at least from the DNA point of view.

Sexual orientation An early twin study of male homosexuality reported remarkable concordance rates of 100 percent for identical twins and 15 percent for fraternal twins (Kallmann, 1952). However, a recent twin study found more

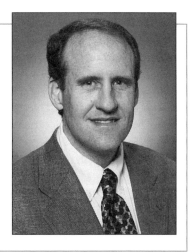

modest concordances of 52 and 22 percent, respectively, and concordance of 22 percent for genetically unrelated adoptive brothers (Bailey & Pillard, 1991). A small twin study of lesbians also yielded evidence for moderate genetic influence (Bailey et al., 1993). This area of research has received considerable attention recently because of reports of linkage between homosexuality and a region at the tip of the long arm of the X chromosome (Hamer et al., 1993; Hu et al., 1995). The X chromosome has been targeted because studies indicate that male homosexuality is more likely to be transmitted from the mother's side of the family. It has been hypothesized that the genetic effect on homosexuality might act indirectly through personality (LeVay & Hamer, 1994).

Self-Esteem

A key variable for adjustment is self-esteem, which is also referred to as a sense of self-worth. Research on the etiology of individual differences in self-esteem has focused on the family environment (Harter, 1983). It is surprising that the possibility of genetic influence had not been considered previously, because it seems likely that genetic influence on personality and psychopathology (especially depression, for which low self-esteem is a core feature) could also affect self-esteem. Twin and adoption studies of self-esteem have recently been reported for teacher and parent ratings in middle childhood (Neiderhiser & McGuire, 1994) and for teacher, parent, and self-ratings in adolescence (McGuire et al., 1994). These studies point to modest genetic influence on self-esteem, but no influence of shared family environment.

Attitudes and Interests

Social psychologists have long been interested in the impact of group processes on change and continuity in attitudes and beliefs. Although it is recognized that social factors are not solely responsible for attitudes, it has been a surprise to find that genetics makes a major contribution to individual differences in attitudes. A core dimension of attitudes is traditionalism, which involves conservative versus liberal views on a wide range of issues. In several twin studies (Eaves et al., 1989), including a study of twins reared apart (Tellegen et al., 1988), identical twin correlations are typically about .65 and fraternal twin correlations are about .50. This pattern of twin correlations suggests heritability of about 30 percent and shared environmental influence of about 35 percent. However, assortative mating is higher for traditionalism than for any other psychological trait, with spouse correlations of about .50. Assortative mating inflates the fraternal twin correlation, thereby lowering estimates of heritability and raising estimates of shared environment (Chapter 8). When assortative mating is taken into account, heritability is estimated to be about 50 percent and shared environmental influence is about 15 percent (Eaves et al., 1989).

Social psychology traditionally uses the experimental approach rather than investigating naturally occurring variation (Chapter 5). There is a need to bring together these two research traditions. For example, Tesser (1993), a social psychologist, separated attitudes into those that were more heritable (such as attitudes about the death penalty) and those that were less heritable (coeducation and the truth of the Bible). In standard social psychology experimental situations, the more heritable items were found to be less susceptible to social influence and more important in interpersonal attraction.

A related area is vocational interests, which involve personality-like dimensions such as realistic, intellectual, social, enterprising, conventional, and artistic. Results from twin studies for vocational interests are similar to results for personality questionnaires, with identical twin correlations of about .50 and fraternal twin correlations of about .25 (Roberts & Johansson, 1974). Moderate genetic influence also emerged in an adoption study of vocational interests (Scarr & Weinberg, 1978b). Evidence for genetic influence was also found in twin studies of work values (Keller et al., 1992) and job satisfaction (Arvey et al., 1989).

These examples of genetic research indicate the potential usefulness of incorporating genetics into social psychology. There are signs that a rapprochement is dawning. For example, a recent text in social development features new insights offered by current genetic research and concludes that "it is the hyphen in the Nature-Nurture formula that now requires our attention above all" (Schaffer, 1996).

SUMMING UP

Genetic research on personality across situations and across time suggests that genetics is largely responsible for continuity and that change is largely due to environmental factors. Some research indicates that heritability increases during development. New directions for research include the role of personality in explaining genetic influence on measures of the environment. Another new direction is the interface between personality and social psychology. Recent research has found evidence for genetic influence on social relationships (parent-offspring relationships and sexual orientation, but not romantic relationships), self-esteem, attitudes, and vocational interests.

Personality Disorders

To what extent is psychopathology the extreme manifestation of normal dimensions of personality? It has long been suggested that this is the case for some psychiatric disorders (e.g., Eysenck, 1952). The few genetic studies that have broached this topic indicate genetic overlap between psychopathology and personality (Carey & DiLalla, 1994). For example, genetic variation in anxiety and depression largely overlaps with neuroticism (Eaves et al., 1989).

Rather than directly investigating the relationship between personality and psychopathology, most genetic research in this area has focused on personality disorders. Unlike the mental disorders described in Chapter 10, personality disorders are personality traits that cause significant impairment or distress. People with personality disorders regard their disorder as part of who they are, their personality, rather than as a condition that can be treated. That is, they do not feel that they were once well and are now ill. For this reason, DSM-IV separates personality disorders from clinical syndromes. This category of disorders (called Axis II), which also includes mental retardation, refers to long-term disorders that date from childhood. Although the reliability, validity, and utility of personality disorders have long been questioned, interest in the genetics of personality disorders and their links to normal personality and to psychopathology is increasing (Nigg & Goldsmith, 1994).

DSM-IV recognizes ten personality disorders, but only four have been investigated in genetic research: schizotypal, obsessive-compulsive, borderline, and antisocial personality disorders. Most genetic research has targeted antisocial personality disorder because of its relevance to criminal behavior. For this reason, antisocial personality disorder is discussed in a separate section that follows a brief presentation of the other three personality disorders.

Schizotypal personality disorder involves less intense schizophrenic-like symptoms and, like schizophrenia, clearly runs in families (e.g., Baron et al., 1985a; Siever et al., 1990). The results of a small twin study suggest genetic influence, yielding 33 percent concordance for identical twins and 4 percent for fraternal twins (Torgersen & Psychol, 1984). Twin studies using dimensional measures of schizotypal symptoms in unselected samples of twins also found evidence for genetic influence (Claridge & Hewitt, 1987; Kendler & Hewitt, 1992).

Genetic research on schizotypal personality disorder focuses on its relationship to schizophrenia and has consistently found an excess of schizotypal personality disorder among first-degree relatives of schizophrenic probands. A summary of such studies found that the risks of schizotypal personality disorder are 11 percent for the first-degree relatives of schizophrenic probands and 2 percent in control families (Nigg & Goldsmith, 1994). Adoption studies have played an important role in showing that schizotypal personality disorder is part of the genetic spectrum of schizophrenia. For example, in the Danish adoption study (see Chapter 10), the rate of schizophrenia in the biological first-degree relatives of schizophrenic adoptees was 5 percent, but 0 percent in their adoptive relatives and relatives of control adoptees (Kety et al., 1994). When schizotypal personality disorder was included in the diagnosis, the rates rose to 24 and 3 percent, respectively, implying greater genetic influence for the spectrum of schizophrenia that includes schizotypal personality disorder (Kendler, Gruenberg, & Kinney, 1994). Twin studies also suggest that schizotypal personality disorder is related genetically to schizophrenia (Farmer, McGuffin, & Gottesman, 1987).

Obsessive-compulsive personality disorder sounds as if it is a milder version of obsessive-compulsive type of anxiety disorder (OCD, described in Chapter 10), and family studies provide some empirical support for this. However, the diagnostic criteria for these two disorders are quite different. The compulsion of OCD is a single sequence of bizarre behaviors, whereas the personality disorder is more pervasive, involving a general preoccupation with trivial details that leads to difficulties in making decisions and getting anything accomplished. Although no family, twin, or adoption studies of diagnosed obsessive-compulsive personality disorder have been reported, twin studies of obsessional symptoms in unselected samples of twins provide some evidence for at least modest heritability (Torgersen & Psychol, 1980; Young et al., 1971). One twin study indicated substantial genetic overlap with neuroticism (Clifford, Murray, & Fulker, 1984). Family studies indicate that obsessional traits are more common (about 15 percent) in relatives of probands with obsessive-compulsive disorder than in controls (5 percent) (Rasmussen & Tsuang, 1984). This finding implies that obsessive-compulsive personality disorder might be part of the spectrum of the obsessive-compulsive type of anxiety disorder.

Another personality disorder that has been the target of genetic research is borderline personality disorder, which comes from the psychoanalytic tradition and includes insecure self-identify, distrust, and self-destructive behavior. Family studies indicate that borderline personality disorder is familial, but it is not specific in that family members are also at increased risk for other personality disorders and major depressive disorder (Nigg & Goldsmith, 1994). Across nine family studies, 12 percent of first-degree relatives of borderline personality disorder probands were affected, in contrast to 2 percent in control families. Family members were also at increased risk for major depressive disorder (14 percent in first-degree relatives of probands and 5 percent in control families) but not for schizophrenia (Baron et al., 1985b; Loranger, Oldham, & Tulis, 1982). This result is another example of genetic research helping to provide an approach to diagnosis based on etiology rather than on symptoms. No twin or adoption studies large enough to yield interpretable results have as yet been reported.

In summary, schizotypal, obsessive-compulsive, and borderline personality disorders appear to be partially heritable. More important, personality

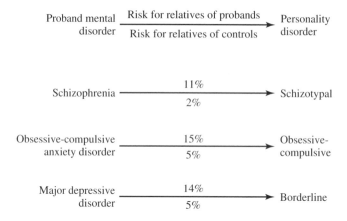

Figure 11.2 Personality disorders are related to psychopathology: Risk for relevant personality disorders is increased for first-degree relatives of probands diagnosed for schizophrenia, obsessive-compulsive disorder, and major depression.

disorders are related etiologically to psychopathology. Figure 11.2 summarizes the risk of personality disorders in first-degree relatives of controls and of probands diagnosed for mental disorders.

Antisocial Personality Disorder and Criminal Behavior

Much more genetic research has focused on antisocial personality disorder than on other personality disorders. Lying, cheating, and stealing are examples of antisocial behavior. At the extreme of antisocial behavior, with chronic indifference to and violation of the rights of others, is antisocial personality (ASP) disorder, a highly heterogeneous disorder. Such individuals were called psychopaths a century ago when the condition was assumed to be a mental illness. Then they were called sociopaths, with the rise of sociology and the assumption that such conditions are due to social conditions. It is now recognized that some, but certainly not all, antisocial behavior can be due to psychological disturbances called antisocial personality disorder. DSM-IV criteria for antisocial personality disorder include a history of illegal or socially disapproved activity beginning before age 15 and continuing into adulthood, irresponsibility, irritability, aggressiveness, recklessness, and disregard for truth. Although antisocial personality disorder shows early roots, the vast majority of juvenile delinquents and children with conduct disorders do not develop antisocial personality disorder (see Chapter 10). For this reason, there is a need to distinguish conduct disorder that is limited to adolescence from antisocial behavior that persists throughout the life span (Caspi & Moffitt, 1995). As diagnosed by DSM–IV criteria, antisocial personality disorder affects about 1 percent of females and 4 percent of males from 13 to 30 years of age (Kessler et al., 1994).

Family studies show that ASP runs in families (Nigg & Goldsmith, 1994), and an adoption study found that familial resemblance is largely due to genetic rather than to shared environmental factors (Schulsinger, 1972). The risk for ASP is increased fivefold for first-degree relatives of ASP males, whether living together or adopted apart. For relatives of ASP females, risk is increased tenfold, a result suggesting that, to be affected for this disproportionately male disorder, females need a greater genetic loading. Although no twin studies of diagnosed ASP are available, a personality questionnaire assessing symptoms of antisocial behavior yielded average correlations of .50 for identical twins and .22 for fraternal twins in three studies of unselected samples (Nigg & Goldsmith, 1994). A small study of identical twins reared apart also found evidence for modest genetic influence for a similar scale of ASP symptoms (Grove et al., 1990), as did an adoption study (Loehlin, Willerman, & Horn, 1987).

A recent twin study of more than 3000 pairs of adult male twins assessed current ASP symptoms as well as a retrospective report of adolescent ASP symptoms (Lyons et al., 1995). For adult ASP symptoms, results similar to other studies were found (correlations were .47 for identical twins and .27 for fraternal twins). For adolescent ASP symptoms, results were similar to studies of adolescent conduct disorder (.39 for identical twins and .33 for fraternal twins), implying little genetic influence and substantial shared environmental influence. Analyses of cross-age twin correlations indicated that both genetic and shared environmental factors contribute to the correlation of about .40 between adolescent and adult ASP symptoms. It is generally accepted that, from adolescence to adulthood, genetic influence increases and shared environmental influence decreases for ASP symptoms (DiLalla & Gottesman, 1989) (see Figure 11.3).

ASP shows interesting genetic relations with other disorders. As mentioned in Chapter 10, in families of ASP probands, males are at increased risk for ASP and drug abuse, whereas females more often have somatization disorder. Of particular interest is the relation between ASP and criminal behavior. For example, two adoption studies of biological parents with criminal records found increased rates of ASP in their adopted-away offspring (Cadoret & Stewart, 1991; Crowe, 1974), suggesting that genetics contributes to the relationship between criminal behavior and ASP. Most genetic research in this area has focused on criminal behavior itself rather than on ASP, because crime can be assessed objectively by using criminal records. However, criminal behavior, although important in its own right, is only moderately associated with ASP. About 40 percent of male criminals and 8 percent of female criminals qualify for a diagnosis of ASP (Robins & Regier, 1991). Clearly, breaking the law cannot be equated with psychopathology (Rutter, 1996a).

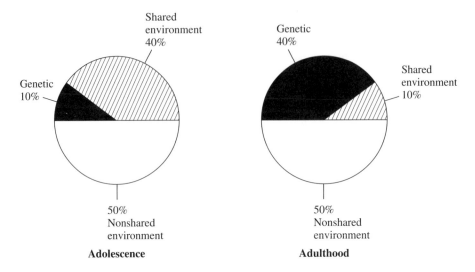

Figure 11.3 The causes of antisocial symptoms change from adolescence to adulthood, with genetics becoming more important and shared environment becoming less important. (Based on Lyons et al., 1995.)

The best twin study of criminal behavior included all male twins born on the Danish Islands from 1881 to 1910 (Christiansen, 1977). Evidence from more than a thousand twin pairs was found for genetic influence for criminal convictions, with an overall concordance of 51 percent for male identical twins and 30 percent for male-male fraternal twins. In 13 twin studies of adult criminality, identical twins are consistently more similar than fraternal twins (Raine, 1993). The average concordances for identical and fraternal twins are 52 and 21 percent.

A recent twin study in the United States of self-reported arrests and criminal behavior involved more than 3000 male twin pairs in which both members served in the Vietnam War (Lyons, 1996). Genetics contributed to self-reported arrests and criminal behavior. However, self-reported criminal behavior before age 15 showed negligible genetic influence. Shared environment made a major contribution to arrests and criminal behavior before age 15, but not later. These results before age 15 are similar to results discussed earlier for ASP symptoms in adolescence and adulthood (Lyons et al., 1995) and for conduct disorder in adolescence (Chapter 10).

Adoption studies are consistent with the hypothesis of significant genetic influence on adult criminality, although adoption studies point to less genetic influence than twin studies do. It has been hypothesized that twin studies overestimate genetic effects because identical twins are more likely to be partners in crime (Carey, 1992). Adoption studies include both the adoptees' study method (Cloninger et al., 1982; Crowe, 1972) and the adoptees' family method (Cadoret et al., 1985b). One of the best studies used the adoptees' study method, beginning with more than 14,000 adoptions in Denmark between 1924 and 1947 (Mednick, Gabrielli, & Hutchings, 1984). Using court convictions as an index of criminal behavior, the researchers found evidence for genetic influence and for genotype-environment interaction, as shown in Figure 11.4. Adoptees were at greater risk for criminal behavior when their biological parents had criminal convictions, a finding implying genetic influence. Unlike the twin study just described, this adoption study (and others) found genetic influence for crimes against property but not for violent crimes (Brennan, Mednick, & Jacobsen, 1996). Genotype-environment interaction is also found. Adoptive parents with criminal convictions had no effect on the criminal behavior of adoptees unless the adoptees' biological parents also had criminal convictions. In other words, the highest rate of criminal behavior was found for adoptees who had both biological parents *and* adoptive parents with criminal records.

A Swedish adoption study of criminality using the adoptees' family method found similar evidence for genotype-environment interaction as well as interesting interactions with alcohol abuse, which greatly increases the likelihood of violent crimes (Bohman, 1996; Bohman et al., 1982). When adoptees' crimes did not involve alcohol abuse, their biological fathers were found to be at increased risk for nonviolent crimes. In contrast, when adoptees' crimes involved alcohol abuse, their biological fathers were *not* at increased risk for crime. These findings suggest that genetics contributes to criminal behavior but not to alcohol-related crimes, which are likely to be more violent crimes.

Research on the genetics of crime has been controversial. For example, a conference on the topic in the United States was postponed and then disrupted (Roush, 1995), although a similar conference in the United Kingdom created little stir (Bock & Goode, 1996). Especially for such sensitive topics, it is

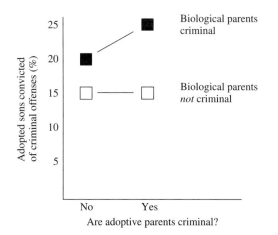

Figure 11.4 Evidence for genetic influence and genotype-environment interaction for criminal behavior in a Danish adoption study. (Adapted from Mednick, Gabrielli, & Hutchings, 1984.)

important to keep in mind the discussion in Chapter 5 concerning the interpretation of genetic effects (see also Raine, 1993; Rutter, 1996b). These issues will become increasingly important as specific genes that contribute to genetic risk are identified.

SUMMING UP

Multivariate genetic research suggests that, at least for some conditions, there may be a continuum of individual differences from normal personality to personality disorders to psychopathology. For example, genetic variation in neuroticism largely accounts for genetic variation in anxiety and depression. Genetic overlap has also been reported between personality disorders and psychopathology: for example, between schizotypal personality disorder and schizophrenia, between obsessive-compulsive personality disorder and obsessive-compulsive anxiety disorder, and between borderline personality disorder and mood disorder. The relationship between antisocial personality disorder and criminal behavior may also be due in part to genetic influences. For symptoms of antisocial personality disorder, genetic influence increases and shared environmental influence decreases from adolescence to adulthood.

Identifying Genes

Recently, an association between a DNA marker and personality has been reported, a sign of things to come. In two separate studies, a DNA marker for a certain neuroreceptor gene (dopamine D-4 receptor, *D4DR*) was reported to be associated with the personality trait of novelty seeking in unselected samples (Benjamin et al., 1996; Ebstein et al., 1995). The DNA marker consists of seven alleles involving 2, 3, 4, 5, 6, 7, or 8 repeats of a 48-base-pair sequence. In both studies, individuals with the longer *D4DR* alleles (6–8 repeats) had significantly higher novelty-seeking scores than individuals with the shorter alleles (2–5 repeats). This association was also found within families, a result indicating that the association is not due to ethnic differences. That is, within the same families, individuals with the longer *D4DR* alleles had significantly higher novelty-seeking scores than their siblings with the shorter *D4DR* alleles. Figure 11.5 shows the distributions of novelty-seeking scores for individuals with the short and the long *D4DR* alleles. The overlap in scores shows that the effect is small, accounting for about 4 percent of the variance in this trait.

As mentioned above, twin studies for sexual orientation point to some genetic influence. Two small linkage studies have reported a linkage between markers on the X chromosome and homosexuality, as discussed in Chapter 6 (Hamer et al., 1993; Hu et al., 1995). A preliminary report of another linkage study has not replicated this result, and other studies are underway (Bailey, 1995).

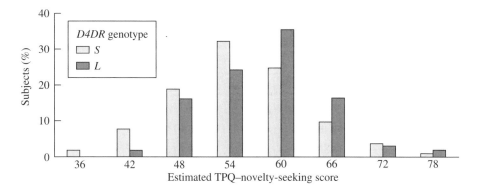

Figure 11.5 Two studies have reported the longer allele for the *D4DR* gene to be associated with increased novelty seeking. The overlap in novelty-seeking scores for those with the shorter allele (*S*) and those with the longer allele (*L*) shows that the effect is modest, accounting for about 4 percent of the variance in novelty seeking. (From Benjamin et al., 1996; used with permission.)

Earlier reports of an association between XYY males and violence were overblown, although there seems to be some increase in hyperactivity and perhaps conduct problems (Ratcliffe, 1994). In a four-generation study of a Dutch family, a deficiency in a gene on the X chromosome that codes for an enzyme (monoamine oxidase A) involved in the breakdown of several neurotransmitters was associated with impulsive aggression and borderline mental retardation in males (Brunner, 1996; Brunner et al., 1993). However, this genetic effect has not yet been found in any other families.

More powerful methods for identifying such QTLs (Chapter 6) for personality are available in research on nonhuman animals (Plomin & Saudino, 1994). For example, as indicated in Chapter 6, several QTLs for fearfulness were identified in mice, as assessed in open-field activity (Flint et al., 1995a). Also, transgenic knock-out gene studies in mice found increased aggression when genes were knocked out for a receptor for an important neurotransmitter (serotonin; Saudou et al., 1994) or an enzyme (neuronal nitric oxide synthase) that plays a basic role in neurotransmission (Nelson et al., 1995).

Summary

More twin data are available for self-report personality questionnaires than for any other domain of psychology, and they consistently yield evidence for moderate genetic influence. Most well studied are extraversion and neuroticism, which yield heritability estimates of about 50 percent for extraversion and about 40 percent for neuroticism across twin and adoption studies. Other personality traits assessed by personality questionnaire also show heritabilities from 30 to 50 percent. There is no replicated example of zero heritability. Environmental influence is almost entirely due to nonshared environmental factors. These surprising findings are not limited to self-report questionnaires. A recent twin study using peer ratings yielded similar results. Although the degree of genetic influence suggested by twin studies using parent ratings of children's personality appears to be inflated by contrast effects, more objective measures, such as behavioral ratings by observers, indicate genetic influence in twin and adoption studies.

New directions for genetic research include looking at personality continuity and change across situations and across time. Results so far indicate that genetics is largely responsible for continuity and that change is largely due to environmental factors. Other new findings include the central role that personality plays in producing genetic influence on measures of the environment. Another new direction for research lies at the border with social psychology. For example, genetic influence has been found for relationships, such as parent-offspring relationships and sexual orientation but not romantic relationships. Other examples include evidence for genetic influence on self-esteem, attitudes, and vocational interests.

Personality disorders, which are at the border between personality and psychopathology, are another growth area for genetic research in personality. It is likely that some personality disorders are part of the genetic continuum of psychopathology: schizotypal personality disorder and schizophrenia, obsessive-compulsive personality disorder and obsessive-compulsive anxiety disorder, and borderline personality disorder and mood disorder. Most genetic research on personality disorders has focused on antisocial personality disorder and its relationship to criminal behavior. From adolescence to adulthood, genetic influence increases and shared environmental influence decreases for symptoms of antisocial personality disorder, including juvenile delinquency and adult criminal behavior.

A QTL association between a dopamine receptor gene and the personality trait of novelty seeking has been reported in two studies. Linkages with the X chromosome have been reported for male sexual orientation and impulsive aggression. Together with powerful mouse models to identify genes for personality, these results signal the dawn of a new era of molecular genetic research on personality.

CHAPTER TWELVE

Health Psychology and Aging

G enetic research in psychology has focused on cognitive disabilities (Chapter 7), general and specific cognitive abilities (Chapters 8 and 9), psychopathology (Chapter 10), and personality (Chapter 11). The reason for this focus is that these are the areas of psychology that have had the longest history of research on individual differences. Much less is known about the genetics of other major domains of psychology that have not traditionally emphasized individual differences, such as perception, learning, and language.

The purpose of this chapter is to provide an overview of genetic research in two new areas of psychology. One of the newest areas is health psychology, sometimes called psychological or behavioral medicine because it lies at the intersection between psychology and medicine. Research in this area focuses on the role of behavior in promoting health and in preventing and treating disease. Although genetic research has just begun in this area, some conclusions can be drawn about relevant topics such as responses to stress, body weight, and addictive behaviors.

The second area is aging. Although genetic research in psychology has neglected the last half of the life span, new research has produced interesting results, especially about issues unique to aging such as quality of life in the later years. The explosion of molecular genetic research on cognitive decline and dementia in the elderly has added momentum to genetic research on psychology and aging.

Health Psychology

Most of the central issues about the role of behavior in promoting health and in preventing and treating disease have not yet been addressed in genetic research (Plomin & Rende, 1991). For example, the first book on genetics and health psychology was not published until 1995 (Turner, Cardon, & Hewitt, 1995).

John K. Hewitt is a faculty fellow in the Institute for Behavioral Genetics and a professor of psychology at the University of Colorado, Boulder. He was educated at the University of Birmingham in England, where he studied psychology and genetics, and then at the Institute of Psychiatry, receiving his doctorate in psychology from the University of London in 1978. His first faculty appointment was in the Department of Psychology at the University of Birmingham. He then moved to the Department of Human Genetics at the Medical College of Virginia. He directed the Virginia Twin Study of Adolescent Behavioral Development until his move to Boulder in 1992. The common theme of Hewitt's research has been the application of biometrical genetics to the elucidation of the development of individual differences in human and animal behavior. This theme has led from the study of learning, activity, and reactivity in wild-trapped and laboratory populations of rats (the subject of his doctoral dissertation) to his current interests in genetic research in behavioral medicine, human behavioral development and developmental psychopathology, and smoking, alcohol, and drug use in relation to personality and motivation factors.

Nonetheless, some conclusions can be drawn about the genetics of three areas relevant to health psychology: the relationship between stress and cardiovascular risk; body weight and obesity; and addictions, including alcoholism, smoking, and other drug abuses.

Stress and Cardiovascular Risk

Individuals differ in their cardiac responses to psychological stress (Turner, 1994). Because such differences in reactivity may be related to cardiovascular disease, the principal cause of death in the United States, this has become one of the major areas of research in health psychology (Manuck, 1994). Attention has begun to turn toward consideration of genetic factors. Ten recent twin studies investigated blood pressure and heart rate reactivity to acute psychological stressors administered in the laboratory (Hewitt & Turner, 1995). These studies show moderate genetic influence and no shared environmental influence on cardiovascular responses to stress.

Attempts to investigate environmental risk in the laboratory in which a standard situation is imposed on all subjects may be thwarted in the real world, where people are free to choose their own situations (Turner, 1994). That is, the laboratory paradigm does not take into account how likely it is that an individual will be exposed to stress outside the laboratory. Some individuals will successfully avoid stressful situations, whereas others actively seek such situations. For this reason, a new direction for research in this area is to monitor stress and cardiovascular responses repeatedly outside the laboratory (Pickering, 1991). For example, one small twin study recorded blood pressure every 20 minutes during a 24-hour period and found heritabilities of about 30 percent for blood pressure and heart rate (Somes et al., in press). A large twin study of blood pressure over 24 years found heritabilities of about 50 percent and evidence for genetic contributions to change from middle to late adulthood (Colletto, Cardon, & Fulker, 1993). Similar results were found in a blood pressure study of young twins, their parents, and adult twins the same age as the parents (Snieder, van Doornen, & Boomsma, 1995). Although these blood pressure studies have not explicitly assessed cardiovascular reactivity to stress, the role of genetics in exposure to stress and other health-related risk factors is increasingly recognized (Plomin, 1995b). For example, genetic factors are involved in lifestyles that lead to risk exposure (Rose, 1992). The topic of genetic contributions to such experiences is discussed further in Chapter 14.

Cardiovascular disease has been the target of much molecular genetic research and has become a model for quantitative trait locus (QTL) research on complex common disorders. Some of the first QTLs were identified in this area (Cambien et al., 1992; Sing & Boerwinkle, 1987). Although a single-gene dominant mutation has been found to be responsible for a rare type of cardiovascular disease called familial hypercholesterolemia, this gene accounts for only a small portion of coronary heart disease (Hobbs et al., 1990). Genes involved in several physiological processes are known to increase risk for cardiovascular disease (Keating & Sanguinetti, 1996). One factor is hypertension. At least ten genes that alter blood pressure have been found, but these genes explain only a small fraction of the total genetic variation in blood pressure in the population (Lifton, 1996).

Psychological research on the relationship between stress and cardiovascular disease has also begun to move in the direction of identifying QTLs responsible for genetic influence on cardiovascular reactivity to stress. For example, a QTL has been reported to be associated with blood pressure changes during stress (Boomsma et al., 1991).

Body Weight and Obesity

Obesity is a major health risk for several medical disorders, especially diabetes and heart disease. Although it is often assumed that individual differences in weight are largely due to environmental factors such as eating habits and exercise, twin and adoption studies consistently lead to the conclusion that genetics accounts for the majority of the variance for weight (Grilo & Pogue-Geile,

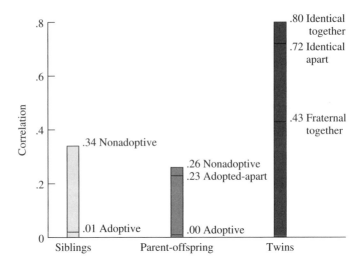

Figure 12.1 Family, adoption, and twin results for body weight. (Derived from Grilo & Pogue-Geile, 1991.)

1991). For example, as illustrated in Figure 12.1, twin correlations for weight based on thousands of pairs of twins are .80 for identical twins and .43 for fraternal twins. Identical twins reared apart correlate .72. Biological parents and their adopted-away offspring are almost as similar in weight (.23) as are nonadoptive parents and their offspring (.26), who share both nature and nurture. Adoptive parents and offspring and adoptive siblings, who share nurture but not nature, do not resemble each other at all for weight.

Together, the results in Figure 12.1 imply a heritability of about 70 percent. Similar results are found for body mass index, which corrects weight for height, and for skinfold thickness, which is an index of fatness (Grilo & Pogue-Geile, 1991). There are few genetic studies of overweight or obesity, in part because weight shows a continuous distribution, which means that diagnostic criteria are somewhat arbitrary (Bray, 1986). Nonetheless, using an obesity cutoff based on body mass index, one twin study reported concordances of 59 percent for identical twins and 34 percent for fraternal twins (Stunkard, Foch, & Hrubec, 1986).

As emphasized in Chapter 5, finding genetic influence does not mean that the environment is unimportant. Everyone can lose weight if they stop eating. The issue is not what *can* happen but rather what *does* happen. That is, to what extent are the obvious differences among people in weight due to genetic and environmental differences that exist in a particular population at a particular time? The answer provided by the research summarized in Figure 12.1 is that genetic differences largely account for individual differences in weight. If everyone ate the same amount and exercised the same amount, people would still differ in weight for genetic reasons.

C L O S E U P

Dorret Boomsma is in the Department of Psychophysiology at the Free University in Amsterdam, the Netherlands. She received M.A. degrees from this department and from the Institute for Behavioral Genetics in Boulder, Colorado. Her doctoral research consisted of a methodological part and an empirical part. The methodological part explored the application of model-fitting analyses to family data. This study led to several extensions of multivariate genetic models that had not been considered before, such as the estimation of single-subject genetic and environmental scores. The empirical twin research provided insight into the genetic architecture of cardiovascular risk factors and also illustrated some of the problems in contacting large samples of twins. Boomsma's research led to the establishment of the Netherlands Twin Register, which each year registers around 50 percent of all newborn twins in the Netherlands. These twins participate in longitudinal studies of behavioral development and child psychopathology. Adolescent twins and their parents take part in a longitudinal study of the genetic and cultural inheritance of health-related behaviors, lifestyle, and personality. Subsamples of adolescent twins participate in psychophysiological studies (electroencephalography, nerve conduction velocity) of neural mechanisms that may mediate the influence of genes on behavior.

This conclusion was illustrated dramatically in an interesting study of dietary intervention in 12 pairs of identical twins (Bouchard et al., 1990b). For three months, the twins were given excess calories and kept in a controlled sedentary environment. Individuals differed greatly in how much weight they gained, but members of identical twin pairs correlated .50 in weight gain. Similar twin studies show that the effects of physical exercise are influenced by genetic factors (Fagard, Bielen, & Amery, 1991).

Such studies do not point to the mechanisms by which genetic effects occur. For example, even though genetic differences occur when calories and exercise are controlled, in the world outside the laboratory, genetic contributions to individual differences might be mediated by individual differences in proximal processes such as food intake and exercise. In other words, individual differences in eating habits and in the tendency to exercise, although typically assumed to be environmental factors responsible for body weight, might be influenced by genetic factors.

Previous chapters have indicated that environmental variance is of the non-shared variety for most areas of psychology. This is also the case for body weight. As noted in relation to Figure 12.1, adoptive parents and their adopted children and adoptive siblings do not resemble each other at all for weight. This finding is surprising, because theories of weight and obesity have largely focused on weight control by means of dieting; yet individuals growing up in the same families do not resemble each other for environmental reasons (Grilo & Pogue-Geile, 1991). Attitudes toward eating and weight also show substantial heritability and no influence of shared family environment (Rutherford et al., 1993). In other words, environmental factors that affect individual differences in weight are factors that make children growing up in the same family different, not similar. The next step in this research is to identify environmental factors that differ for children growing up in the same family. For example, although it is reasonable to assume that children in the same family share similar diets, this may not be the case.

Genetic factors that affect body weight begin to have their effects in early childhood (Meyer, 1995). Three longitudinal genetic studies are especially informative. A longitudinal twin study from birth through adolescence found no heritability for birth weight, increasing heritability during the first year of life, and stable heritabilities of 60 to 70 percent thereafter (Matheny, 1990) (see Figure 12.2).

The second study, a longitudinal adoption study from infancy to childhood, also found substantial heritability of weight throughout childhood (Cardon, 1994b, 1995). Moreover, longitudinal genetic analyses of sibling adoption data showed that there is substantial genetic continuity from year to year during childhood but there is some genetic change as well, especially during infancy.

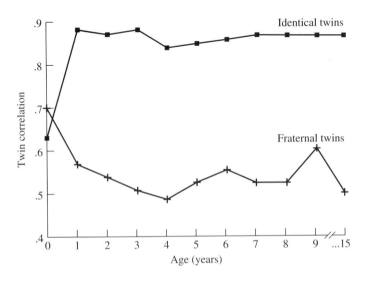

Figure 12.2 Identical and fraternal twin correlations for weight from birth to 15 years of age. (Derived from Matheny, 1990.)

Parent-offspring adoption data indicated little genetic continuity from infancy to adulthood but a surprising degree of continuity from childhood to adulthood. These data were also used to examine the rapid growth in body fat that typically begins at about six years of age. When this growth spurt occurs earlier in development, it is predictive of obesity in adulthood. Heritability of the timing of this growth spurt is about 40 percent, and the growth spurt is related genetically to adult body mass index.

The third study is a longitudinal study of about 4000 pairs of twins at 20 years of age and again at 45 years (Stunkard et al., 1986). The heritabilities of weight and body mass index were about 80 percent at both ages. Although longitudinal genetic analyses indicated that genetic effects largely contribute to continuity from 20 to 45 years, about 25 percent of the genetic effects at 45 years are different from genetic effects at 20 years. In other words, genetic processes contribute to some change as well as to substantial continuity from young adulthood to middle adulthood.

Obesity is the target of intense molecular genetic research because of the so-called *obese* gene in mice. In the 1950s, a recessive mutation that caused obesity in the homozygous condition was discovered in mice. When these obese mice were given blood from a normal mouse, they lost weight, a result suggesting that the obese mice were missing some factor important in control of weight. The gene was cloned and was found to be similar to a human gene (Zhang et al., 1994). The gene's product, a hormone called *leptin*, was shown to reduce weight in mice by decreasing appetite and increasing energy use (Halaas et al., 1995). However, obese humans do not appear to have defects in the leptin gene. The gene that codes for the leptin receptor in the brain has also been cloned from another mouse mutant (Chua et al., 1996). Mutations in this gene might contribute to genetic risk for obesity. However, like the leptin gene, it is also possible that there are no mutations in the leptin receptor gene. Although both genes appear to be critical mechanisms in weight control, it is possible that no genetic variation exists in either gene, which would mean that these genes are not responsible for the substantial genetic contribution to individual differences in weight.

Like most complex traits, there is no evidence for major single-gene effects on human obesity. For this reason, it is likely that QTLs, multiple genes of various effect sizes, are responsible for the substantial genetic contribution to obesity.

SUMMING UP

Recent genetic research on health psychology shows moderate genetic influence for cardiovascular responses to stress and no influence of shared family environment. Body weight shows high heritabilities, about 70 percent, and little influence of shared environment. Genetic effects on weight are largely stable after infancy, although there is some evidence for genetic change. Obesity is the target of much molecular genetic research because of the recent discovery of two genes involved in obesity in mice.

Addictions

Alcohol abuse, smoking, and abuse of other drugs are major health-related behaviors. Most research in this area has focused on alcoholism.

Alcoholism Clearly there are many steps in the path to alcoholism. For example, there is choice in whether or not to drink alcohol at all, in the amount one drinks, in the way that one drinks, and in the development of tolerance and dependence. Each of these steps might involve different genetic mechanisms. For this reason, alcoholism is likely to be highly heterogeneous.

Nonetheless, over 30 family studies have shown that alcoholism runs in families, although the studies vary widely in the size of the effect and in diagnostic criteria (Cotton, 1979). For males, alcoholism in a first-degree relative is by far the single best predictor of alcoholism. For example, a family study of 300 alcohol-abusing probands found an average risk of about 40 percent in first-degree male relatives and 20 percent in female relatives. The risk rates in the general population are about 20 percent for males and 5 percent for females, using the same diagnostic procedures for assessment (Reich & Cloninger, 1990).

Results of twin and adoption studies of alcoholism vary greatly, but taken together they indicate moderate heritability for males and modest or even negligible heritability for females (McGue, 1994). For example, a recent twin study of male alcoholics reported concordances of 77 percent for identical twins and 54 percent for fraternal twins (McGue, Pickens, & Svikis, 1992). For females, concordances were 39 percent for identical twins and 42 percent for fraternal twins, results suggesting no genetic influence and substantial influence of shared family environment. Similar results were found in the largest adoption study of alcoholism, which included more than 600 reared-away offspring of alcoholic biological parents (Cloninger, Bohman, & Sigvardsson, 1981). The rates of alcoholism were 23 percent for adopted males and 15 percent for control males. For females, the rates were 5 and 3 percent, respectively. As is often found in psychopathology (Chapter 10), earlier onset and more severe alcoholism appears to be more heritable (McGuffin et al., 1994). For example, in the twin study just mentioned (McGue et al., 1992), heritability for males was found to be twice as high for onset of alcoholism before age 21 relative to alcoholism that begins after age 21.

Depression often co-occurs with alcoholism. Multivariate genetic research indicates that alcoholism and depression are largely due to different genes (McGuffin et al., 1994; Merikangas, 1990). Other possible mediators of genetic influence on alcoholism have also been explored, such as personality, alcohol sensitivity, and cognitive factors (Cloninger, 1987b; McGue, 1993).

An influential classification based on the adoption study mentioned earlier (Cloninger et al., 1981) suggests that early-onset alcoholism in males associated with alcohol-related aggression, called Type II alcoholism, is especially heritable.

One of the strongest areas of behavioral genetic research in rodents is called *psychopharmacogenetics*, genetic effects on behavioral responses to drugs, and much of this research involves alcohol (Bloom & Kupfer, 1995; Broadhurst, 1978; Crabbe & Harris, 1991). In 1959, it was shown that inbred strains of mice differ markedly in their preference for drinking alcohol, an observation that implies genetic influence (McClearn & Rodgers, 1959). Inbred strain differences have subsequently been found for many behavioral responses to alcohol (Phillips & Crabbe, 1991).

Selection studies provide especially powerful demonstrations of genetic influence. For example, one study successfully selected for sensitivity to the effects of alcohol (McClearn, 1976). When mice are injected with the mouse equivalent of several drinks, they will "sleep it off" for various lengths of time. "Sleep time" in response to alcohol injections was measured by the time it took mice to right themselves after being placed on their backs in a cradle (Figure 12.3). Selection for this measure of alcohol sensitivity was successful, providing a

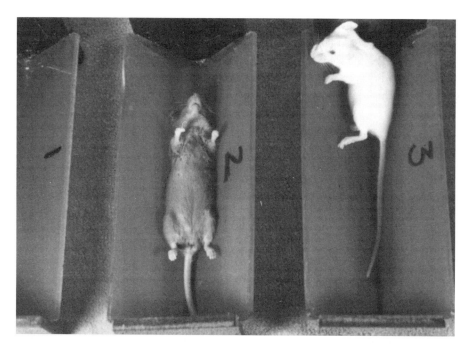

Figure 12.3 The "sleep cradle" for measuring time interval for loss of righting response after alcohol injections in mice. In cradle 2, a long-sleep mouse is still on its back, sleeping off the alcohol injection. In cradle 3, a short-sleep mouse has just begun to right itself. (Courtesy of E. A. Thomas.)

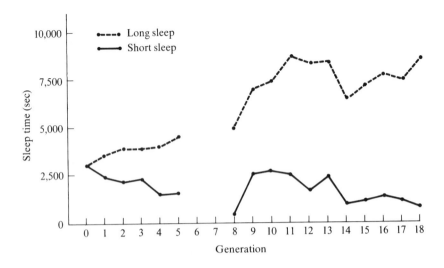

Figure 12.4 Results of alcohol sleep-time selection study. Selection was suspended during generations 6 through 8. (From G. E. McClearn, unpublished data.)

powerful demonstration of the importance of genetic factors (Figure 12.4). After 18 generations of selective breeding, the long-sleep (LS) animals "slept" for an average of two hours. Many of the short-sleep (SS) mice were not even knocked out, and their average "sleep time" was only about ten minutes. By generation 15, there was no overlap between the LS and SS lines (Figure 12.5). That is, every mouse in the LS line slept longer than any mouse in the SS line.

The steady divergence of the lines over 18 generations indicates that many genes affect this measure. If just one or two genes were involved, the lines would

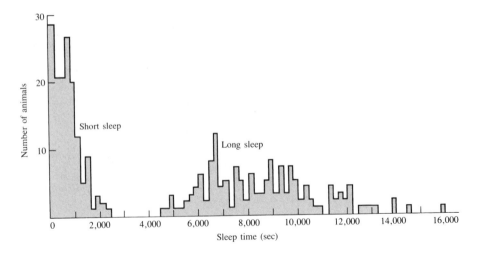

Figure 12.5 Distributions of alcohol sleep time after 15 generations of selection. (From G. E. McClearn, unpublished data.)

John Crabbe is a research career scientist at the V.A. Medical Center and a professor in the Department of Behavioral Neuroscience at the Oregon Health Sciences University, Portland, Oregon. He is director of an NIH Alcohol Research Center, whose goals are to identify the genetic map locations of genes affecting neuroadaptation to alcohol and to develop new genetic animal models to explore alcohol's effects. He received his doctorate in biopsychology from the University of Colorado in 1973 and was trained there at the Institute for Behavioral Genetics. After postdoctoral work at the University of Wisconsin-Madison, he was a lecturer in psychology at San Jose State University and the University of California, Santa Barbara, and then held a two-year research position at a Dutch pharmaceutical company. Crabbe moved to his current location in 1979. His research has concentrated on the analysis of drug-related responses in mice. He has developed several mouse lines selectively bred to be sensitive or resistant to different effects of alcohol. Most recently, he has been studying special mouse populations with behavioral, neurochemical, and molecular biological techniques designed to identify QTLs affecting drug-related behavior. These mouse genes can then be located in the human genome.

completely diverge in a few generations. Selected lines provide important animal models for additional research on pathways between genes and behavior. For example, the LS and SS lines have been extensively used as mouse models of alcohol sensitivity (Collins, 1981). Other selection studies include successful selection for susceptibility to seizures during withdrawal from alcohol dependence in mice and for voluntary alcohol consumption in rats (Crabbe et al., 1985). These are powerful genetic effects. For example, mice in the line selected for susceptibility to seizures are so sensitive to withdrawal that they show symptoms after a single injection of alcohol.

Psychopharmacogenetic studies of mice have become increasingly important in identifying QTLs associated with drug-related behavior (Crabbe, Belknap, & Buck, 1994). Several QTLs for alcohol-related behaviors have been reported, using the F_2 and the RI QTL methods described in Chapter 6 (Crabbe et al., 1994; Plomin & McClearn, 1993b). As also discussed in Chapter 6, mouse QTL research is especially exciting because it can nominate candidate QTLs that can then be tested in human QTL research.

Molecular genetic studies of human alcoholism are also underway. Much is known about genes involved in the metabolism of alcohol. For example, about half of all Chinese and Japanese individuals have an allele that, when homozygous, leads to inactivity of a key enzyme in the metabolism of alcohol. The resulting buildup in this step of alcohol metabolism leads to unpleasant symptoms such as flushing and nausea when alcohol is consumed. This genetic variant results in reduced alcohol consumption and has been implicated as the reason why rates of alcoholism are much lower in Asian populations than in Caucasian populations (Hodgkinson, Mullan, & Murray, 1991). Within Caucasian populations, several studies have found an allelic association between a DNA marker for a dopamine receptor (*DRD2*) and alcoholism (reviewed by Stone & Gottesman, 1993) and other drug abuse (Uhl et al., 1993). Although other studies have failed to find this association (Gelernter, Goldman, & Risch, 1993), a recent analysis of all studies supports it (Neiswanger, Hill, & Kaplan, 1995). Large-scale studies scanning the genome for other linkages are also in progress.

SUMMING UP

Results of twin and adoption studies of alcoholism vary greatly, but all together they suggest moderate heritability for males and modest heritability for females. Early-onset alcoholism and more severe alcoholism appear to be more highly heritable. Psychopharmacogenetics has been a very active area of research using mouse models of drug use and abuse, especially for alcohol. For example, selection studies have documented genetic influence on many behavioral responses to drugs. QTLs for alcohol-related behavior in mice have been identified. An association between a DNA marker for a dopamine receptor and alcoholism in humans has also been reported.

Smoking Nicotine is a highly addictive drug and, in a sense, is the most lethal drug, because tobacco use is associated with the death of hundreds of thousands of people each year in the United States alone (Peto et al., 1992). Although nicotine is an environmental agent, individual differences in susceptibility to its addictive properties are influenced by genetic factors. Five twin studies with more than a thousand twin pairs each from four countries all point to genetic influence on smoking (Heath & Madden, 1995). For example, the largest study includes 12,000 pairs from Sweden of whom half smoked (Medlund et al., 1977). If one twin currently smoked, the probability that the co-twin smoked was 75 percent for identical twins and 63 percent for fraternal twins. Across these twin studies, analyses based on the liability-threshold model (see Chapter 3) suggest heritabilities of liability to smoking of about 60 percent and some shared environmental influence (Heath & Madden, 1995). In a study of 42 pairs of identical twins reared apart, their concordance for smoking was 79 percent (Shields, 1962).

   The reasons why people start to smoke appear to differ from the reasons why people persist in smoking and in the amount they smoke (Heath &

Martin, 1993). For example, shared environment, probably due to peers rather than parents, plays a much larger role in smoking initiation than in smoking persistence (Rowe & Linver, 1995). An interesting developmental side to this issue is that adolescence is a critical period for smoking initiation. Few people start smoking after adolescence. In addition, multivariate genetic analyses show that most of the genetic variance in smoking initiation can be accounted for by personality traits such as novelty seeking (Heath & Madden, 1995) and by depression (Kendler et al., 1993b). However, personality is not related to persistence or quantity of smoking. The same genetic factors appear to be involved in these two aspects of smoking.

Other drugs Inbred strain and selection studies in mice have documented genetic influence on sensitivity to almost all drugs subject to abuse (Crabbe & Harris, 1991). Human studies are difficult because drugs such as amphetamines and cocaine are illegal (Seale, 1991). An interesting recent finding is that *exposure* to drugs shows genetic influence. In a study of more than a thousand male twin pairs from the Vietnam Era Twin Registry, significant heritability was found, not just for use of marijuana, stimulants, sedatives, cocaine, opiates, and psychedelics, but also for exposure to each drug (Tsuang et al., 1992). Exposure to drugs is usually, and reasonably, thought to be an environmental risk factor. However, results such as these raise the possibility that genetic factors contribute to experience, a topic discussed in Chapter 14.

S U M M I N G UP

Moderate genetic influence has been found for persistence and quantity of smoking. Shared environmental influence plays a larger role for initiation of smoking. Mouse research shows genetic influence on sensitivity to nearly all drugs subject to abuse. In humans, even *exposure* to drugs may be influenced by genetic factors.

Psychology and Aging

Aging is another example of a new area in psychology that is being introduced to genetic research. Like health psychology, aging is an area of great social significance. The average age of most societies is increasing, primarily as a result of improvements in health care. For example, in the United States, the number of people age 65 and older will double from 10 to 20 percent during the next 30 years (U.S. Bureau of the Census, 1995). The fastest growing group of adults is those over age 85. Worldwide, this group is growing nearly twice as fast as the population as a whole (Chawla, 1993). Although obvious changes occur later in life, it is not possible to lump these older individuals into a category of "the elderly" because older adults differ greatly biologically and psychologically. The question for genetics is the extent to which genetic factors contribute to individual differences in functioning later in life.

Surprisingly little genetic research in psychology has been directed toward the last half of the life span (Bergeman & Plomin, in press). Chapter 7 described genetic research on dementia for which a few twin studies found moderate genetic influence. Dementia is a focal area for molecular genetic research. Several genes have been described that together account for most cases of a rare form of dementia that occurs in middle adulthood. The best example of a QTL in psychology is the association between apolipoprotein E and typical late-onset dementia.

Another interesting finding about genetics and cognitive aging was described in Chapter 8: The heritability of general cognitive ability increases throughout the life span. In later life, heritability estimates reach 80 percent, one of the highest heritabilities reported for behavioral traits.

Not enough research has been conducted on specific cognitive abilities throughout the life span to be able to conclude whether heritabilities of specific cognitive abilities also increase during development. However, this conclusion seems likely, at least as a general rule, because genetic influence on specific cognitive abilities largely overlaps with genetic influence on general cognitive ability (Chapter 9). Not mentioned in this discussion of multivariate genetic analysis in Chapter 9 is a distinction made in the field of cognition and aging between "fluid" abilities such as spatial ability that decline with age and "crystallized" abilities such as vocabulary that increase with age (Baltes, 1993). Although it has been assumed that fluid abilities are more biologically based and crystallized abilities more culturally based, genetic research so far has found that fluid and crystallized abilities are equally heritable (Pedersen, 1996).

For psychopathology and personality, the few genetic studies in later life yield results similar to those described in Chapters 10 and 11 for research earlier in life (Bergeman, in press). One twin study of nonclinical depression in older twins found only modest genetic influence (about 15 percent heritability) for depressive symptoms, although heritability was greater for the oldest twins in this sample (Gatz et al., 1992). Shared environmental influence was surprisingly strong in this study, accounting for about 25 percent of the variance.

For personality, a few traits that have been subjected to genetic research in later life have not been studied earlier in the life span. So-called Type A behavior, hard-driving and competitive behavior that is of special interest because of its reputed link with heart attacks, shows moderate heritability typical of other personality measures in older twins (Pedersen et al., 1989b). Another interesting personality domain is locus of control, which refers to the extent that outcomes are believed to be due to one's own behavior or chance. For some older individuals, this sense of control declines, and the decline is linked to declines in psychological functioning and poor health. A twin study later in life found moderate genetic influence for two aspects of locus of control, sense of responsibility and life direction (Pedersen et al., 1989a). However, the key variable of the perceived role of luck in determining life's outcomes showed no genetic influence and substantial shared environmental influence. This finding, although in need of replication, stands out from the usual finding in personality research of moderate genetic influence and no shared environmental influence.

The famous U.S. Supreme Court Justice Oliver Wendell Holmes quipped that "those wishing long lives should advertise for a couple of parents, both belonging to long-lived families" (Cohen, 1964, p. 133). Genetic research, however, indicates only modest genetic influence on longevity. For example, a study of more than 500 pairs of twins reported correlations of .23 for identical twins and .00 for fraternal twins for longevity (McGue et al., 1993d). Similar results suggesting modest genetic influence on longevity have been found in other twin, family, and adoption studies (Bergeman, in press).

Psychologists are especially interested in how well we live, the quality of life, not just how long we live. Health and functioning in daily life show moderate genetic influence later in life, as does the relationship between health and psychological well-being (Harris et al., 1992) and life satisfaction (Plomin & McClearn, 1990). Another aspect of quality of life is self-perceived competence. One study of older twins found that six dimensions of self-perceived competence—including interpersonal skills, intellectual abilities, and domestic skills—show heritabilities of about 50 percent (McGue, Hirsch, & Lykken, 1993c).

SUMMING UP

Surprisingly few twin and adoption studies in psychology have been directed toward the last half of the life span. Nonetheless, dementia is one of the most intense areas of molecular genetic research. The best example of a QTL in psychology is the association between apolipoprotein E and late-onset dementia. Longevity shows only modest genetic influence. The few twin and adoption studies of psychological traits in the later years yield results that are generally similar to those found earlier in the life span. Quality of life indicators later in life also show some genetic influence.

As discussed in Chapter 14, a study of older twins was the first to show genetic influence on life events, especially life events such as conflict with spouse over which the individual has some control (Plomin et al., 1990b). Additional analyses indicated that genetic influence on life events primarily emerges for women and can be explained by genetic influence on personality traits (Saudino et al., in press-b). Social support later in life also shows genetic influence in terms of satisfaction with relationships but not number of relationships (Bergeman et al., in press). Over a six-year period, genetic factors were largely responsible for continuity of social support (Bergeman et al., submitted). Genetic factors also primarily accounted for the association between social support and psychological well-being (Bergeman et al., 1991). Another study of elderly twins indicated that genetic factors contribute to the association between education and cognitive functioning later in life (Carmelli, Swan, & Cardon, 1995).

Summary

Two new areas of psychology from which interesting genetic results are emerging are health psychology and aging. Within health psychology, one example of genetic research is stress and cardiovascular risk. Several twin studies show moderate genetic influence on cardiovascular responses to stress in the laboratory as well as outside the laboratory.

A second example of genetic research on health psychology is body weight and obesity. Although most theories of weight gain are environmental, genetic research consistently shows substantial genetic influence on individual differences in body weight, with heritabilities of about 70 percent. Also interesting in light of environmental theories is the consistent finding that shared family environment does not affect weight. Longitudinal genetic studies indicate that genetic influences on weight are surprisingly stable after infancy, although there is some evidence for genetic change even during adulthood. Much current interest in molecular genetic research focuses on two genes for obesity originally found in mice.

A third example from health psychology is addictions. Alcoholism in males shows moderate genetic influence, with stronger genetic influence for alcoholism that is early in onset, severe, and associated with aggression. For females, genetic influence on alcoholism is modest. Selection studies of alcohol-related behaviors in mice demonstrate genetic influence, provide animal models for research, and yield QTLs. Molecular genetic studies are underway for human alcoholism. Persistence and quantity of smoking also show moderate genetic influence; initiation of smoking shows a larger role for shared environment, probably due to peers rather than parents.

Genetic research has only recently addressed the last half of the life span. Dementia and cognitive decline in later life is one of the most intense areas of molecular genetic research in psychology. For general cognitive ability, twin and adoption studies indicate that heritability increases during adulthood. Some research suggests that heritability of depression also increases in later life. Personality traits generally show results similar to those for younger ages, moderate heritability and no shared family environment. Several quality of life measures also show moderate genetic influence in studies of elderly individuals.

CHAPTER THIRTEEN

Evolutionary Psychology

Although its roots lie firmly with Darwin's ideas of more than a century ago, evolutionary thinking has only recently established itself in psychology. This chapter offers an overview of evolutionary theory and two related fields. Population genetics provides a quantitative basis for investigating forces, especially evolutionary forces, that change gene and genotype frequencies. The second related field is sociobiology, which looks at evolution from the perspective of the gene rather than from that of the individual. Finally, the chapter considers the relationship between genetics viewed on an evolutionary time scale and contemporary genetic variation, which has been the focus of research in behavioral genetics.

Charles Darwin

One of the most influential books ever written is Charles Darwin's 1859 *On the Origin of Species* (Figure 13.1). Darwin's famous trip around the world on the *Beagle* led him to observe the remarkable adaptation of species to their environments. For example, he made particularly compelling observations about 14 species of finches found in a small area on the Galápagos Islands. The principal differences among these finches were in their beaks, and each beak was exactly appropriate for the particular eating habits of the species (Figure 13.2).

Theology of the time proposed an "argument from design," which viewed the adaptation of animals and plants to the circumstances of their lives as evidence of the Creator's wisdom. Such exquisite design, so the argument went, implied a "Designer." Darwin was asked to serve as naturalist on the surveying voyage of the *Beagle* in order to provide more examples for the "argument from design." However, during his voyage, Darwin began to realize that species, such as the Galápagos finches, were not designed once and for all. This realization led to his heretical theory that species evolve one from another: "Seeing this gradation and diversity of structure in one small, intimately related group of

birds, one might really fancy that from an original paucity of birds in this archipelago, one species had been taken and modified for different ends" (Darwin, 1896, p. 380). For 20 years after his voyage, Darwin gradually and systematically marshalled evidence for his theory of evolution.

Darwin's theory of evolution begins with variation within a population. Variation exists among individuals in a population due, at least in part, to heredity. If the likelihood of surviving to maturity and reproducing is influenced even to a slight degree by a particular trait, offspring of the survivors will show slightly more of the trait than their parents' generation. In this way, generation after generation, the characteristics of a population can gradually change. Over a sufficiently long period, the cumulative changes can be so great that populations become different species, no longer capable of interbreeding successfully

For example, the different species of finches that Darwin saw on the Galápagos Islands may have evolved because individuals in a progenitor species differed slightly in the size and shape of their beaks. Certain individuals with slightly more powerful beaks may have been more able to break open hard seeds. Such individuals could survive and reproduce when seeds were the main source of food. The beaks of other individuals may have been better at catching insects and this gave them a selective advantage at certain times. Generation after generation, these slight differences led to other differences, such as different habitats. For example, seed eaters made their living on the ground and insect eaters lived in the trees. Eventually, the differences became so great that offspring of the seed

Figure 13.2 The 14 species of finches in the Galápagos Islands and Cocos Island. (a) A woodpecker-like finch that uses a twig or cactus spine instead of its tongue to dislodge insects from tree-bark crevices. (b–e) Insect eaters. (f,g) Vegetarians. (h) The Cocos Island finch. (i–n) The birds on the ground eat seeds. Note the powerful beak of (i), which lives on hard seeds. (From "Darwin's finches" by D. Lack. Copyright © 1953 by Scientific American, Inc. All rights reserved.)

eaters and insect eaters rarely interbred. Different species were born. A Pulitzer prize–winning account of 25 years of repeated observations of Darwin's finches, *The Beak of the Finch* (Weiner, 1994), shows natural selection in action.

Darwin's most notable contribution to the theory of evolution was his principle of *natural selection*:

> Owing to this struggle [for life], variations, however slight and from whatever cause proceeding, if they be in any degree profitable to the individuals of a species, in their infinitely complex relations to other organic beings and to their physical conditions of life, will tend to the preservation of such individuals, and will generally be inherited by the offspring. The offspring, also, will thus have a better chance of surviving, for, of the many individuals of any species which are periodically born, but a small number can survive. (Darwin, 1859, pp. 51–52)

Although Darwin used the phrase "survival of the fittest" to characterize this principle of natural selection, it could more appropriately be called reproduction of the fittest. Mere survival is necessary, but it is not sufficient. The key to the spread of alleles in a population is the relative number of offspring produced who themselves survive and reproduce.

Darwin convinced the world that species evolved by means of natural selection, but his theory had serious gaps, mainly because the mechanism for heredity, the gene, was not yet understood. Gregor Mendel's work was not published until seven years after the publication of the *Origin of Species*, and even then it was ignored until the turn of the century. Ironically, a copy of Mendel's manuscript was found unopened in Darwin's files (Allen, 1975). Mendel provided the answer to the riddle of inheritance, which led to an understanding of how variability arises through mutations and how genetic variability is maintained generation after generation (Chapter 2).

Darwin considered behavioral traits to be just as subject to natural selection as physical traits were. In *Origin of Species*, an entire chapter is devoted to instinctive behavior patterns. In a later book, *The Descent of Man and Selection in Relation to Sex*, Darwin discussed intellectual and moral traits in animals and humans, concluding that the difference between the mind of a human being and the mind of an animal "is certainly one of degree and not of kind" (1871, p. 101).

A century later, evolutionary thinking is making major inroads in psychology (Buss, 1991). Three of many examples of this trend in psychology are language, fears, and strategies of human mating. Theories about the origins of language have moved sharply toward universal instincts that include "modules" in the brain thought to be responsible for surprisingly specific aspects of language (Pinker, 1994). Concerning fears, it is thought that humans and other primates are much more afraid of snakes and spiders than of automobiles and guns, even though the latter are much more likely to be harmful, because fear of snakes and spiders was adaptive in our evolutionary past (Marks & Nesse,

1994; Mineka et al., 1984). In relation to strategies of human mating, why do males prefer younger females and females prefer older males as mates in nearly all cultures? It has been suggested that younger women are more able to bear and care for children, whereas older men have more resources to provide for their offspring (Buss, 1994).

It should be emphasized that just because a particular behavior is adaptive in an evolutionary sense, by no means does this imply that it is necessarily morally acceptable or desirable.

SUMMING UP

Darwin proposed that species evolve one from another. Hereditary variation among individuals results in differences in reproductive fitness. This process of natural selection changes species and can lead to new species that rarely interbreed. Gaps in the theory of evolution occurred because the mechanism of heredity, the gene, was not understood in Darwin's time. Natural selection affects behavior just as much as it affects anatomy. Evolutionary thinking is playing an increasingly important role in psychology.

Population Genetics

Darwin's evidence for the evolution of species, such as the beaks of the Galápagos finches, relied on qualitative descriptions. Population genetics provides evolution with a quantitative basis. Its unique contribution is to describe allelic and genotypic frequencies in populations and to study the forces that change these frequencies, such as natural selection.

In the absence of opposing forces, the frequencies of alleles and genotypes remain the same generation after generation. As explained in Box 2.2, this stability is called Hardy-Weinberg equilibrium. Population geneticists investigate the forces that change this equilibrium (e.g., Hartl & Clark, 1989). For example, selection against a rare recessive allele is very slow, and it is for this reason that most deleterious alleles are recessive. Suppose that a recessive allele were lethal in the homozygous condition, when two such alleles are inherited. Further suppose that the frequency of the allele were 2 percent in the original population. If no homozygous recessive individuals were to reproduce for 50 generations, the frequency for this undesirable allele would only change from 2 to 1 percent. In contrast, complete selection against a dominant allele would wipe out the allele in a single generation. As mentioned in Chapter 2, the dominant allele responsible for Huntington's disease persists because its lethal effect is not expressed until after the reproductive years.

Natural selection is often discussed in terms of *directional selection* of this sort, a process in which a deleterious allele is selected against. If successful, directional

selection would remove genetic variability. Another type of selection maintains different alleles rather than favoring one allele over another, a process that is especially interesting because genetic variability within a species is the focus of behavioral genetics. In contrast to directional selection, this type of selection is called *stabilizing selection* because it leads to a *balanced polymorphism*. Suppose that selection operated against both dominant and recessive homozygotes. In this process, heterozygotes would reproduce relatively more than the two homozygous genotypes. However, heterozygotes always produce homozygotes as well as heterozygotes (Box 2.2). Genetic variability is thus maintained.

Sickle-cell anemia in humans is a specific example of this kind of balanced polymorphism. Although few individuals afflicted with this serious disease (recessive homozygotes) survive to reproduce, the allele is nonetheless maintained in relatively high frequency in some African populations and among African-Americans. This high frequency of an essentially lethal recessive allele is due to the higher relative fitness of heterozygotes (carriers). Heterozygotes are more resistant than normal homozygotes to a form of malaria prevalent in certain parts of Africa.

Another sort of stabilizing selection involves environmental diversity. As noted in relation to Darwin's finches, if environments encountered by a species are diverse, selection pressures can differ and foster genetic variability. Box 13.1 describes an example of stabilizing selection in which shell markings of snails differ between woodlands and grasslands (Clarke, 1975).

A balanced polymorphism can also occur if selection depends on the frequency of a genotype. For example, selection that favors rare alleles produces genetic variability. Individuals with a rare genotype might use resources that are not used by other members of the species and thus gain a selective edge.

BOX **13.1**

Maintaining Genetic Variability in Snails ·

Shell markings of a single species of land snail display great variation in the pattern of stripes and colors (see figure). Such genetic variability has been around for a long time; fossil snails tens of thousands of years old have similar varieties of shell. This genetic variability is maintained by selection. When snails are found in woodlands, their shells are likely to be without bands. However, snails in grasslands are likely to have banded shells. The fact that shell banding is correlated with habitat suggests that selection is at work.

Direct evidence for selection comes from an examination of the shells of snails captured by thrushes, who smash the snails on stones to break them open (Clarke, 1975). The most conspicuous snails in a particular habitat—that is, banded snails in woodlands, unbanded snails in grasslands—are most often preyed on by the thrushes. Given that this species of land snail occupies both woodlands and grasslands, genetic variability for shell banding will continue as the result of such selection.

Variation in the shell markings of a single species of land snail, *Cepaea nemoralis*.
(From "The causes of biological diversity" by B. Clarke. Copyright © 1975 by Scientific American, Inc. All rights reserved.)

Predator-prey relationships can also be frequency dependent. For example, predatory birds and mammals tend to attack more common types of prey.

Another type of frequency-dependent selection involves mate selection in which rare genotypes have an edge. For example, in fruit flies, females are more likely to mate with a rare male (Ehrman & Seiger, 1987). Like the other types of stabilizing selection, frequency-dependent sexual selection maintains genetic variability in a species.

In addition to considering forces that change allelic frequency, population genetics also investigates systems of mating—inbreeding and assortative mating— that change genotypic frequencies without changing allelic frequencies. Inbreeding involves matings between genetically related individuals. If inbreeding occurs, offspring are more likely than average to have the same alleles at any locus, which means that recessive traits are more likely to be expressed. Inbreeding reduces heterozygosity and increases homozygosity. In relation to the deriva-tion of inbred strains mentioned in Chapter 5, population genetics shows that, after 20 generations of brother-sister matings, at least 98 percent of all loci are homozygous. Inbreeding often leads to a reduction in viability and fertility, called *inbreeding depression*. Inbreeding depression is caused by the increase in homozygosity for deleterious recessive alleles. Although inbreeding reduces genetic variability, its overall effect on genetic variability in natural populations is negligible because it is relatively rare. The other side of the coin is *hybrid vigor*, or *heterosis*. These terms refer to an increase in viability and performance when different inbred strains are crossed. Outbreeding reintroduces heterozy-gosity and masks the effects of deleterious recessive alleles.

Assortative mating, phenotypic similarity between mates, is another system of mating that changes genotypic, not allelic, frequency. As discussed in Chapter 8, assortative mating for a particular trait increases genotypic variance for that trait in a population. Although assortative mating is modest for most behavioral traits, increases in genetic variance due to assortative mating accumulate over generations. In other words, even a small amount of assortative mating can greatly increase genetic variability after many generations.

SUMMING UP

Population genetics is the study of allelic and genotypic frequencies in popula-tions and of the forces that change these frequencies, such as natural selection. Stabilizing selection such as frequency-dependent selection increases genetic variability in a population. Inbreeding has little overall effect on genotypic variability because it is relatively rare at a population level. Assortative mating increases genotypic variability cumulatively over generations.

Sociobiology

The trend toward more evolutionary thinking in psychology has been fueled by *sociobiology* (Wilson, 1975), which extended Darwin's theory of individual fitness to consider what is called *inclusive fitness*. Inclusive fitness is the fitness of an individual plus part of the fitness of kin that is genetically shared by the individual (Hamilton, 1968). Inclusive fitness and kin selection explain altruistic acts that do not directly benefit the individual. If the net result of an altruistic act helps more of that individual's genes to survive and to be transmitted to future generations, the act is adaptive even if it results in the death of the individual.

For example, the founder of quantitative genetics, R. A. Fisher, long ago suggested an example of kin selection and inclusive fitness that involves distastefulness of some butterfly larvae (Fisher, 1930). A bird will learn that certain larvae taste bad, but the lesson costs the larva its life. However, sibling eggs are laid in a cluster, and inclusive fitness is served by the sacrifice of one larva if two siblings (the genetic equivalent of the sacrificed larva) are saved. Inclusive fitness switches the focus from the individual to the gene, which explains the title of a classic book in this area, *The Selfish Gene* (Dawkins, 1976). Acts that appear to be altruistic can be interpreted in terms of "selfish" genes that are maximizing their reproduction through inclusive fitness.

Sociobiology has offered novel and interesting hypotheses that stem from the simple principle of inclusive fitness and kin selection. For example, why do mothers provide most of the care of offspring in the vast majority of mammalian species, including humans? Unless a species is completely monogamous (as eagles are, for example), males have less invested in their offspring. Males can have many offspring by many females, but each female must devote large amounts of energy to each pregnancy and, in mammals, provide sustenance after birth. In terms of inclusive fitness, the fitness of females is better served by increased care of each offspring, because females must make a substantial investment in each one of them. In many cases, however, the male's investment is little more than copulation, and he can maximize his inclusive fitness by having more offspring by different females. A related reason for the relative investments of mothers and fathers in the care of their offspring is that females can always be sure that they share half of their genes with their young. Males, however, cannot be sure that offspring are theirs. According to sociobiology, the greater altruism of mothers to their offspring is no less selfish from the point of view of genes than that of fathers. Although there are many hypotheses of this sort in which "selfish altruism" of genes evolved through kin selection, it is also likely that some positive social behaviors evolved through the less devious mechanism of individual selection (de Waal, 1996).

What about individual differences, the focus of behavioral genetics? For example, why do some mothers provide better care to their offspring than others? Although sociobiology usually focuses on species-typical behavior, some

Figure 13.3 Ground squirrel in position for alarm call. (Photograph by George D. Lepp; courtesy of Paul W. Sherman.)

research considers individual differences within a species. For example, an early empirical study asked why some ground squirrels make alarm calls in response to predators even though the squirrel that makes the alarm call is more likely to be killed by the predator (Sherman, 1977) (see Figure 13.3). Kin selection predicts that alarm calls will occur more often among individuals with more relatives. Observations showed that females with kin, especially females with young, were more likely to make alarm calls. However, more research is required to determine whether individual differences in alarm calling has a genetic basis.

Evolutionary Psychology and Behavioral Genetics

Evolutionary thinking is essential to psychology in order to paint the broad strokes in the portrait of our species that show the similarities to and differences from other species. For example, the fact that we are mammals, defined in terms of the mammary gland, means that we have evolved a system in which mothers care for their young after birth. The fact that we are primates has many evolutionary implications, such as extremely slow postnatal development, which requires long-term care by parents. Also fundamental for understanding our species are facts such as these: Our species uses language naturally, walks upright on two feet, and has eyes in the front of the head that permit depth perception.

In contrast to thinking about average differences between species on an evolutionary time scale, behavioral genetics focuses on genetic variability that exists today among individuals within a population. Although both perspectives are useful, it is important to recognize the distinctions between these perspectives, because the causes of average differences between species are not necessarily related to the causes of individual differences within a species. For example, although use of language evolved in our species, this does not mean that individual differences within our species in the ability to use language are due to genetic differences. The origins of individual differences is an empirical question that can be addressed by genetic research. Conversely, finding genetic influence on individual differences for a particular trait has no necessary evolutionary implications for that trait.

In summary, investigating the causes of average differences between species, which is the focus of evolutionary psychology, is a different level of analysis that asks questions and gets answers different from those asked and received in investigations of the causes of individual differences within species, which is the focus of behavioral genetics. There is increasing interest in attempting to integrate these perspectives (Buss, 1991; Fuller, 1983).

SUMMING UP

Sociobiology is an extension of evolutionary theory that focuses on inclusive fitness and kin selection. Sociobiology and other evolutionary approaches usually concentrate on average differences between species considered on an evolutionary time scale. This level of analysis is different from that of most behavioral genetic research, which focuses on the genetic and environmental origins of individual differences within species in current populations.

Summary

Charles Darwin's 1859 book on the origin of species convinced the world that species evolved one from the other rather than being created once and for all. Reproductive fitness is the key to natural selection. Gaps in Darwin's theory of evolution occurred because the mechanism for heredity, the gene, was not understood at that time. Although Darwin noted that natural selection affected behavior as much as bones, evolutionary thinking has only entered the mainstream of psychology in recent years.

Population genetics investigates forces that change allelic and genotypic frequencies. Because behavioral genetics focuses on genetic variability within a species, types of natural selection that increase genetic variation in a population are especially interesting, such as balanced polymorphisms due to heterozygote advantage or frequency-dependent selection. Inbreeding and assortative mating change genotypic but not allelic frequencies. Although inbreeding can reduce genetic variability, its rarity in the human species makes its effects on populations negligible. Assortative mating, on the other hand, increases genotypic variability for many behavioral traits.

This trend toward evolutionary thinking in psychology has in part been fueled by sociobiology with its emphasis on inclusive fitness and kin selection that goes beyond Darwin's focus on individual reproductive fitness. Most evolutionary psychology considers average differences between species on an evolutionary time scale. This level of analysis is different from that of most behavioral genetics, which focuses on contemporary individual differences.

CHAPTER FOURTEEN

Environment

G enetic research is changing the way we think about the environment. For example, two of the most important discoveries from genetic research in psychology are about nurture rather than nature. The first discovery is that environmental influences tend to make children growing up in the same family different, not similar. Because environmental influences that affect psychological development are not shared by children in the same family, they are called *nonshared environment*. The second discovery is equally surprising: Many environmental measures widely used in psychology show genetic influence. This research suggests that people create their own experiences in part for genetic reasons. This topic has been called the *nature of nurture*, although in genetics it is known as *genotype-environment correlation* because it refers to experiences that are correlated with genetic propensities. Another important concept at the interface between nature and nurture is *genotype-environment interaction*, genetic sensitivity to environments.

Nonshared environment, genotype-environment correlation, and genotype-environment interaction are the topics of this chapter. The chapter's goal is to show that some of the most important questions in genetic research involve the environment and some of the most important questions for environmental research involve genetics (Rutter et al., in press-c). Genetic research will profit if it includes sophisticated measures of the environment, environmental research will benefit from the use of genetic designs, and psychology will be advanced by collaboration between geneticists and environmentalists. These are ways in which some psychologists are putting the nature-nurture controversy behind them and bringing nature and nurture together in the study of development in order to understand the processes by which genotypes eventuate in phenotypes.

Three reminders about the environment are warranted. First, genetic research provides the best available evidence for the importance of environmental factors. The surprise from genetic research in psychology has been the discovery that genetic factors are so important throughout psychology, sometimes accounting for as much as half of the variance. However, the excitement about this discovery should not overshadow the fact that environmental factors are at least as important. Heritability rarely exceeds 50 percent and thus "environmentality" is rarely less than 50 percent.

Second, in quantitative genetic theory, the word *environment* includes all influences other than inheritance, a much broader use of the word than is usual in psychology. By this definition, environment includes, for instance, prenatal events and biological events such as nutrition and illness, not just family socialization factors.

Third, as explained in Chapter 5, genetic research describes *what is* rather than predicts *what could be*. For example, high heritability for height means that height differences among individuals are largely due to genetic differences, given the genetic and environmental influences that exist in a particular population at a particular time (*what is*). Even for a highly heritable trait such as height, an environmental intervention such as improving children's diet or preventing illness could affect height (*what could be*). Such environmental factors are thought to be responsible for the average increase in height across generations, for example, even though individual differences in height are highly heritable in each generation.

Nonshared Environment

From Freud onward, most theories about how the environment works in development implicitly assume that offspring resemble their parents because parents provide the family environment for their offspring and that siblings resemble each other because they share that family environment. Twin and adoption research during the past two decades has dramatically altered this view. The reason why genetic designs such as twin and adoption methods were devised was to address the possibility that some of this widespread familial resemblance may be due to shared heredity rather than to shared family environment. The surprise is that genetic research consistently shows that family resemblance is almost entirely due to shared heredity rather than to shared family environment. As indicated in Chapters 10 and 11, shared family environment plays a negligible role in much psychopathology and personality. Chapter 12 showed that shared family environment also has little effect on alcoholism in males and, most surprisingly, on body weight. Only a few possible exceptions to this rule have been found, such as conduct disorder in adolescence. Results for cognitive abilities are more complex (Chapters 8 and 9). The evidence is clear that, in childhood, about a quarter of the variance of general cognitive ability is due to shared family environment. However, after adolescence, the role of shared family environment is negligible. Although twin studies suggest some influence of shared family environment for specific cognitive abilities and especially for school achievement, studies of adoptive relatives are needed to provide direct tests of the importance of shared family environment in this context.

Environmental influence is important, accounting for at least half of the variance for most domains of psychology, but it is generally not shared family environment that causes family members to resemble each other. The salient environmental influences are not shared by family members (Figure 14.1). This remarkable finding means that environmental influences that affect development operate to make children growing up in the same family no more similar than children growing up in different families. Shared and nonshared environment are not limited to family environments. Experiences outside the family can also be shared or not shared by siblings, such as peer groups and educational and occupational experiences.

The nonshared environment component of variance refers to variance not explained by heredity or by shared family environment, and it includes error of measurement. For example, for self-report personality questionnaires (Chapter 11), genetics accounts for about 40 percent of the variance, shared family environment for 0 percent, and nonshared environment for 60 percent of the variance. Such questionnaires are usually at least 80 percent reliable, which means that about 20 percent of the variance is due to error of measurement. In other words, systematic nonshared environmental variance excluding error of measurement accounts for about 40 percent of the variance (i.e., 60 percent – 20 percent).

It is important that the specific processes responsible for nonshared environmental effects be identified, because it is only by identifying the details of causal mechanisms that policy and practice implications can be determined. Nevertheless, genetic designs provide an essential starting point in their quantification of the net effect of genetic and environmental influences in the populations studied. Provided the limitations of such quantifications are taken into account, the findings can provide very valuable pointers. For example, if the net effect of genetic factors is substantial, then there may be value in seeking to identify the specific genes responsible for that strong genetic effect, as is

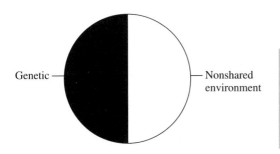

Genetic — Nonshared environment

Figure 14.1 Phenotypic variance for most psychological traits is caused by environmental variance as well as genetic variance, but the environmental variance is largely of the nonshared variety.

the case with respect to autism. Similarly, if environmental influences are largely nonshared rather than shared, this finding should deter researchers from relying solely on family-wide risk factors that pay no attention to the ways in which these influences impinge differentially on different children in the same family. An important caveat, however, is that it may be misleading to dismiss the possibility of shared environmental effects if there is no measurement of specific environmental risk factors. For example, although a twin study of depression and alcoholism found no evidence of shared environmental influence, when a specific shared environmental factor—parental loss—was included in the genetic model, it showed a significant effect (Kendler et al., 1992a, 1996). Still, genetic designs have been helpful in leading researchers in the right direction. Current research is trying to identify specific sources of nonshared environment and to investigate associations between nonshared environment and psychological traits, as discussed later.

Estimating Nonshared Environment

How do genetic designs estimate the net effect of nonshared environment? Chapter 5 focused on heritability, which is estimated, for example, by comparing identical and fraternal twin resemblance or by using adoption designs. Environmental variance is variance not explained by genetics. Shared family environment is estimated as family resemblance not explained by genetics. Nonshared environment is the rest of the variance, variance not explained by genetics or by shared family environment. The conclusion that environmental variance is largely nonshared refers to this residual component of variance, usually estimated by model-fitting analyses. However, more direct tests of shared and nonshared environments make it easier to understand how they can be estimated.

A direct test of shared family environment is resemblance among adoptive relatives. Why do genetically unrelated adoptive "siblings" correlate about .25 for general cognitive ability in childhood? The answer must be shared family environment because adoptive siblings are unrelated genetically. This result fits with the conclusion in Chapter 8 that about a quarter of the variance of general cognitive ability in childhood is due to shared family environment. By adolescence, the correlation for adoptive siblings plummets to zero and is the basis for the conclusion that shared family environment has a negligible impact in the long run. For personality and much psychopathology, adoptive siblings correlate near zero, a value implying that shared environment is unimportant and that environmental influences, which are substantial, are of the nonshared variety.

Just as genetically unrelated adoptive siblings provide a direct test of shared family environment, identical twins provide a direct test of nonshared environment. Because they are identical genetically, differences within pairs of identical twins can only be due to nonshared environment. For example, for self-report personality questionnaires, identical twins typically correlate about .45. This value means that about 55 percent of the variance is due to non-shared environment plus error of measurement. Identical twin resemblance is also only moderate for most mental disorders, an observation implying that nonshared environmental influences play a major role.

Differences within pairs of identical twins is a conservative estimate of nonshared environment, because twins often share special environments that increase their resemblance but do not contribute to similarity among "normal" siblings. For example, for general cognitive ability, identical twins correlate about .85, a result that does not seem to leave much room for nonshared environment (i.e., $1 - .85 = .15$). However, fraternal twins correlate about .60, whereas nontwin siblings correlate about .40, implying that twins have a special shared twin environment that accounts for as much as 20 percent of the variance. For this reason, the identical twin correlation of .85 may be inflated by .20 because of this special shared twin environment. In other words, about a third of the variance of general cognitive ability may be due to nonshared environment [i.e., $1 - (.85 - .20) = .35$]. Another example of underestimating the importance of nonshared environment in twin studies is attitudes. As explained in Chapter 11, in twin studies, attitudes appear to show substantial influence from shared family environment, but most of this is due to high assortative mating for attitudes. That is, when the high assortative mating for attitudes is taken into account, estimates of shared family environmental influence for attitudes are modest and most of the environmental variance is nonshared.

Greater detail about estimating nonshared environment and evidence for the importance of nonshared environment in psychology can be found elsewhere (Plomin, Chipuer, & Neiderhiser, 1994b). The next step in research on nonshared environment is to identify specific factors that make children growing up in the same family so different.

Identifying Specific Nonshared Environment

To identify nonshared environmental factors, it is necessary to begin by assessing aspects of the environment specific to each child, rather than aspects shared by siblings. Many measures of the environment used in studies of psychological development are general to a family rather than specific to a child. For example, whether or not their parents have been divorced is the same for two children in the family. Assessed in this family-general way, divorce cannot be a source of differences in siblings' outcomes, because it does not differ for two children in the same family. However, research on divorce has shown that divorce affects children in a family differently (Hetherington & Clingempeel, 1992). If the divorce is assessed in a child-specific way (for example, by the children's different perceptions about the stress caused by the divorce so that it may, in fact, differ among siblings), it could well be a source of differential sibling outcome.

Even when environmental measures are specific to a child, they can be shared by two children in a family. Research on siblings' experiences is needed to assess the extent to which aspects of the environment are shared. For example, to what extent are maternal vocalizing and maternal affection shared by siblings in the same family? Observational research on maternal interactions with siblings assessed when each child was one and two years old indicates that mothers' spontaneous vocalizing correlates substantially across the siblings (Chipuer & Plomin, 1992). This research implies that maternal vocalizing is an experience shared by siblings. In contrast, mothers' affection yields negligible correlations across siblings, indicating that maternal affection is not shared.

Some family structure variables, such as birth order and sibling age spacing, are, by definition, nonshared environmental factors. However, these have generally been found to account for only a small portion of variance in psychological outcomes. Research on more dynamic aspects of nonshared environment has found that children growing up in the same family lead surprisingly separate lives (Dunn & Plomin, 1990). Siblings perceive their parents' treatment of them quite differently, although parents report that they treat their children similarly. Observational studies tend to back up the children's perspective.

Table 14.1 shows sibling correlations for measures of family environment in a study focused on these issues and called the Nonshared Environment and

TABLE 14.1

Sibling Correlations for Measures of Family Environment

Type of Data	Sibling Correlation
Child reports	
Parenting	.25
Sibling relationship	.40
Parent reports	
Parenting	.70
Sibling relationship	.80
Observational data	
Child to parent	.20
Parent to child	.30

SOURCE: *Adapted from Plomin (1994b).*

Adolescent Development (NEAD) project (Reiss et al., 1994, 1995). During two, two-hour visits to 720 families with sibling offspring ranging in age from 10 to 18 years, a large battery of questionnaire and interview measures of the family environment was administered to both parents and offspring, and parent-child interactions were videotaped during a session when problems in family relationships were discussed. Sibling correlations for children's reports of their family interactions (for example, children's reports of their parents' negativity) were modest, as they were for observational ratings of child-to-parent interactions and parent-to-child interactions. This finding suggests that these experiences are largely nonshared. In contrast, parent reports yielded high sibling correlations, for example, when parents reported on their own negativity toward each of the children. Although this may be due to a rater effect in that the parent rates both children, the high sibling correlations indicate that parent reports of children's environments are not good sources of candidate variables assessing nonshared environmental factors.

Nonshared environment is not limited to measures of the family environment. Indeed, experiences outside the family as siblings make their own way in the world are even more likely candidates for nonshared environmental influence. For example, how similarly do siblings experience peers, social support, and life events? The answer is "only to a limited extent"; correlations across siblings for these experiences range from about .10 to .40 (Plomin, 1994a). It is also possible that nonsystematic factors, such as accidents and illnesses, initiate differences between siblings. Compounded over time, small differences in experience might lead to large differences in outcome.

Identifying Specific Nonshared Environment That Predicts Psychological Outcomes

Once child-specific factors are identified, the next question is whether these nonshared experiences relate to psychological outcomes. For example, to what extent do differences in parental treatment account for the nonshared environmental variance known to be important for personality and psychopathology? Although research in this area has only just begun, some success has been achieved in predicting differences in adjustment from sibling differences in their experiences (Hetherington, Reiss, & Plomin, 1994). The NEAD project mentioned earlier provides an example in that negative parental behavior directed specifically to one adolescent sibling (controlling for parental treatment of the other sibling) relates strongly to that child's antisocial behavior and, to a lesser extent, to that child's depression (Reiss et al., 1995). It appears that most of these associations involve negative aspects of parenting such as conflict and negative outcomes such as antisocial behavior. Associations are generally weaker for positive parenting such as affection.

When associations are found between nonshared environment and outcome, the question of direction of effects is raised. That is, is differential parental negativity the cause or the effect of sibling differences in antisocial behavior? One way to begin to disentangle cause and effect is by means of longitudinal analyses. In the NEAD project, the adolescent siblings and their families were studied again three years later, yielding similar associations. If differential parental treatment causes sibling differences, differential parental treatment at the first time of assessment would be expected to predict changes in siblings' adjustment over the three-year period. Conversely, a child effect would predict that the child's behavior should be associated with changes in parental behavior. Although either longitudinal direction of effects was found, this was not a strong test because both parental treatment and sibling adjustment were quite stable over the three years, leaving little room to find change in the siblings' adjustment.

Genetic research is beginning to suggest that some differential parental treatment of siblings is in fact the effect rather than the cause of sibling differences. One of the reasons why siblings differ is genetics. Siblings are 50 percent similar genetically, but that means that siblings are also 50 percent different. Research on nonshared environment needs to be embedded in genetically sensitive designs in order to distinguish true nonshared environmental effects from sibling differences due to genetics. For this reason, the NEAD project included identical and fraternal twins, full siblings, half siblings, and genetically unrelated siblings. Multivariate genetic analysis of associations between parental negativity and adolescent adjustment yielded an unexpected finding: Most of these associations were mediated by genetic factors (Pike et al., 1996b). This finding implies that differential parental treatment of siblings to a substantial extent reflects genetically influenced differences between the siblings, such as differences in personality. The role of genetics on environmental influences is given detailed consideration in the next section.

Nonetheless, in these multivariate genetic analyses, some modest nonshared environmental effects were found that are not mediated by genetics, especially for the adolescents' own ratings of their parents' negativity and of their adjustment (Pike et al., 1996a). Correlating identical twin differences in experience with identical twin differences in outcome is a direct test of nonshared environment independent of genetics. As shown in Table 14.2, significant correlations were found in the NEAD project between identical twin differences in parental negativity and their differences in depression and antisocial behavior as rated by the adolescent and by the mother (Pike et al., 1996c). It is interesting that videotaped observations of parent-adolescent interactions yielded evidence for nonshared environment for adolescent antisocial behavior but not for depression (last column of Table 14.2).

These nonshared environmental effects emerged only when the same source was used to assess environment and outcome, for instance, when the adolescents rated their parents' negativity and also rated their own depression. Nonshared environmental effects were not found across raters, for example, between adolescents' ratings of their parents' negativity and parents' ratings of the adolescents'

TABLE **14.2**

Correlations Between Adolescent Identical Twin Differences in Parental Negativity and Differences in Adolescent Outcome

Measure	Correlation Between Identical Twin Differences in Parental Negativity and Adolescent Outcome		
	Adolescent Report	Mother Report	Videotaped Observation
Outcome: adolescent depressed			
Mother negativity	.33*	.23*	.02
Father negativity	.28*	.25*	.02
Outcome: adolescent antisocial			
Mother negativity	.33*	.27*	.29*
Father negativity	.39*	.54*	.29*

*$p < .05$.
SOURCE: *Pike et al. (1996c).*

depression. Although this finding might be due to rater bias, it is also possible that the source-specific differences are real. That is, the experiences and behaviors measured may be different across the situations experienced by the different informants (Pike et al., 1996c).

SUMMING UP

Environmental influences largely operate in a nonshared manner, making children growing up in the same family different from one another. Differences within pairs of identical twins provide a direct test of nonshared environment. Resemblance for adoptive siblings directly tests the importance of shared environment. Attempts to identify specific sources of nonshared environment indicate that many sibling experiences differ. Some of these sibling differences in experience relate to psychological outcomes. However, the sources of nonshared environment remain unclear because associations between sibling differences in experience and sibling differences in outcome are in part mediated genetically.

No matter how difficult it may be to find specific nonshared environmental factors within the family, it should be emphasized that nonshared environment is generally the way the environment works in psychology. It seems reasonable that experiences outside the family, experiences with peers or life events, for example, might be richer sources of nonshared environment. It is also possible that chance contributes to nonshared environment in the sense of idiosyncratic experiences or the subtle interplay of a concatenation of events (Dunn & Plomin, 1990). Francis Galton, the founder of behavioral genetics, suggested that nonshared environment is largely due to chance:

> The whimsical effects of chance in producing stable results are common enough. Tangled strings variously twitched, soon get themselves into tight knots. (Galton, 1889, p. 195)

However, before we conclude that nonshared environment is due to chance, possible systematic sources of nonshared environment need more investigation.

Implications

The discovery of the major importance of nonshared environment has far-reaching implications for understanding how the environment works in psychological development. Whatever the salient factors might be, they operate to make two children growing up in the same family different from one another. Environmental influences that affect psychological development do not operate on a family-by-family basis but rather on an individual-by-individual basis. That is, their effects are relatively specific to each child rather than general for all children in a family.

Theories of socialization and much psychological research focus on environmental factors at a level of analysis that does not consider differences between children growing up in the same family. For example, when viewed in this way, parental education, parental attitudes about child-rearing, and parents' marital relationships are shared by siblings. Shared environmental factors that do not differ between children growing up in the same family cannot explain why children growing up in the same family are different. However, the effects of such factors might not be shared, as mentioned earlier. For example, the effects of variables such as parental divorce may be nonshared because the events affect children in the family differently. The message is not that family experiences are unimportant but that the effects of environmental influences are specific to each child, not general to an entire family.

Nonshared environmental effects may come about in several ways, and research is only just beginning to differentiate among the possibilities. Thus, nonshared environmental effects may derive from experiences within or outside the family that apply to only one child in the family. These experiences might involve, for example, physical abuse, hospitalization, bullying at school, or the influence of a delinquent peer group. In addition, influences that are apparently shared by members of a family can impinge in a child-specific fashion. Thus, when there is family-wide discord or quarreling, one child may be particularly likely to be embroiled in the conflict or may be the focus of scapegoating.

Alternatively, the key influence may lie in the contrasting ways in which the parents treat the children (e.g., one being favored over the other), or in the relationships among the siblings (e.g., the dominance of one leading to dependency in the other). Finally, even when all the influences impinge entirely equally on children in the same family, the environmental effects may nevertheless be nonshared simply because children differ in their susceptibilities, as discussed in the following sections on genotype-environment correlation and interaction. Genetic research strategies are needed for the effective study of environmental risk mechanisms.

The critical question for understanding how the environment influences psychological development is why children in the same family are so different. To address this question, it is obviously necessary to study more than one child per family in order to identify sibling differences in experience and to investigate the relationship between these different experiences and differences in their psychological outcomes. Answers to the question why children in the same family are so different pertain not only to sibling differences. This is a key to unlocking the environmental origins of psychological development for all children.

Genotype-Environment Correlation

In addition to showing that environmental influences in psychology are largely of the nonshared variety, genetic research is also changing the way we think about the environment by showing that we create our experiences in part for genetic reasons. That is, genetic propensities are correlated with individual differences in experiences, an example of a phenomenon known as genotype-environment correlation. In other words, what seems to be environmental effects can reflect genetic influence because these experiences are influenced by genetic differences among individuals. This is just what genetic research during the past decade has found: When environmental measures are used as outcome measures in twin and adoption studies, the results consistently point to some genetic influence, as discussed later. For this reason, genotype-environment correlation has been described as genetic control of exposure to the environment (Kendler & Eaves, 1986).

Genotype-environment correlation adds to phenotypic variance for a trait (Appendix B), but it is difficult to detect the overall extent to which phenotypic variance is due to the correlation between genetic and environmental effects (Plomin, DeFries, & Loehlin, 1977). For this reason, these discussions focus on detection of specific genotype-environment correlations rather than on estimating their overall contribution to phenotypic variation.

What are the processes by which genetic factors contribute to variations in environmental exposure? For example, to what extent are traditional traits such as cognitive abilities, personality, and psychopathology mediators of this genetic contribution? What is even more important, does genetic influence on environmental measures contribute to the prediction of psychological outcomes from environmental measures? That is, does heredity contribute to the prediction of adolescent adjustment from measures of family environment? These questions can be viewed as questions about genotype-environment correlation for specific measures of environment.

Three Types of Genotype-Environment Correlation

There are three types of genotype-environment correlation: passive, evocative, and active (Plomin et al., 1977). The passive type occurs when children passively inherit from their parents family environments that are correlated with their genetic propensities. The evocative, or reactive, type occurs when individuals evoke reactions from other people on the basis of their genetic propensities. The active type occurs when individuals select, modify, construct, or reconstruct experiences that are correlated with their genetic propensities (see Table 14.3).

For example, consider musical ability. If musical ability is heritable, musically gifted children are likely to have musically gifted parents who provide them with both genes and an environment conducive to the development of musical ability (passive genotype-environment correlation). Musically talented children might also be picked out at school and given special opportunities (evocative). Even if no one does anything about their musical talent, gifted children might seek out their own musical environments by selecting musical friends or otherwise creating musical experiences (active).

Passive genotype-environment correlation requires interactions between

TABLE 14.3

Three Types of Genotype-Environment Correlations

Type	Description	Source of Environmental Influence
Passive	Children receive genotypes correlated with their family environment	Parents and siblings
Evocative	Individuals are reacted to on the basis of their genetic propensities	Anybody
Active	Individuals seek or create environments correlated with their genetic proclivities	Anybody or anything

SOURCE: *Plomin et al. (1977).*

genetically related individuals. The evocative type can be induced by anyone who reacts to individuals on the basis of their genetic proclivities. The active type can involve anybody or anything in the environment. Genotype-environment correlation can also be negative. As an example of negative genotype-environment correlation, slow learners might be given special attention to boost their performance.

Three Methods to Detect Genotype-Environment Correlation

Three methods are available to investigate the contribution of genetic factors to the correlation between an environmental measure and a psychological trait. These methods differ in the type of genotype-environment correlation that they detect. The first method is limited to detecting the passive type. The second method detects the evocative and active types. The third method detects all three types.

The first method compares correlations between environmental measures and traits in nonadoptive and adoptive families (Figure 14.2). In nonadoptive families, a correlation between a measure of family environment and a psychological trait of children could be environmental in origin, as is usually assumed. However, genetic factors might also contribute to the correlation. Genetic mediation would occur if genetically influenced traits of parents are correlated with the environmental measure and with the children's trait. For example, a correlation between the Home Observation for Measurement of the Environment (HOME) and children's cognitive abilities could be mediated by genetic factors that affect both the cognitive abilities of parents and also their scores on the HOME. In contrast, in adoptive families, this indirect genetic path between family environment and children's traits is not present, because adoptive parents are not genetically related to their adopted children. For this reason, a genetic

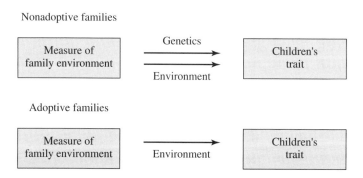

Figure 14.2 Passive genotype-environment correlation can be detected by comparing correlations between family environment and children's traits in nonadoptive and adoptive families.

contribution to the covariation between family environment and children's traits is implied if the correlation is greater in nonadoptive families than in adoptive families. The genetic contribution reflects passive genotype-environment correlation, because children in nonadoptive families passively inherit from their parents both genes and environment that are correlated with the trait.

This method has uncovered significant genetic contributions to associations between family environment and children's psychological development in the Colorado Adoption Project (Table 14.4). For example, the correlation between the HOME and cognitive development of two-year-olds is higher in nonadoptive families than in adoptive families (Plomin, Loehlin, & DeFries, 1985). The same pattern of results was found for correlations between the HOME and language development. Longitudinal predictions from family environment to children's outcomes also suggest long-term genetic mediation. For example, the correlation between the HOME at two years and general cognitive ability at seven years is greater in nonadoptive families than in adoptive families, as shown in Table 14.4 (Coon et al., 1990). Although the HOME is correlated genetically with children's cognitive development, it is not correlated genetically with parents' general cognitive ability (Bergeman & Plomin, 1988). In other words,

TABLE 14.4

Correlations Between Family Environment and Children's Traits in Nonadoptive and Adoptive Families

Measure of Family Environment*	Children's Trait	Environment-Trait Correlation	
		Nonadoptive Families	Adoptive Families
2-year HOME	2-year general cognitive ability	.42	.27
2-year HOME	2-year language	.50	.32
2-year HOME	7-year general cognitive ability	.31	.08
1-year FES	7-year behavior problems		
	Parent ratings	.36	.06
	Teacher ratings	.24	.08

*HOME, Home Observation for Measurement of Environment; FES, Family Environment Scales.
SOURCE: Plomin (1994a).

parental cognitive ability does not appear to be responsible for this passive genotype-environment correlation.

Evidence for passive genotype-environment correlation is not limited to cognitive ability. Indeed, when measures of family environment predict children's development, the correlation is usually mediated in part by genetic factors (Plomin, DeFries, & Fulker, 1988a). For example, a measure of family environment when children were one year old was correlated with parents' and teachers' ratings of children's behavior problems at seven years (Neiderhiser, 1994). Again, correlations in nonadoptive families were higher than in adoptive families, a difference suggesting genetic mediation.

The second method for finding specific genotype-environment correlations involves correlations between biological parents' traits and adoptive families' environment (Figure 14.3). This method addresses the other two types of genotype-environment correlation, evocative and active. Traits of biological parents can be used as an index of adopted children's genotype, and this index can be correlated with any measure of the adopted children's environment. Although biological parents' traits are a weak index of adopted children's genotype, a finding that biological parents' traits correlate with the environment of their adopted-away children suggests that the environmental measure reflects genetically influenced characteristics of the adopted children. That is, adopted children's genetic propensities evoke reactions from adoptive parents.

Attempts to use this method in the Colorado Adoption Project yielded only meager evidence for evocative and active genotype-environment correlation. For example, biological mothers' general cognitive ability did not correlate significantly with HOME scores in the adoptive families of their adopted-away children. Nonetheless, a few possible examples of evocative and active genotype-environment correlation emerged, as shown in Table 14.5. For example, adopted children whose biological mothers rated themselves as more active were in adoptive homes with higher HOME scores. This finding might mean that adoptive mothers are more responsive when their adopted children are genetically disposed toward higher activity.

A developmental theory of genetics and experience predicts that the evocative and active forms of genotype-environment correlation become more

Figure 14.3 Evocative and active genotype-environment correlation can be detected by the correlation between biological parents' traits (as an index of adopted children's genotype) and the environment of adoptive families.

TABLE 14.5

Correlations Between Biological Parents' Traits (as an Index of Adopted Children's Genotype) and the Environment of Adoptive Families

Biological Mothers' Trait	Measure of Adoptive Family Environment	Correlation
Activity	4-year HOME	.16
Impulsivity	4-year HOME	.17
Depression	4-year HOME	−.19
Fearfulness	1-year FES control	−.15

SOURCE: *Plomin (1994a).*

important as children experience environments outside the family and begin to play a more active role in the selection and construction of their experiences (Scarr & McCartney, 1983).

This hypothesis can be tested as the Colorado Adoption Project sample is followed into adolescence. A recent adoption study using this approach found evidence for genotype-environment correlation for antisocial behavior in adolescence (Ge et al., 1996). Genetic risk for the adoptees was indexed by anti-social personality disorder or drug abuse in their biological parents. Adoptees at genetic risk had adoptive parents who were more negative in their parenting than adoptive parents of control adoptees. Moreover, this effect was shown to be mediated by the adolescent adoptees' own antisocial behavior, an observation suggesting evocative genotype-environment correlation.

The third method to detect genotype-environment correlation involves multivariate genetic analysis of the correlation between an environmental measure and a trait (Figure 14.4). This method is the most general in the sense that it detects genotype-environment correlation of any kind—passive, evocative, or active. As explained in Appendix B, multivariate genetic analysis estimates the extent to which genetic effects on one measure overlap with genetic effects on another measure. In this case, genotype-environment correlation is implied if

Figure 14.4 Passive, evocative, and active genotype-environment correlation can be detected using multivariate genetic analysis of the correlation between environmental measures and traits.

genetic effects on an environmental measure overlap with genetic effects on a trait measure.

Multivariate genetic analysis can be used with any genetic design and with any type of environmental measure, not just measures of the family environment. For example, a sibling adoption design was used to compare cross-correlations between one sibling's HOME score and the other sibling's general cognitive ability for nonadoptive and adoptive siblings at two years of age in the Colorado Adoption Project (Braungart, Fulker, & Plomin, 1992b). As shown in Table 14.6, the phenotypic correlation between the HOME and children's cognitive ability was .42. The sibling cross-correlations between the HOME and cognitive ability were .37 for nonadoptive siblings and .12 for adoptive siblings, suggesting substantial genetic mediation. Multivariate genetic model fitting indicated that about half of the phenotypic correlation between the HOME and children's cognitive ability is mediated genetically. A subsequent analysis indicated that

TABLE 14.6

Examples of Genotype-Environment Correlation from Multivariate Genetic Analyses

Environmental Measure	Trait Measure	Phenotypic Correlation	Genetic Contribution to the Phenotypic Correlation
2-year HOME	2-year-old general cognitive ability	.42	.19
2-year HOME	2-year-old tester rating of attention	.28	.28
Mothers' negativity	Adolescent depression	.33	.24
Mothers' negativity	Adolescent antisocial behavior	.60	.40
Social support	Adult depression	−.20	−.13

SOURCE: *Plomin (1994a).*

tester-rated attention of the children accounted for the rest of the genetic influence on the HOME (Saudino & Plomin, in press).

In adolescence, multivariate genetic analyses also found substantial genetic mediation of correlations between measures of family environment and adolescents' depression and antisocial behavior in the NEAD project mentioned earlier (Pike et al., in press-a). For each of these correlations, more than half of the correlation is mediated genetically, as shown in rows 3 and 4 of Table 14.6. Evidence for genetic mediation has also been found in adulthood in correlations between parenting and parents' personality (Chipuer et al., 1992), between social support and depression and well-being (Bergeman et al., 1991; Kessler et al., 1992), between life events and personality (Saudino et al., in press-b), between socioeconomic status and health (Lichtenstein et al., 1992a), between socioeconomic status and general cognitive ability (Lichtenstein, Pedersen, & McClearn, 1992b; Tambs et al., 1989), and between education and cognitive functioning in elderly individuals (Carmelli, Swan, & Cardon, 1995). Despite the evidence for genetic mediation, these studies may also identify "pure" environmental effects independent of genetic mediation. For example, an analysis using an extended twin-family design found a pure environmental effect of childhood parental loss on alcoholism in women (Kendler et al., 1996).

The Nature of Nurture

Even though the first research on this topic was published just a decade ago, at least two dozen studies using various genetic designs and measures have converged on the conclusion that measures of the environment show genetic influence. Most of this genetic research has focused on the family environment (Plomin, 1995c). How can a measure of the family environment show genetic influence? Genetic influence can arise in three ways that carry different implications. First, many environmental measures are ratings based on the perceptions of an individual family member. Such ratings can be open to biases deriving from genetically influenced characteristics of that person. For example, some people tend to be hypersensitive and suspicious, reading hostile intentions into the quite innocent behaviors of other people. In other words, the supposed environmental measure actually reflects personality characteristics rather than objective experiences. In reality, the measure is not functioning as an environmental measure at all. One solution lies in the use of observations by external raters or the use of ratings by different people in order to investigate experiences beyond such possible perceptual biases.

Second, even with measures that are free of perceptual bias, genetic factors may influence individual differences in environmental risk exposure. That is, people may act in ways that make it more likely that they will be involved in conflict, will fall out with friends or neighbors, will divorce, lose their job, or be without emotional support. In this case, there is nothing genetic in the environmental measure as such, but the extent to which individuals experience these environments may be genetically influenced. Such findings are important because effective means of reducing environmental risks need to involve knowledge about the origins of these risks. However, the origins of a risk factor have no necessary correlation with the way the risk factor operates to affect outcomes, that is, the mode of risk mediation. For example, people choose to smoke for reasons deriving from personality characteristics, social mores, and the availability of cigarettes. But the risks to physical health have nothing to do with these mechanisms—the risks to health stem from carcinogenic tars, carbon monoxide in smoke, and the effects of nicotine on the blood vessels. It is important to move beyond the origins of a risk factor to investigate the way the risk operates; different research strategies are needed to examine risk mechanisms. As indicated earlier, multivariate genetic analyses of the correlation between environmental measures and outcomes can be helpful, but they are insufficient in themselves for a thorough study of environmental risk mechanisms.

Third, genotype-environment correlation may mean that the risk is truly genetically mediated. Thus, parental personality disorder is associated with a considerable increase in family conflict and hostile parental behavior (an environmental risk), but it also involves the transmission to the children of susceptibility genes for this and related disorders. The issue is whether the antisocial behavior or conduct disorder in the offspring derives from genetic risk or environmental risk or a combination of the two. Genetic findings have been important in showing that, even when there is an environment that carries real risks, part of the risk is actually genetically mediated. The issue here is not the measures or the origins of the risk factors but the mode of risk mediation.

One further caveat is necessary. When there is a correlation between genotype and environment, this result is not a genetic effect or an environmental effect—it is both. That is, genotype-environment correlation could reflect either an environmental effect on the individual or the effect of the individual on the environment. Longitudinal data within genetically sensitive designs provide one way to dissect the direction of effects. Another way is to identify specific genes that are correlated with experiences.

Observations A widely used measure of the home environment that combines observations and interviews is the Home Observation for Measurement of the Environment (Caldwell & Bradley, 1978). The HOME assesses aspects of the home environment such as parental responsivity, encouraging developmental advance, and provision of toys. In an adoption study of the HOME, correlations for nonadoptive and adoptive siblings were compared when each child was one year old and again when each child was two years old (Braungart et al., 1992b). As shown in Table 14.7, HOME scores are more similar for nonadoptive siblings than for adoptive siblings at both one and two years, results suggesting genetic influence on the HOME. Genetic factors were estimated to account for about 40 percent of the variance of HOME scores. Although an attempt to extend this measure to three and four years of age did not show genetic influence (Braungart, 1994), videotaped interactions of the mother and siblings together showed some genetic influence on maternal attention and intrusiveness (Rende et al., 1992).

Other observational studies of mother-infant interaction in infancy, using the adoption design (Dunn & Plomin, 1986) and the twin design (Lytton, 1977, 1980), show genetic influence. An interesting aspect of the twin study was that

TABLE 14.7

Correlations for Nonadoptive and Adoptive Siblings for HOME Scores at Ages 1 and 2

Environmental Measure	Sibling Correlation	
	Nonadoptive	Adoptive
1 year	.58	.35
2 years	.57	.40

SOURCE: *Braungart et al. (1992).*

TABLE 14.8

**Model-Fitting Heritability Estimates
for Videotaped Observations of Parent-Initiated
and Child-Initiated Interactions**

Initiator	Target	Measure	Heritability
Adolescent	Mother	Positivity	.59
		Negativity	.48
Adolescent	Father	Positivity	.64
		Negativity	.52
Mother	Adolescent	Positivity	.18
		Negativity	.38
Father	Adolescent	Positivity	.18
		Negativity	.24

SOURCE: *O'Connor et al. (1995).*

parent-initiated interactions and child-initiated interactions were rated separately. Genetic effects emerged primarily for child-initiated interactions, as would be expected because the twins were the children rather than the parents. That is, a child-based genetic design can only detect genetic factors in parents' behavior that reflect genetically based behavioral differences among children. If a similar study using a parent-based genetic design were conducted, such as adult twins interacting with their children, greater genetic influence would be expected for parental contributions to family interactions and less genetic influence for children's contributions.

The NEAD project, mentioned earlier, included six groups of adolescent siblings (identical twins, fraternal twins, and full siblings in nondivorced families; and full siblings, half siblings, and unrelated siblings in stepfamilies). Videotaped observations were obtained for each parent interacting with each child when the parent-child dyad was engaged in ten-minute discussions around problems and conflict relevant to the dyad. Table 14.8 lists heritability estimates for parent-initiated and adolescent-initiated positive and negative interactions from model-fitting analyses of the six sibling groups (O'Connor et al., 1995). Significant heritability was found for all measures, and greater heritability was found for adolescent-initiated interactions than for parent-initiated interactions.

Questionnaires These few observational studies suggest that genetic effects on family interactions are not solely in the eye of the beholder. Most genetic research on the nature of nurture has used questionnaires rather than observations. Questionnaires add another source of possible genetic influence: the subjective processes involved in perceptions of the family environment.

TABLE **14.9**

Model-Fitting Heritability Estimates for Questionnaire Assessments of Parenting

Rater	Ratee	Measure	Heritability
Adolescent	Mother	Positivity	.30
		Negativity	.40
Adolescent	Father	Positivity	.56
		Negativity	.23
Mother	Mother	Positivity	.38
		Negativity	.53
Father	Father	Positivity	.22
		Negativity	.30

SOURCE: *Plomin et al. (1994c).*

The pioneering research in this area was two twin studies of adolescents' perceptions of their family environment (Rowe, 1981, 1983b). Both studies found substantial genetic influence on adolescents' perceptions of their parents' acceptance and no genetic influence on perceptions of parents' control.

The NEAD project was designed in part to investigate genetic contributions to diverse measures of family environment. As shown in Table 14.9, significant genetic influence was found for adolescents' ratings of composite variables of their parents' positivity and negativity (Plomin et al., 1994c). The highest heritability of the 12 scales that contributed to these composites was for a measure of closeness (e.g., intimacy, supportiveness), which yielded heritabilities of about 50 percent for both mothers' closeness and fathers' closeness as rated by the adolescents. As found in Rowe's original studies and in several other studies, measures of parental control showed the lowest heritabilities. The NEAD project also assessed parents' perceptions of their parenting behavior toward the adolescents (lower half of Table 14.9). Parents' ratings of their own behavior yielded heritability estimates similar to those for the adolescents' ratings of their parents' behavior.

More than a dozen other studies of twins and adoptees have reported genetic influence on questionnaires of family environment (Plomin, 1994a). These include studies of adult twins reared apart, who were asked to rate the family environment in which they were reared (Plomin et al., 1988b). Even though members of each twin pair were reared in different families, identical twins reared apart rated their family environments more similarly than did fraternal twins reared apart in relation to their family's warmth (cohesion, expressiveness) and the family's emphasis on personal growth (achievement, culture), but not for control (Table 14.10). How can genetic effects emerge when the reared-apart twins rate

TABLE 14.10

Twin Correlations and Heritabilities for Adult Twins' Retrospective Ratings of Their Childhood Family Environment

Measure	Twins Reared Apart		Twins Reared Together		Model-Fitting Heritability
	Identical	Fraternal	Identical	Fraternal	
Warmth	.37	.29	.66	.42	.38*
Control	.00	.17	.60	.31	.11
Personal growth	.42	.26	.53	.45	.19*

*$p < .05$.

SOURCE: *Plomin et al. (1988b).*

different families? Although subjective processes of perception are likely to be involved, it is also possible that members of the two families were responding similarly to genetically influenced characteristics of the separated twins.

CLOSE UP

David C. Rowe is a professor of family studies, with joint appointments in genetics and psychology, at the University of Arizona. He is one of six principal investigators for the National Longitudinal Study of Adolescent Health. He received his A.B. from Harvard University in 1972 and his Ph.D. in psychology in 1977 from the University of Colorado, where he received special training in behavioral genetics at the Institute for Behavioral Genetics. Rowe taught at Oberlin College and the University of Oklahoma before moving to the University of Arizona in 1988. His interests range across the field of behavioral genetics. His studies of delinquency in siblings and twins have highlighted both environmental and genetic influences on these socially troubling behaviors. His twin studies of perception of family environment brought into focus genetic variation in measures of "environmental" variables. Rowe's research increasingly uses quantitative genetic models that estimate genetic and environmental effects on both individual variation and group means (e.g., on sex differences). His most recent interest is in the molecular genetics of behavior. His research in this area is a molecular genetic study of childhood hyperactivity/conduct disorder.

Not surprisingly, twins reared apart rated their families as less similar than did twins reared together in the same family, but both twins reared apart and twins reared together yielded evidence for genetic influence on these retrospective reports of family environment. In model-fitting analyses based on the data for all four twin groups, heritability was about 40 percent for warmth, 20 percent for personal growth, and 10 percent for control.

Genetic influence on environmental measures also extends beyond the family environment. For example, genetic influence has been found for characteristics of children's peer groups (e.g., Manke et al., 1995), television viewing (Plomin et al., 1990c), classroom environments (Jang, 1993), work environments (Hershberger, Lichtenstein, & Knox, 1994), social support (Bergeman et al., 1990; Kessler et al., 1992), and life events (Kendler et al., 1993c; McGuffin, Katz, & Rutherford, 1991; Plomin et al., 1990b), including accidents in childhood (Phillips & Matheny, 1995), divorce (McGue & Lykken, 1992), exposure to drugs (Tsuang et al., 1992), and exposure to trauma (Lyons et al., 1993).

In summary, diverse genetic designs and measures converge on the conclusion that genetic factors contribute to experience. A major direction for research in this area is to investigate the causes and consequences of genetic influence on measures of the environment.

SUMMING UP

We create our experiences in part for genetic reasons. Three types of genotype-environment correlation are involved: passive, evocative, and active. Results of three methods to detect genotype-environment correlation suggest that the passive type is most important in childhood. A developmental theory predicts that evocative and active forms of genotype-environment correlation become more important later in development. Twin and adoption studies indicate that genetic factors may be involved in perceptions and even observations of exposure to environments and may in part mediate the effect of environmental risk.

Implications

Research using diverse genetic designs and measures leads to the conclusion that genetic factors often contribute substantially to measures of the environment, especially the family environment. The most important implication of finding genetic contributions to measures of the environment is that the correlation between an environmental measure and a psychological trait does not necessarily imply environmental causation. Genetic research often shows that genetic factors are importantly involved in correlations between environmental measures and psychological traits. In other words, what appears to be an environmental risk might actually reflect genetic factors. Three independent methods that can assess passive, evocative, and active genetic contributions to correlations between environmental measures and psychological traits were described.

This research does not mean that experience is entirely driven by genes. Widely used environmental measures show some significant genetic influence, but most of the variance in these measures is not genetic. Nonetheless, environmental measures cannot be assumed to be environmental just because they are called environmental. Indeed, research to date suggests that it is safer to assume that measures of the environment include some genetic effects. Especially in families of genetically related individuals, associations between measures of the family environment and children's developmental outcomes cannot be assumed to be purely environmental in origin. Taking this argument to the extreme, a recent book concluded that socialization research is fundamentally flawed because it has not considered the role of genetics (Rowe, 1994).

These findings support a current shift from thinking about passive models of how the environment affects individuals toward models that recognize the active role we play in selecting, modifying, and creating our own environments. Progress in this field depends on developing measures of the environment that reflect the active role we play in constructing our experience.

Genotype-Environment Interaction

The previous section focused on correlations between genotype and environment. Genotype-environment correlation refers to the role of genetics in *exposure* to environments. Genotype-environment interaction involves genetic *sensitivity*, or susceptibility, to environments. There are many ways of thinking about genotype-environment interaction (Kendler & Eaves, 1986; Plomin, DeFries, & Loehlin, 1977; Rutter & Pickles, 1991; Wahlsten, 1990). Although the statistical interaction between genetic and environmental effects is useful for detecting some forms of genotype-environment interaction, genetic sensitivity to environmental factors is not limited to statistical interactions; a broader range of research strategies is needed.

PKU is one of the best examples of genetic sensitivity to a particular environmental factor (see Chapter 7). Phenylalanine in food has a very different effect on children who are homozygous for the PKU allele than on other children. Because the homozygous children cannot metabolize this amino acid, its metabolic products build up and damage the developing brain. Phenylalanine has no harmful effect on other children. Conversely, a diet low in phenylalanine has a major effect on children homozygous for the PKU allele—it prevents mental retardation. But the low phenylalanine diet has no harmful or beneficial effect on other children. One of the goals of molecular genetic research is to find genotype-environment interactions like this in which environmental interventions prevent the negative effects of genetic disorders.

For complex traits influenced by many genes, it is more difficult to identify genotype-environment interaction than single-gene disorders like PKU. Chapter 11 described an example of genotype-environment interaction for criminal behavior found in two adoption studies (Bohman, 1996; Brennan, Mednick, & Jacobsen, 1996). The highest rate of criminal behavior was found for adoptees who had both biological parents *and* adoptive parents with criminal records. That is, criminal convictions of adoptive parents led to increased criminal convictions of their adopted children mainly when the adoptees' biological parents also had criminal convictions. A recent example of a similar type of genotype-environment interaction has been reported for adolescent conduct disorder (Cadoret et al., 1995). Genetic risk was indexed by biological parents' antisocial personality diagnosis or drug abuse, and environmental risk was assessed by marital, legal, or psychiatric problems in the adoptive family. Adoptees at genetic risk were more sensitive to the environmental effects of stress in the adoptive family. Adoptees at low genetic risk were unaffected by stress in the adoptive family. This result confirms previous research that also showed interactions between genetic risk and family environment in the development of adolescent antisocial behavior (Cadoret, Cain, & Crowe, 1983; Crowe, 1974).

The twin method has also been used to identify genotype-environment interaction. One twin's phenotype can be used as an index of the co-twin's genetic risk to explore interactions with measured environments. For example, the effect of stressful life events on depression is greater for individuals at genetic risk for depression (Kendler et al., 1995). The approach is stronger when twins reared apart are studied, an approach that has also yielded some evidence for genotype-environment interaction (Bergeman et al., 1988). Another use of the twin method to study interaction simply asks whether heritability differs in two environments. For example, this approach was used to show that heritability of alcoholism is greater for unmarried women than for married women (Heath, Jardine, & Martin, 1989).

Similar attempts to find genotype-environment interaction for cognitive ability have been less successful, although it is always possible that the wrong environmental factors have been assessed. Using data from the classic adoption study of Skodak and Skeels (1949), general cognitive ability scores were compared for adopted children whose biological parents were high or low in level of education (as an index of genotype) and whose adoptive parents were high or low in level of education (as an index of environment) (Plomin et al., 1977). Although the level of education of the biological parents showed a significant effect on the adopted children's general cognitive ability, no environmental effect was found for adoptive parents' education and no genotype-environment interaction was found. A similar adoption analysis using more extreme groups found both genetic and environmental effects but, again, no evidence for genotype-environment interaction (Capron & Duyme, 1989, in press). Other attempts to find genotype-environment interaction for cognitive ability in infancy and childhood have not been successful in adoption analyses (Plomin et al., 1988a).

Genotype-environment interaction is easier to study in animals in the laboratory, where both genotype and environment can be manipulated. Chapter 8 described one of the best-known examples of genotype-environment interaction. Maze-bright and maze-dull selected lines of rats responded differently to "enriched" and "restricted" rearing environments (Cooper & Zubek, 1958). The enriched condition had no effect on the maze-bright selected line, but it improved the maze-running performance of the maze-dull rats. The restricted environment was detrimental to the performance of the maze-bright rats but had little effect on the maze-dull rats. This result is an interaction in that the effect of restricted versus enriched environments depends on the genotype of the animals. Other examples from animal research in which environmental effects on behavior differ as a function of genotype have been found (Erlenmeyer-Kimling, 1972; Fuller & Thompson, 1978; Mather & Jinks, 1982), although a series of learning studies in mice failed to find replicable genotype-environment interactions (Henderson, 1972).

More examples of genotype-environment interactions are likely to be found as measures of the environment are included in genetic research, as researchers examine the extremes of genotypes and environments, and especially as specific genes whose effects can be studied in interaction with experience are identified. Examples of genotype-environment interaction that have been found so far are often of the type suggested by the *diathesis-stress* model of psychopathology (Gottesman, 1991): Individuals at genetic risk for psychopathology (diathesis) are especially sensitive to the effects of stressful environments.

The recognition through behavioral genetic research of genotype-environment correlations and interactions emphasizes the power of genetic research to elucidate environmental risk mechanisms. It also serves to demonstrate that genetic findings contradict the assumptions of many critics of genetics that genetic factors are deterministic in their effects or neglectful of environmental influences.

SUMMING UP

Animal studies have provided examples in which environmental effects on behavior differ as a function of genotype, which is genotype-environment interaction. It is more difficult to identify genotype-environment interaction for human behavior. A few examples have been found in which stressful environments primarily affect individuals who are at genetic risk.

Identifying Genes for Experience

A surprising implication of finding genetic influence on environmental measures used in psychology is that it ought to be possible to find DNA markers associated with these measures, genes for experience. Finding genes for experience would greatly facilitate research on the mechanisms by which genetic dispositions guide individuals in their active creation of their own experiences. For example, do these genes correlate more with passive experience early in development and with evocative experience and especially with active experience later in development? Are the genes correlated with genetically influenced characteristics of children or of their parents? Indeed, it might help to find genes associated with behavior if we looked for them, not independent of experience, but rather in their interplay with measured aspects of the environment.

Conversely, identifying genes associated with behavior would revolutionize research on genotype-environment correlation and interaction. Even if such genes accounted for just a small amount of variance, they would make it possible to study how such genes correlate with and interact with specific aspects of experience.

Summary

Genetic research can tell us as much about nurture as it can about nature. Two of the most important findings from genetic research in psychology involve the environment. First, genetic research has shown that environmental influences work in a nonshared manner, making children growing up in the same family different from one another. Second, genetic factors often contribute to measures of the environment that are widely used in psychological research. Genetic factors are also responsible in part for the correlation between environmental measures and psychological traits.

In addition to providing the best available evidence for the importance of the environment in psychology, genetic research shows that environmental effects tend not to be shared by family members. For example, resemblance between adoptive siblings is negligible for many traits, indicating that shared environment is not important. Differences within pairs of identical twins also suggest the importance of nonshared environment. Attempts to identify specific sources of nonshared environment have found that family environments are experienced differently by children growing up in the same family. Experiences outside the family, such as those with peers, are likely to be even more important sources of nonshared environment. Nonshared experiences are related to psychological outcomes, especially for negative aspects of parenting and negative outcomes. However, cause and effect in these associations is not clear. Recent research suggests that genetic factors largely mediate the association between nonshared family experiences and differences in siblings' outcomes.

This suggestion leads into the second finding at the interface between nature and nurture: Our experiences are influenced in part by genetic factors. This is the topic of genotype-environment correlation. Genotype-environment correlations are of three types: passive, evocative, and active. Three methods are available to assess specific genotype-environment correlations between psychological traits and measures of the environment. These three methods have identified several examples of genotype-environment correlation. Dozens of studies using various genetic designs and measures of the environment converge on the conclusion that genetic factors contribute to the variance of measures of the environment.

Another aspect of the interface between nature and nurture is genotype-environment interaction. Animal studies, in which both genotype and environment can be controlled, have yielded examples in which environmental effects on behavior differ as a function of genotype. Although it is more difficult to identify genotype-environment interaction for human behavior, some examples have been found. The general form of these interactions is that stressful environments primarily have their effect on individuals who are genetically at risk.

Understanding how nature and nurture correlate and interact will be greatly facilitated when specific genes that are associated with behavior and with experience are identified.

CHAPTER FIFTEEN

Behavioral Genetics in the Twenty-first Century

Predicting the future of behavioral genetics is not a matter of crystal ball gazing because the momentum of recent developments makes the field certain to thrive, especially as behavioral genetics continues to flow into the mainstream of psychological research. This momentum is propelled by new findings, methods, and projects both in quantitative genetics and in molecular genetics. Another reason for optimism about the continued growth of genetics in psychology is that leading psychologists have begun to incorporate genetic strategies in their research (Plomin, 1993). This trend is important because the best behavioral genetic research is likely to be done by psychologists who are not primarily geneticists. Experts from behavioral domains will focus on traits and theories that are pivotal to those domains and interpret their research findings in ways that will achieve the most important advances. For this reason, a major motivation for writing this book is to enlist the aid of the next generation of psychologists in studies of important behavioral phenomena, using the theory and methods of behavioral genetics.

Quantitative Genetics

The future will no doubt witness the application of twin and adoptee strategies, as well as genetic research using nonhuman animal models, to other psychological traits. Behavioral genetics has only scratched the surface of possible applications, even within the domains of cognitive disabilities (Chapter 7), cognitive abilities (Chapters 8 and 9), psychopathology (Chapter 10), and personality (Chapter 11). For example, for cognitive abilities, most research has focused on general cognitive ability and major group factors of specific cognitive abilities. The future of quantitative genetic research in this area lies in more fine-grained analyses of cognitive abilities and in the use of information-processing and biological approaches to cognition. For psychopathology, genetic research has just begun to consider disorders other than schizophrenia and the major mood disorders. Much remains to be learned about disorders in childhood, for example. Personality is so complex that it can keep researchers busy for decades, especially as they go beyond self-report questionnaires to use other measures such as observations. A rich territory for future exploration is the link between psychopathology and personality.

Cognitive disabilities and abilities, psychopathology, and personality have been the targets for the vast majority of genetic research in psychology because these areas have traditionally considered individual differences. Three new areas of psychology that are beginning to be explored genetically were described in Chapters 12 and 13: health psychology, psychology and aging, and evolutionary psychology. Some of the oldest areas of psychology—perception, learning, and language, for example—have not emphasized individual differences and as a result have yet to be explored systematically from a genetic perspective. Entire disciplines within the social and behavioral sciences, such as economics, education, and sociology, are still essentially untouched by genetic research.

Genetic research in psychology will continue to move beyond simply demonstrating that genetic factors are important or estimating heritabilities. The questions *whether* and *how much* genetic factors affect psychological dimensions and disorders represent important first steps in understanding the origins of individual differences. But these are only first steps. The next steps involve the question *how*, the mechanisms by which genes have their effect. How do genetic effects unfold developmentally? What are the biological pathways between genes and behavior? How do nature and nurture interact and correlate? Examples of these three directions for genetic research in psychology—developmental genetics, multivariate genetics, and "environmental" genetics—have been seen throughout the preceding chapters. The future will see more research of this type as behavioral genetics continues to move beyond merely documenting genetic influence. Large-scale, collaborative research projects that focus on these issues are underway.

Developmental genetic analysis considers change as well as continuity during development throughout the human life span. Two types of developmental questions can be asked. First, do genetic and environmental components of variance change during development? The most striking example to date involves general cognitive ability (Chapter 8). Genetic effects become increasingly important throughout the life span. Shared family environment is important in childhood but its influence becomes negligible after adolescence. The second question concerns the role of genetic and environmental factors in age-to-age change and continuity during development. Using general cognitive ability again as an example, we find a surprising degree of genetic continuity from childhood to adulthood. However, some evidence has been found for genetic change as well, for example, during the transition from early to middle childhood when formal schooling begins. Interesting developmental discoveries are not likely to be limited to cognitive development—it just so happens that most developmental genetic research so far has focused on cognitive development.

Multivariate genetic research addresses the covariance between traits rather than the variance of each trait considered by itself. A surprising finding in relation to specific cognitive abilities is that the same genetic factors affect most cognitive abilities (Chapter 9). For psychopathology, a key question is why so many disorders co-occur. Multivariate genetic research suggests that genetic overlap between disorders may be responsible for this comorbidity (Chapter 10). Another basic question in psychopathology is heterogeneity. Are there subtypes of disorders that are genetically distinct? Multivariate genetic research is critical for investigating the causes of comorbidity and heterogeneity and for identifying the most heritable constellations (comorbidity) and components (heterogeneity) of psychopathology.

Two other general directions for multivariate genetic research are links between the normal and the abnormal and between behavior and biology. A fundamental question is the extent to which genetic and environmental effects on disorders are merely the quantitative extremes of the same genetic and environmental factors that affect the rest of the distribution. Or are disorders different in kind, not just in quantity, from the normal range of behavior? Multivariate genetic analysis also can be used to explicate the mechanisms by which genetic factors influence behavior by identifying genetic correlations between behavior and biological processes such as neurotransmitter systems. It cannot be assumed that the nexus of associations between biology and behavior is necessarily genetic in origin. Multivariate genetic analysis is needed to investigate the extent to which genetic factors mediate these associations.

"Environmental" genetics will continue to explore the interface between nature and nurture. As described in Chapter 14, genetic research has made some of the most important discoveries about the environment in recent decades, especially nonshared environment and the role of genetics in experience. More discoveries about environmental mechanisms can be predicted as the environment continues to be investigated in the context of genetically sensitive designs. Much remains to be learned about interactions and correlations between nature and nurture.

In summary, no crystal ball is needed to predict that behavioral genetic research will continue to flourish as it turns to other areas of psychology and, especially, as it goes beyond the rudimentary questions of *whether* and *how much* to ask the question *how*. Such research will become increasingly important as it guides molecular genetic research to the most heritable components and constellations throughout the human life span as they interact and correlate with the environment. In return, developmental, multivariate, and "environmental" genetics will be transformed by molecular genetics.

Molecular Genetics

Psychology is at the dawn of a new era in which molecular genetic techniques will revolutionize genetic research in psychology by identifying specific genes that contribute to genetic variance for complex dimensions and disorders. The quest is to find, not *the* gene for a trait, but the multiple genes that affect the trait in a probabilistic rather than predetermined manner. The breathtaking pace of molecular genetics (Chapter 6) leads us to predict that psychologists will routinely use DNA markers as a tool in their research to identify some of the relevant genetic differences among individuals. This is a safe prediction, because it is already happening in research on dementia and cognitive decline in the elderly. It is now standard practice for research in this area to take advantage of the genetic risk information provided by the DNA marker for apolipoprotein E (Chapter 7), even when researchers are interested primarily in psychosocial risk mechanisms.

To answer questions about how genes influence behavior, nothing can be more important than identifying specific genes involved and characterizing the genes' products. As specific genes are found that begin to account for some of the widespread genetic influence in psychology, more precise questions can be asked, using measured genotypes. Do the effects of the genes change during development? Do the genes correlate with some aspects of a trait but not others (heterogeneity) or do their effects extend across several traits (comorbidity)? Are genes for disorders also associated with normal dimensions and vice versa? Do the genetic effects interact or correlate with the environment?

Nature and Nurture

The controversy that swirled around behavioral genetics research in psychology during the 1970s (Chapter 8) has largely faded. One of many signs of the increasing acceptance of genetics is that behavioral genetics was identified by the American Psychological Association at its centennial celebration in 1992 as one of the two themes that best represent the future of psychological research (Plomin & McClearn, 1993a). This is one of the most dramatic shifts in the modern history of psychology. Indeed, the wave of acceptance of genetic influence in psychology is growing into a tidal wave that threatens to engulf the second message coming from behavioral genetic research. The first message is that genes play a surprisingly important role throughout psychology. The second message is just as important: Individual differences in complex psychological traits are due at least as much to environmental influences as they are to genetic influences.

The first message will become more prominent during the next decade as more genes are identified that are responsible for the widespread influence of genetics in psychology. As explained in Chapter 5, it should be emphasized that genetic effects on complex traits describe *what is*. Such findings do not predict *what could be* or prescribe *what should be*. Genes are not destiny. Genetic effects represent probabilistic propensities, not predetermined programming.

A related point is that, for complex traits such as behavioral traits, genetic effects refer to average effects in a population, not to a particular individual. For example, one of the strongest DNA associations with a complex behavioral disorder is the association between allele 4 of the apolipoprotein E gene and late-onset dementia (Chapter 7). Unlike simple single-gene disorders, this association does not mean that allele 4 is necessary or sufficient for the development of dementia. Many people with dementia do not have the allele and many people with the allele do not have dementia. A particular gene may be associated with a large average increase in risk for a disorder, but it is likely to be a weak predictor at an individual level. The importance of this point concerns the dangers of labeling individuals on the basis of population averages.

Genetic influence does not mean that the environment is unimportant. To the contrary, genetic research provides the best available evidence for the importance of environmental influences and has produced some of the most important findings in psychology about the environment and how it affects development (Chapter 14).

The relationship between genetics and equality, an issue that lurks in the shadows, causing a sense of unease about genetics in psychology, was discussed in Chapter 5. The main point is that finding genetic differences among individuals does not compromise the value of social equality. But knowledge alone by no means accounts for societal and political decisions. Values are just as important in the decision-making process. Decisions, both good and bad, can be made with or without knowledge. Nonetheless, scientific findings are often misused and scientists, like the rest of the population, need to be concerned to diminish misuse. We believe firmly, however, that better decisions can be made with knowledge than without.

Finding widespread genetic influence creates new problems to consider in psychology. For example, could evidence for genetic influence be used to justify the status quo? Will people at genetic risk be labeled and discriminated against? When genes are found for psychological traits, will parents use them prenatally to select "designer" children? New knowledge also provides new opportunities. For example, finding genes associated with a particular disorder makes it more likely that environmental preventions and interventions that are especially effective for the disorder can be found. Knowing that certain children have increased genetic risk for a disorder could make it possible to prevent the disorder before it appears, rather than trying to treat the disorder after it appears and causes other problems. Moreover, it should not be assumed that once a gene associated with some psychopathological disorder is found, the logical next step is to get rid of it. For example, genes that persist in the population may be the result of stabilizing selection (Chapter 13), which might mean that the genes have both good and bad outcomes.

Two other points should be made in this regard. First, most powerful scientific advances create new problems. For example, consider prenatal screening for genetic defects. This advance has obvious benefits in terms of detecting chromosomal and genetic disorders before birth. Combined with abortion, prenatal screening can relieve parents and society of the tremendous burden of severe birth defects. However, it also raises ethical problems concerning abortion and creates the possibility of abuses, such as compulsory screening and mandatory abortion. Despite the problems created by advances in science, we would not want to cut off the flow of knowledge and its benefits in order to avoid having to confront such problems.

The second point is that it is wrong to assume that environmental explanations are good and that genetic explanations are dangerous. Tremendous harm was done by the environmentalism that prevailed until the 1960s, when the pendulum swung back to a more balanced view that recognized genetic as well as environmental influences. For example, environmentalism led to blaming children's problems on what their parents did to them in the first few years of life. Imagine that, in the 1950s, you are among the 1 percent of parents who had a child who became schizophrenic in late adolescence. You face a lifetime of concern. And then you are told that the schizophrenia was caused by what you did to the child in the first few years. The sense of guilt would be overwhelming. Worst of all, such parent blaming was not correct. There is no evidence that early parental treatment causes schizophrenia. Although the environment is important, whatever the salient environmental factors might be, they are not shared family environmental factors. Most important, we now know that schizophrenia is substantially influenced by genetic factors.

Our hope for the future is that the next generation of psychologists will wonder what the nature-nurture fuss was all about. We hope they will say, "Of course, we need to consider nature as well as nurture in understanding psychology." The conjunction between nature and nurture is truly *and*, not *versus*.

The basic message of behavioral genetics is that each of us is an individual. Recognition of, and respect for, individual differences is essential to the ethic of individual worth. Proper attention to individual needs, including provision of the environmental circumstances that will optimize the development of each person, is a utopian ideal and no more attainable than other utopias. Nevertheless, we can approach this ideal more closely if we recognize, rather than ignore, individuality. Acquiring the requisite knowledge warrants a high priority because human individuality is the fundamental natural resource of our species.

APPENDIX A

Brief Overview of Statistics

Variance and covariance are central to behavioral genetics research. The purpose of this appendix is to provide an overview of these statistics.

Statistics Describing Distributions

Table A.1 lists scores of a small sample of two inbred strains of mice for activity in an open field (Chapter 5). Figure A.1 illustrates the distribution of these scores in the form of frequency histograms. How can we describe these two distributions? We first calculate an average score (or some measure of *central tendency*) and then describe the variability of the scores. The average score is not very useful in itself, because it adequately represents the scores only if there is little variability. The distributions in Figure A.1 show substantial variability, as do distributions for most behavioral characteristics. The *average*, or *mean*, is simply the sum of scores divided by the total number of scores:

$$\overline{X} = \frac{\Sigma X}{N}$$

where $\overline{X}$ refers to mean, ΣX is the sum of scores, and N is the number of scores. The sum of the scores obtained from the six A subjects is 306. Thus,

$$\overline{X}_A = \frac{306}{6} = 51$$

The mean score of C57BL subjects in the sample is

$$\overline{X}_C = \frac{1092}{6} = 182$$

These means are indicated in Figure A.1.

TABLE A.1

Activity Scores of Two Inbred Strains of Mice

Strain A	Strain C57BL
29	155
29	157
44	161
58	199
63	202
83	218

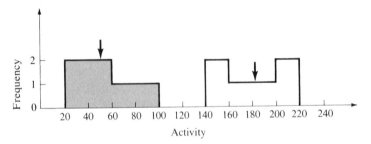

Figure A.1 Frequency histograms of the activity scores of two inbred strains of mice: A (shaded) and C57BL. The means are indicated by arrows.

As the name implies, *variance* is a measure of variability or dispersion. The more spread out the distribution, the greater the variance. Variance is described relative to the mean of the sample. The difference between each subject's score and the mean is computed (i.e., $X - \overline{X}$). Some of these deviations are above the mean and are thus positive numbers; those below the mean are negative numbers. We would like to obtain an average deviation from the mean in order to describe the variability in the distribution. However, if we simply summed the deviations from the mean, the positive deviations would balance the negative deviations and the sum would always be zero. The solution is to square the deviations from the mean and then calculate an average squared deviation. This is the definition of variance. However, the sum of the squared deviations from the mean is divided by $N - 1$ for technical reasons (in order to obtain an unbiased estimate of the variance). In short, the variance of a sample (V) is

$$V = \frac{\Sigma(X - \overline{X})^2}{N - 1}$$

To illustrate the calculation of V, the data given in Table A.1 are presented again in Table A.2, along with corresponding deviations from means and squared deviations. As you can see, the variance of activity scores in the C57BL sample is somewhat larger than that in the A sample.

Because the variance is the average of the *squared* deviations from the mean, the values obtained are expressed in squared units, rather than in the actual units of measure. Despite this, variance has many important applications in genetics. But another measure of variability is provided by the square root of the variance, the so-called *standard deviation*, which is more easily interpreted.

If a sample has been drawn at random from a population with a normal distribution of a trait (see Figure A.2), the sample standard deviation (s) provides a useful estimate of dispersion of the trait within that population. Approximately two-thirds of the population (68 percent) fall within one standard deviation above and below the mean, and about 96 percent of the observations fall within two standard deviations. Thus, in a large population of mice of the A strain with a mean of 51 and a standard deviation of 21.14, approximately two-thirds of their activity scores would fall within the range of 51 ± 21.14, that is, between 29.86 and 72.14. The precision of such estimates increases along with the sample size.

TABLE A.2

**Examples of Variance Estimation from Activity Scores
of Two Inbred Strains of Mice**

	Strain A	
X_i	$X_i - \overline{X}$	$(X_i - \overline{X})^2$
29	−22	484
29	−22	484
44	−7	49
58	+7	49
63	+12	144
83	+32	1024
$\Sigma X_i = 306$	$\Sigma(X_i - \overline{X}) = 0$	$\Sigma(X_i - \overline{X})^2 = 2234$

$$\overline{X}_A = 51 \qquad\qquad V_A = \frac{2234}{5} = 446.8$$

$$s_A = \sqrt{V_A} = \sqrt{446.8} = 21.14$$

	Strain C57BL	
X_i	$X_i - \overline{X}$	$(X_i - \overline{X})^2$
155	−27	729
157	−25	625
161	−21	441
199	+17	289
202	+20	400
218	+36	1296
$\Sigma X_i = 1092$	$\Sigma(X_i - \overline{X}) = 0$	$\Sigma(X_i - \overline{X})^2 = 3780$

$$\overline{X}_C = 182 \qquad\qquad V_C = \frac{3780}{5} = 756.0$$

$$s_C = \sqrt{V_C} = \sqrt{756.0} = 27.50$$

Statistics Describing the Relationship
Between Two Variables

When two variables are measured for each subject, or when the same variable is measured on pairs of subjects (for example, pairs of twins or parents and their offspring), we can analyze the relationship between the two measures. The question is usually phrased in terms of *covariance*, which literally means "shared variance." It tells us the extent to which the measures relate to one another. If there is substantial covariance between two

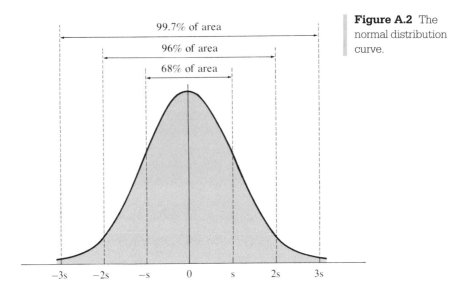

Figure A.2 The normal distribution curve.

variables (X and Y), then a subject above the mean on X is also likely to be above the mean on Y. Like variance, the covariance statistic is based on deviations from the mean of each variable. It is computed by multiplying each subject's deviation from the mean of X by the subject's deviation from the mean of Y. *Cross products* of these deviations are summed across subjects and divided by the size of the sample (actually, $N - 1$). In short, the sample covariance between two variables (Cov_{XY}) is

$$\text{Cov}_{XY} = \frac{\Sigma[(X-\overline{X})(Y-\overline{Y})]}{N-1}$$

Consider the hypothetical data presented in Table A.3 and plotted in Figure A.3. Note that Y tends to increase as X increases. The variance of X is 2.5, and the variance of Y is 10. The covariance between X and Y is 3.

Covariances are easier to interpret if we divide them by a product of standard deviations or by the appropriate variance. The two major statistics are *correlation* and *regression*. Sometimes one of these methods is more appropriate for certain quantitative genetic analyses, and sometimes the other is more suitable. A correlation coefficient standardizes the covariance by dividing it by the product of the standard deviations of X and Y. This is known as standardization because it results in equal units of X and Y. The regression coefficient is the covariance divided by the variance of just one of the variables. For example, if we are predicting offspring scores (Y) from parental scores (X), the regression Y on X (b_{YX}) is the covariance divided by the variance of X. Thus, the regression coefficient is not standardized. It is expressed in terms of observed units of measure. It expresses the average number of units that Y will change for each unit change of X.

In summary, the formulas for a correlation coefficient (r_{XY}) and a regression coefficient (b_{YX}) are

$$r_{XY} = \frac{\text{Cov}_{XY}}{\sqrt{(V_X)\,(V_Y)}} \quad \text{and} \quad b_{YX} = \frac{\text{Cov}_{XY}}{(V_X)}$$

Note that if the standard deviations of X and Y are equal, then the correlation coefficient and the regression coefficient are the same. If $\sqrt{V_X} = \sqrt{V_Y}$ then $\sqrt{(V_X)}\,(V_Y) = V_X$, so that both the denominator and the numerator are identical for the correlation and the regression.

Correlation

Table A.3 illustrates the computation of a correlation. The covariance (3) divided by the product of the standard deviations is 0.6. A correlation of zero (or near zero) indicates that the two variables are independent: Scores on one variable tell us nothing about scores on the other. A high positive or negative correlation (close to +1 or –1) indicates a close relationship. Because the correlations are standardized, they are easily related to variances. Squaring the correlations yields the percentage of variance in one variable related to the variance with the other. The correlation of 0.6 in Table A.3 indicates that 36 percent of the variance in Y is related to the variance of X (and vice versa). Because the variance of Y is 10.0, 0.36 × 10 = 3.6 is the variance of Y related to the variance of X. This statement means that the rest of the variance of Y, 6.4, is not related to the variance of X.

We have been using the phrase "related to" rather than "caused by" because correlations do not in themselves prove the existence of a causal relationship. In genetics, however, causal associations between genotype and phenotype can be inferred. When a causal relationship between two variables (X and Y) has been established, the correlation coefficient can be used to estimate the variance in Y caused by the variation in X. In the previous example, this would mean that, if X is held constant, 64 percent of the variance in Y will remain.

TABLE A.3

Sample Calculation of a Correlation Coefficient, r_{XY}

X	$X-\overline{X}$	$(X-\overline{X})^2$	Y	$Y-\overline{Y}$	$(Y-\overline{Y})^2$	$(X-\overline{X})(Y-\overline{Y})$
1	−2	4	2	−4	16	+8
2	−1	1	8	+2	4	−2
3	0	0	6	0	0	0
4	+1	1	4	−2	4	2
5	+2	4	10	+4	16	+8
Σ 15	0	10	30	0	40	12

$$\overline{X} = 3 \qquad V_X = \frac{10}{4} \qquad \overline{Y} = 6 \qquad V_Y = \frac{40}{4} \qquad \mathrm{Cov}_{XY} = \frac{12}{4}$$

$$= 2.5 \qquad\qquad = 10 \qquad\qquad = 3$$

$$r_{XY} = \frac{3}{\sqrt{(2.5)(10)}} = 0.6$$

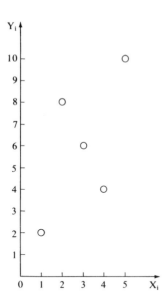

Figure A.3 Plot of hypothetical data presented in Table A.3.

Regression

The regression of Y on X for the same data (Table A.3) is 1.2:

$$b_{YX} = \frac{\text{Cov}_{XY}}{V_X} = \frac{3.0}{2.5} = 1.2$$

Thus, for every unit of change in X, Y changes an average of 1.2 units. This regression coefficient can be used to show how the variance of Y can be partitioned into two parts—one due to variation in X and one that is independent of X. In overview, we will use an equation to predict scores on Y, given scores on X. Then we will obtain the variance of the Y scores as they were predicted by X scores. The deviation of the actual Y scores from the predicted Y scores can be squared and averaged to produce the variance of Y that is independent of X.

The regression coefficient describes the change in Y predicted by a unit of change in X. Such a prediction may seem unnecessary, given that we already have information regarding both variables. However, from the sample regression, we may estimate Y for other members of the population for whom we have information regarding only variable X. More important, for our present purpose, the regression can be used to draw a straight line through the observed points, like that in Figure A.4. This line is called a *"least squares" regression line* because the sum of the squared deviations from the predicted points is at a minimum. This prediction equation is

$$\hat{Y} = \overline{Y} + b_{YX}(X - \overline{X})$$

where $\hat{Y}$ is the predicted value of Y, given information on X. Thus, the predicted value of Y is derived from the deviation of the X score from its mean, weighted by the regression coefficient. From the data of Table A.3,

$$\hat{Y} = 6 + 1.2(X - 3) = 6 + 1.2X - 3.6 = 2.4 + 1.2X$$

Using this equation, we can calculate the expected value of Y corresponding to each observed value of X in Table A.3. These observed and expected values are presented in Table A.4 and graphed in Figure A.4. For example, the X score of 2 predicts a Y score of 4.8, because $2.4 + 1.2(2) = 4.8$. This predicted value has been entered as a point on the straight line in Figure A.4.

The variance of these predicted scores of Y is the variance of Y due to variation in X. As calculated in column 4 of Table A.5, the variance of the predicted Y scores is 3.6, the same answer obtained by using the correlation coefficient ($r^2 V_Y$). Of course, the variance of Y not predicted by X is the rest of the variance of Y (that is, $10 - 3.6 = 6.4$). However, we can directly calculate the variance of Y not predicted by X by obtaining the deviation of each Y value from its predicted value and then deriving the variance of these deviations as in column 6 of Table A.5.

Concordance

When traits are not continuous—most notably, for disorders that are diagnosed as either-or dichotomies—correlation is less appropriate as an index of familial resemblance. So, a simpler statistic called concordance is used as an index of familial resemblance for dichotomous traits. *Concordance* is the percentage of family members of probands who also have the proband's disorder. There are two types of concordance, pairwise and probandwise. Pairwise concordance should be computed if only one member of a pair (e.g., the first-born member of a pair) could be ascertained as a proband. In the more typical case in which both members of the pair could be ascertained as probands, probandwise concordance should be estimated (DeFries & Alarcón, 1996). The following examples of the two types of concordance assume that 100 pairs of identical twins were identified in which at least one member of each pair is affected.

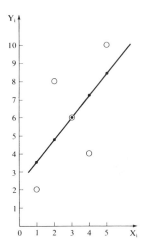

Figure A.4 Plot of observed values (open circles) and expected values (small black dots) of Y, corresponding to observed values of X. Expected values are obtained from the regression equation, $\hat{Y} = 2.4 + 1.2X$.

TABLE A.4

Observed Values of X and Y and Expected Values of Y

X	Y	$\hat{Y}$
1	2	3.6
2	8	4.8
3	6	6.0
4	4	7.2
5	10	8.4

TABLE A.5

Calculation of the Variance in Y Due to Both Regression and Deviations from Regression

Y	$\hat{Y}$	$\hat{Y} - \overline{\hat{Y}}$	$(\hat{Y} - \overline{\hat{Y}})^2$	$Y - \hat{Y}$	$(Y - \hat{Y})^2$
2	3.6	−2.4	5.76	−1.6	2.56
8	4.8	−1.2	1.44	+3.2	10.24
6	6.0	0.0	0.00	0.0	0.00
4	7.2	+1.2	1.44	−3.2	10.24
10	8.4	+2.4	5.76	+1.6	2.56
Σ 30	30.0	0.0	14.40	0.0	25.60

$$V_Y = 10 \qquad V_{\hat{Y}} = \frac{14.4}{4} = 3.6 \qquad V_{Y-\hat{Y}} = \frac{25.6}{4} = 6.4$$

Of these 100 pairs, 30 pairs are concordant (both members of the pair are affected) and 70 pairs are discordant (only one member of the pair is affected). In other words, these 100 twin pairs include 130 probands (60 probands from the 30 concordant pairs and 70 probands from the 70 discordant pairs).

As illustrated in Figure A.5a, pairwise concordance is 30 percent (i.e., 30 of 100 pairs are concordant). Although the simplicity of pairwise concordance is appealing, proband-wise concordance provides a better index of resemblance because it yields a risk estimate that can be compared with other risk estimates, such as the population risk. Using the same example, we calculate probandwise concordance as the number of probands in concordant pairs (60) divided by the total number of probands (130). Thus, probandwise concordance in this example is 46 percent, as shown in Figure A.5b. Probandwise concordance can be interpreted as risk. If you had an identical twin with this disorder, you would have a 46 percent chance of developing the disorder.

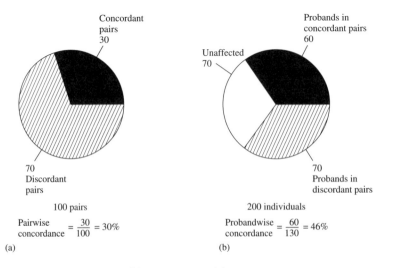

Figure A.5 Calculation of (a) pairwise and (b) probandwise concordance.

APPENDIX B

Quantitative Genetic Theory and Model Fitting

Quantitative genetics provides the basis for a general theory of the etiology of individual differences of a scope and power rarely seen in psychology. The theory is unlike most contemporary theories in psychology in that it is not limited to a particular substantive domain. It is analogous in this respect to learning theory, which does not predict what is learned but rather the processes by which learning occurs. Quantitative genetics is similar to learning theory in another respect, in that it does not predict which neurotransmitters are involved in the learning process. Although quantitative genetics is based on the proposition that variation in DNA leads to phenotypic variation, it does not specify which genes are responsible for phenotypic variance.

At its most general level, quantitative genetic theory provides an expectation that individuals will differ in complex behavioral traits. In this way, quantitative genetics organizes a welter of data on individual differences so that they are no longer viewed as imperfections in the species type or as nuisance error in analysis of variance, but rather as the quintessence of evolution.

An attractive feature of quantitative genetic theory is that, in philosophy of science jargon, the theory is progressive. That is, it leads to new predictions that can be verified empirically. In addition, potential problems with the theory are examined rather than ignored. Newer views of the philosophy of science attempt to return empirical evidence to its role as judge of scientific truth (e.g., Gholson & Barker, 1985). According to this view, successful theory maximizes empirical successes and minimizes conceptual liabilities. In this sense, quantitative genetics is a powerful theory of the origin of individual differences.

The Single-Gene Model

Although quantitative genetics was developed for application to traits influenced by genes at many loci, the underlying model is based on segregation at only a single locus. Once we have described gene action at a single locus, we can generalize to the polygenic case.

Genotypic Value

Genotypic values are expressed as deviations from the mid-homozygote point, as indicated in Figure B.1. The homozygote with the higher value will be referred to as A_1A_1. The genotypic value for A_1A_1 will be +a. The genotypic value of the other homozygote, A_2A_2, is –a. The values +a and –a are equidistant from the mid-homozygote point. However, the genotypic value of the heterozygote, A_1A_2 (symbolized by d), is dependent on the gene action at the locus. If there is no dominance, d will equal zero and will fall at the mid-homozygote point. If A_1 is partially dominant to A_2, d will be closer to a, as in the example in Figure B.1. If dominance is complete, that is, if the observed value for A_1A_2 equals that of A_1A_1, then d = +a. If A_2 is dominant to A_1, d will be negative.

Figure B.1 Assigned genotypic values. (After *Introduction to Quantitative Genetics* by D. S. Falconer. Copyright © D. S. Falconer, 1960, p. 113, Longman Group Ltd., London and New York.)

Additive Genetic Value

The additive effect of genes is merely the extent to which they "add up" or sum according to gene dosage. More specifically, the additive genetic value is the genotypic value expected from gene dosage, as illustrated in Figure B.2, where $p = q = \frac{1}{2}$. Gene dosage is the number of a particular allele (say, the A_1 allele) present in a genotype. As gene dosage increases by one (for example, from the A_2A_2 genotype to the A_1A_2), the expected genotypic value increases by a constant unit. Although additive genetic values depend on allele frequencies in the population, this will not affect our example. If there is no dominance, gene dosage will predict genotypic values exactly. However, dominance can cause the actual genotypic values to deviate from expected values.

Another way of thinking about additive genetic values is to consider that every allele in the genotype has some average effect. In this sense, the additive genetic value is the sum of these average effects of alleles across the genotype. Additive genetic value is a fundamental component of genetic influence, because it represents the extent to which

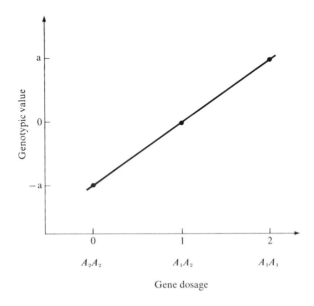

Figure B.2 Additive genetic values predicted by gene dosage, when d = 0. Because there is no dominance and $p = q = \frac{1}{2}$, the genotypic values are the same as the additive genetic values.

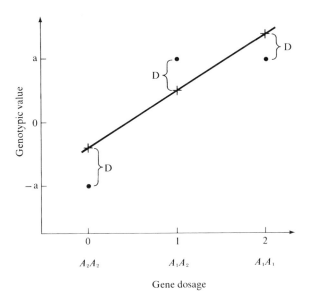

Figure B.3 Genotypic values (black dots) when dominance is complete. Regression line predicts additive genetic values (crosses) based on gene dosage. Dominance deviations (Ds) are the difference between the additive genetic values and the actual genotypic values.

genotypes "breed true" from parents to offspring. If a parent has one "dose" of a particular allele, each offspring of that parent has a 50 percent chance of receiving that allele. If the offspring receive that allele, its effect will be added in to the same extent that it was added into the total genotype of the parent. It does not matter how many alleles are involved at a locus (or as we shall see in the next section, how many loci are involved). Additive genetic values are simply the extent to which the effects of the alleles add up according to gene dosage.

Dominance Deviation

If there is dominance, average effects of alleles do not simply add up according to gene dosage as shown in Figure B.2. Dominance deviations are the difference between the expected (or additive) genotypic value and the actual genotypic value. Dominance allows for the fact that alleles at a given locus can interact with each other, rather than simply adding up in linear fashion. For example, if there is complete dominance, genotypic values will fall at the points plotted in Figure B.3. In Figure B.2, where there was no dominance and $p = q = \frac{1}{2}$, the genotypic values are the same as the additive genetic values, and they fall on a straight line. In Figure B.3, however, the genotypic values do not fall on a straight line through these points. Regression of genotypic value on gene dosage yields the genotypic values predicted by gene dosage. This, of course, is our definition of additive genetic values. Thus, the crosses on the regression line in Figure B.3 are additive genetic values. If there is dominance, this prediction of genotypic values from gene dosage will be slightly off. Dominance, as represented by the Ds in Figure B.3, is thus the deviation of the actual genotypic value from the regression line, which represents the predicted genotypic value based on gene dosage.

Dominance is important because it represents genetic influence that does not "breed true." If dominance occurs, a parent's genotypic value is due to some particular combination of alleles at a locus. Offspring cannot receive both of those alleles from the parent. Therefore, offspring will be genetically different from the parent to some extent if alleles do not add up in their effect. In summary, we have partitioned the genotypic value into two parts: one that is predicted by gene dosage and one that is not. Additive genetic values are the extent to which genotypic values add up according to gene dosage; dominance is the extent to which they do not add up.

The Polygenic Model

Not only can we consider the additive and nonadditive effects at a single locus, we can also sum these effects across loci. This concept is the essence of the polygenic extension of the single-gene model. Just as additive genetic values are the summation of the average effects of two alleles at a single locus, they can also be summed across the many loci that may influence a particular phenotypic character. Similarly, dominance deviations from additive genetic values can also be summed for all the loci influencing a character. Thus, it is relatively easy to generalize the single-gene model to a polygenic one with many loci, each with its own additive and nonadditive effects. However, we need to introduce one more concept: *epistatic interaction.*

Epistatic Interaction Deviation

Dominance is the nonadditive interaction of alleles at a single locus. When we consider several loci, we need to consider the possibility that a particular allele interacts not only with the allele at the same locus on the homologous chromosome, but also with alleles at other loci. This type of interaction is called *epistasis*. In other words, dominance is *intralocus* interaction, and epistasis is *interlocus* interaction. For example, consider two loci (*A* and *B*) that affect a phenotypic character. Both the additive genetic values and the dominance deviations are summed across the two loci. However, a particular combination of a certain allele at locus *A* and another at locus *B* may influence the phenotype in ways not explainable by the additive and dominance effects. Epistasis refers to this sort of effect.

In summary, we can partition genetic effects into three components: additive, dominance, and epistatic. At a single locus, the genotypic value includes additive and dominance effects. When we consider the effects across two or more loci, the additive and dominance effects are summed but may not yield the joint genotypic value, because of epistatic interaction among alleles at different loci. In symbolic terms,

$$G = A + D + I$$

where G is the genotypic value due to all loci, A is the sum of the additive genetic values across all loci, D is the sum of the dominance deviations across all loci, and I symbolizes the deviations due to epistatic interactions. Epistatic interactions may be of several types. They may involve interactions between additive genetic values at different loci, between dominance deviations at different loci, between additive genetic values at one locus and dominance deviations at another locus, and so on.

Phenotypic Value

Up to this point, we have considered only genetic influences. Although that is a work-able approach for traits such as those that Mendel studied in pea plants in a controlled environment, environmental influences are so important for other traits that analyses must consider both genetic and environmental factors. The basic model of quantitative genetic theory simply says that the phenotype of an individual is due to a genotypic value (including A, D, and I) and an environmental effect due to all nongenetic causes. However, science seldom studies a single individual. Our focus is on phenotypic differences in a population and on the genetic and environmental differences that create those differences. So, instead of thinking of P as an individual's phenotypic value, we will consider it as the individual's deviation from the population mean.

Thus, quantitative genetic theory begins with a model in which observed (phenotypic) deviations from the mean for some character in a population are a function of environmental (E) and genetic (G) deviations, which combine in an additive (linear) manner. However, this model may also include a nonadditive, or interaction, term (G × E) to deal with possible nonadditive combinations of genetic and environmental effects, just as dominance and epistasis allow for the possibility of nonadditive effects for single and multiple loci. Symbolically,

$$P = G + E + (G \times E)$$

The symbol G × E does not necessarily refer to multiplication of G and E. It designates the contribution of some nonadditive function of G and E to the phenotype, independent of the main effects of G and E. That is, an environmental factor may have a greater effect on some genotypes than on others, and a genotype may be expressed differently in various environments.

Variance

Each of the components is expressed as a deviation from the mean, but we want to express them in terms of variance. As described in Appendix A, variance is the sum of individuals' squared deviations from the mean, divided by the number of individuals. Let us take the G deviation and express it as variance. The variance of G simply involves squaring the genetic deviations, summing the squared deviations, and dividing by the sample size. Let us also partition the variance of G into parts due to the variance of each of its components, G = A + D + I. The variance of G (V_G) can be expressed as the covariance of G with itself. (The covariance of a variable with itself is the same as its variance.) Therefore,

$$V_G = \text{Cov}(G)(G)$$
$$= \text{Cov}(A + D + I)(A + D + I)$$
$$= V_A + V_D + V_I + 2\text{Cov}(A)(I) + 2\text{Cov}(D)(I) + 2\text{Cov}(A)(D)$$

Because A, D, and I are not correlated, the covariance terms drop out and we are left with

$$V_G = V_A + V_D + V_I$$

In other words, genetic variance is due to additive genetic variance, dominance variance, and variance resulting from epistatic interactions. Additive genetic values are equivalent to genotypic values expected from gene dosage. Thus, additive genetic variance may be thought of as genetic variance due to variation in gene dosage. In the same way, dominance and epistatic variance (or nonadditive genetic variance) are the genetic variance that is not predicted by gene dosage. It should be noted that, even if dominance is complete (e.g., d = +a), genetic variance may still have a substantial component due to additive genetic variance.

In a similar manner, we can determine the variance for the general model, P = G + E + (G × E). The symbol G × E is defined as being uncorrelated with either G or E; however, G and E may themselves be correlated. The variance of the phenotypic deviations (V_P) is a function of the squared deviations for the other components, as follows:

$$V_P = \text{Cov}(P)(P)$$

$$= \text{Cov}[G + E + (G \times E)][G + E + (G \times E)]$$

$$= V_G + V_E + 2\text{Cov}(G)(E) + V_{G \times E}$$

In other words, observed variance in a population includes components due to genetic variance (V_G) and those due to environmental variance (V_E). However, phenotypic variance also contains components added by the correlation between genetic and environmental effects [$2\text{Cov}(G)(E)$], as well as by the interaction between G and E. Although error of measurement is also likely in the variance of a phenotype, we will ignore it for now.

An Example of the Polygenic Model

An example illustrating this model may be helpful. The example is hypothetical, because we cannot often measure genotypic values and because we do not know the environmental values.

Suppose that we know the genetic, environmental, and phenotypic deviations from the mean for a number of individuals, as indicated in Table B.1. Because these values are expressed as deviations from the mean, the mean in all cases is zero. In this example, the genetic variance is 2.0, the environmental variance is 2.0, and the phenotypic variance is 4.0. Thus, $V_P = V_G + V_E$. There is no variance added by the covariance between G and E because there is no covariance between G and E in this example. (Satisfy yourself that this is true by multiplying the deviations of G by the deviations of E and then summing the cross products.)

Now let us suppose that genes and environment are perfectly correlated, as in Table B.2. The genetic and environmental variances remain the same (2.0), but the phenotypic variance is now 8.0 instead of 4.0. The added variance is due to the correlation between genetic and environmental deviations: $V_P = V_G + V_E + 2\text{Cov}(G)(E) = 2 + 2 + 4 = 8$. It should be noted that, even if we somehow removed variance due to the correlation between G and E, V_G and V_E would remain unchanged. In fact, correlation between G and E will contribute substantially to V_P only when both V_G and V_E are substantial (Jensen, 1974). Our example illustrates a positive correlation between G and E, in that large deviations in G correspond to large deviations in the same direction in E. Negative correlation between G and E would decrease rather than increase V_P.

Let us now add the G × E interaction to the example, retaining the positive correlation between G and E. We said that G × E refers to any nonadditive effect of G and

TABLE B.1

Hypothetical Genetic, Environmental, and Phenotypic Deviations from the Mean for Five Individuals

Individual	G	+	E	=	P
1	−2		+1		−1
2	−1		−2		−3
3	0		0		0
4	+1		+2		+3
5	+2		−1		−1
	$V_G = 2.0$		$V_E = 2.0$		$V_P = 4.0$

Note: To keep this example as simple as possible, we will consider these individuals as constituting a population rather than a sample, thus ignoring problems of sampling. As a result, variances are obtained by dividing by N, rather than by $N-1$.

SOURCE: *After Plomin, DeFries, and Loehlin (1977).*

E. In our example, however, we will assume that the nonadditive function is, in fact, G multiplied by E (see Table B.3). The variance of the G × E values around their mean of 2.0 is 6.8. Genetic variance, environmental variance, and variance due to the correlation between G and E [2Cov(G)(E)] remain 2.0, 2.0, and 4.0, respectively. Adding the $V_{G \times E}$ term yields 14.8, which is the phenotypic variance.

Although we cannot often measure genetic variance, environmental variance, or genotype-environment interaction directly, this hypothetical example indicates that all four components can contribute to phenotypic variance for a character. Because we cannot measure these components directly, we estimate them indirectly from the resemblance of relatives.

TABLE B.2

Hypothetical Genetic, Environmental, and Phenotypic Deviations from the Mean for Five Individuals When Genetic and Environmental Deviations Are Perfectly Correlated

Individual	G	+	E	=	P
1	−2		−2		−4
2	−1		−1		−2
3	0		0		0
4	+1		+1		+2
5	+2		+2		+4
	$V_G = 2.0$		$V_E = 2.0$		$V_P = 8.0$

SOURCE: *After Plomin, DeFries, and Loehlin (1977).*

TABLE B.3

Hypothetical Genetic and Environmental Deviations from the Mean and Phenotypic Values for Five Individuals When Genetic and Environmental Deviations Are Perfectly Correlated and When There Is an Interaction Between G and E

Individual	G	+	E	+	G × E	=	P
1	−2		−2		+4		0
2	−1		−1		+1		−1
3	0		0		0		0
4	+1		+1		+1		+3
5	+2		+2		+4		+8
	$V_G = 2.0$		$V_E = 2.0$		$V_{G\times E} = 6.8$		$V_P = 14.8$

SOURCE: *After Plomin, DeFries, and Loehlin (1977).*

Covariance of Relatives

If we could measure genetic and environmental effects for individual subjects, we could directly estimate V_G and V_E in populations. Instead, quantitative genetic analyses proceed indirectly, estimating the various genetic and environmental components of variance from relationships that differ in genetic or environmental relatedness (Chapter 5). For example, full siblings who have both parents in common are twice as similar genetically as half siblings with only one parent in common. If genes influence a particular behavior, then the double genetic similarity of full siblings should make them more similar for that behavior than half siblings are. Quantitative behavioral genetic methods involve comparisons of several such relationships, in which genetic similarity is varied while environmental similarity is held constant, or vice versa. The purpose of this section is to provide the theoretical background for behavioral genetic studies of familial resemblance.

Covariance

Appendix A described covariance, correlation, and regression. Covariance between X and Y in a sample is the sum of the cross products of the deviations from the mean of X and the corresponding deviations from the mean of Y, divided by $N - 1$. Correlation and regression express covariance as a proportion of variance. For now, we will focus on covariance. Appendix A considered the covariation between two variables, X and Y, for several individuals—that is, the extent to which individuals' scores on X covaried with scores on Y. Now we will consider covariance between relatives rather than between variables. That is, instead of considering the covariance between two traits, X and Y, for individuals measured on both traits, we will consider the covariance between twins or between parents and their offspring for a single variable. If members of a family are more similar than individuals picked at random from the population (i.e., if their deviations from the mean are in the same direction), there is covariance.

Both genetic and environmental hypotheses predict similarities between relatives. Relatives share genes to some extent and thus should be similar if genes affect the particular behavior under study. Environmental hypotheses also predict that members of the same family should be similar because they are subject to much the same environmental influences. For example, if certain child-rearing practices are thought to be important influences on the development of personality, then children in the same family subjected to similar child-rearing practices should be similar in those aspects of personality. Later, we shall see how the knowledge that certain family relationships are not as similar genetically or environmentally as others provides the basis for untangling genetic and environmental influences. However, the point here is that both genetic and environmental hypotheses predict covariance among relatives living together.

There is zero covariance between pairs of unrelated individuals picked at random. Because such individuals share neither genes nor environment, their scores do not covary. Other family relationships, however, share both genes and environment. We can describe covariance between relatives as $Cov(P_1)(P_2)$, where P_1 is the phenotype of one relative and P_2 is the phenotype of the other. In the previous section, we noted that $P = G + E$, and we can substitute that for $Cov(P_1)(P_2)$:

$$Cov(P_1)(P_2) = Cov(G_1 + E_1)(G_2 + E_2)$$

Remember that quantitative genetics always addresses differences rather than universals (Chapter 5). That is, when we say that genetically unrelated individuals do not share genes, we mean that they are uncorrelated for polymorphic loci. Much DNA is identical for all humans, and much DNA is even identical for all mammals. Similarly, when we say that individuals do not share environment, we mean that their experiences are uncorrelated. Many environmental factors, from oxygen to caregivers, can be viewed as important environmental constants shared by all members of our species. To the extent that such genetic and environmental factors are constants, they do not contribute to differences among individuals. However, even factors such as these could be viewed as contributing to individual differences. In terms of oxygen, do children living at high altitudes differ developmentally from those at sea level? A major concern of developmental psychologists is the extent to which different styles of caregiving affect development. However, our point is that quantitative genetics only addresses genetic and environmental factors that make a difference, not the genes and environments that are the same for everyone.

Shared and Nonshared Influences

Not all genetic, nor environmental, influences for a particular behavior make family members similar to one another. Identical twins are, of course, identical genetically and thus share all genetic influences. However, for other family relationships, there are both shared and nonshared genetic influences. Genetic theory predicts differences between genetically related individuals other than identical twins. In contrast, environmental theories have not typically predicted differences between members of the same family.

Although V_A, V_D, and V_I contribute in various ways to different familial relationships, for the moment we shall consider only genetic variance that the relatives have in common. Parents and offspring are first-degree relatives, as are full siblings. Consider a single locus with two alleles. An offspring has a fifty-fifty chance of inheriting one particular allele rather than the other from the parent. For this reason, first-degree relatives are 50 percent similar genetically; in other words, half of the genetic variance is shared between them. The other half of the genetic variance does not covary between them, so it makes them different from each other. Such reshuffling of genes is the consequence of meiosis and the source of genetic variability. Thus, we can divide the genetic contribution to the phenotype of an individual into two parts—that part that the individual shares, or has *in common*, with the relative (G_C) and the part that is not shared with the relative.

Similarly, environmental influences may be shared by relatives. Other aspects of the environment make family members different from one another; these aspects refer to nonshared environments, discussed in Chapter 14. Thus, we can also divide the environmental contribution to the phenotype into influences shared with the relative (E_C) and those independent of the relative.

In the previous equation, by definition, only G_C and E_C can contribute to the phenotypic covariance between relatives:

$$\text{Cov}(P_1)(P_2) = \text{Cov}(G_C + E_C)(G_C + E_C)$$

The covariance of G_C with G_C is equivalent to the variance of G_C (that is, V_{G_C}). As we indicated earlier, a variable completely covaries with itself, meaning that the covariance of a variable with itself is the same as its variance. In the same way, $\text{Cov}(E_C)(E_C) = V_{E_C}$.

Now we can express the phenotypic covariance between relatives in terms of components of variance:

$$\text{Cov}(P_1)(P_2) = V_{G_C} + V_{E_C}$$

In other words, for a particular character, the covariance between relatives includes the genetic variance and the environmental variance resulting from shared genetic and shared environmental influences.

Genotype-Environment Correlation and Interaction

The model we have used up to this point is oversimplified. Earlier, we mentioned the correlation and interaction between genetic and environmental factors. These components of variance also enter the picture when we consider the covariance among relatives. $\text{Cov}(G_C + E_C)(G_C + E_C)$ includes the covariance between G_C and E_C, in that it is equivalent to $V_{G_C} + V_{E_C} + 2\text{Cov}(G_C)(E_C)$. Covariance between genetic and environmental deviations can add to phenotypic variance. It can also add to the covariance between relatives. In addition, when we substituted G + E for P, we did not consider the G × E interaction. The G × E interaction shared by relatives will also contribute to their phenotypic covariance.

Genetic Covariance Among Relatives

Our general model for the covariance of relatives is also too simple, because it treats only shared genetic variance rather than distinguishing between V_A, V_D, and V_I. These components of genetic variance contribute variously to different types of family relationships (see Table B.4). Parents and their offspring share one-half of their additive genetic variance, as discussed in the previous section. (For this reason, additive genetic variance provides a measure of the extent to which characters "breed true.") However, parents and offspring do not share genetic variance due to dominance. Remember that dominance is the result of nonadditive combinations of alleles at loci. Offspring cannot obtain a chromosome *pair* from one parent. Thus, although dominance may contribute to the phenotypes of parent and offspring, this genetic factor will not be shared by them.

TABLE B.4

Contribution of Additive Genetic (V_A), Dominance (V_D), and Common Environmental (V_{E_C}) Influences to the Phenotypic Covariance of Relatives

Phenotypic Covariance Between:	V_A		V_D		V_{E_C}
Parents and offspring (PO)	$\frac{1}{2}$	+	0	+	$V_{E_{C(PO)}}$
Half siblings (HS)	$\frac{1}{4}$	+	0	+	$V_{E_{C(HS)}}$
Full siblings (FS)	$\frac{1}{2}$	+	$\frac{1}{4}$	+	$V_{E_{C(FS)}}$
Fraternal twins (DZ)	$\frac{1}{2}$	+	$\frac{1}{4}$	+	$V_{E_{C(DZ)}}$
Identical twins (MZ)	1	+	1	+	$V_{E_{C(MZ)}}$

Another factor that contributes to genetic covariance among relatives is assortative mating (Chapter 8). Assortative mating adds to the genetic similarity between parents and their offspring, as well as to that between siblings (Jensen, 1978). For example, if assortative mating exists, a correlation between mothers and their children will include not only the genetic similarity between the mothers and their children but also some part of the genetic similarity between the children and their fathers.

Siblings, like parents and their offspring, share half of the additive genetic variance that influences a character. However, siblings also share one-fourth of the dominance variance, because full siblings can be expected to receive the same alleles from both parents one-fourth of the time and thus have the same dominance deviation.

Fraternal twins are siblings who happen to be born at the same time. Two eggs are fertilized by different sperm. For this reason, they are sometimes referred to as dizygotic (two-zygote) twins. Like other siblings, dizygotic (DZ) twins can be either the same sex or different, and they share half of the additive genetic variance and one-fourth of the variance due to dominance. Monozygotic (MZ) twins begin life as a single zygote that splits sometime during the first few weeks of life. Because they are genetically identical, identical twins are always of the same sex. They share all genetic variance—V_A, V_D, and V_I.

Finally, half siblings who share only one parent thus share only one-fourth of the additive genetic variance (half as much as full siblings). However, unlike full siblings, half siblings do not share any dominance variance. Because half siblings have only one parent in common, they cannot inherit the same chromosome pairs and thus cannot share in allelic interactions at a given locus.

Sometimes we need to consider the covariance of behavioral measures for one relative with the average measures for a number of other relatives. For example, we might consider the covariance between offspring and the average parental scores, rather than scores for a single parent. Or we could turn it around and look at the covariance between a single parent and the average of all of that parent's offspring. In general, the expected covariances for such averaged relationships are the same as those discussed above for relatives considered one at a time.

What about epistasis? We noted earlier that, in addition to additive effects of alleles across loci (V_A), there is also nonadditive genetic variance. Although some of this nonadditive variance is due to interactions between alleles at a locus (V_D), the rest is due to nonadditive interactions between alleles at different loci (V_I). Because identical twins are genetically identical, their phenotypic covariance includes all additive and nonadditive genetic variance. However, phenotypic covariance for other familial relationships includes only some of the variance due to nonadditive interactions (Falconer & Mackay, 1996). Fortunately, this complexity turns out empirically to be less important than it might seem. Additive genetic variance accounts for the majority of genetic variance in most behavioral characters for which such information is available.

Table B.4 summarizes the additive, dominance, and environmental components of variance responsible for the phenotypic covariance of relatives. For example, the phenotypic covariance between fraternal twins includes half of the additive genetic variance ($\frac{1}{2}V_A$), one-fourth of the nonadditive genetic variance due to dominance ($\frac{1}{4}V_D$), and environmental influences common to members of fraternal twin pairs ($V_{EC(DZ)}$). In contrast, identical twins' covariance includes all additive and nonadditive genetic variance, as well as environmental influences common to members of identical twin pairs ($V_{EC(MZ)}$). Such differences in the components of covariance are used to estimate the various components of genetic and environmental variance.

Heritability

As explained in Chapter 5, heritability is the proportion of phenotypic variance that is attributable to genotypic variance:

$$\text{Heritability} = \frac{V_G}{V_P}$$

In the numerical example presented in Table B.1, both the genetic and environmental variances are equal to 2.0. In the simplest case, when there is no correlation or interaction between genetic and environmental factors, the phenotypic variance is 4.0. In this case, heritability is 0.5, meaning that 50 percent of the phenotypic variance is explained by genetic variance. The other 50 percent of the phenotypic variance is caused by environmental variance. Tables B.2 and B.3 indicate that correlations or interactions between genetic and environmental factors will increase phenotypic variance. These effects can also contribute to the phenotypic resemblance between relatives and thus can have an effect on behavioral genetic analyses.

There are two types of heritability. *Broad-sense heritability* (h_B^2) is the type of heritability we have been discussing. It is the proportion of phenotypic differences due to all sources of genetic variance, regardless of whether the genes operate in an additive or nonadditive manner. *Narrow-sense heritability* (h^2), on the other hand, is the proportion of phenotypic variance due solely to additive genetic variance. If V_G refers to all genetic variance and V_A refers to additive genetic variance, then

$$h_B^2 = \frac{V_G}{V_P} \text{ and } h^2 = \frac{V_A}{V_P}$$

Narrow-sense heritability is particularly interesting in the context of selective breeding studies, where the important question is the extent to which offspring will resemble their parents. As we noted earlier, additive genetic variance involves the extent to which characters "breed true." On the other hand, broad-sense heritability is important in many other contexts. The most important situation involves the relative extent to which individual differences are due to genetic differences of any kind. We can obtain the appropriate answer by assessing broad-sense heritability.

Path Analysis

The concept of heritability can also be illustrated by the analysis of *paths*—the statistical effect of one variable on another, independent of other variables. For some, it is easier to understand the concept of heritability visually in a path model rather than strictly in algebraic terms.

We can construct a path model of the effects of genetic and environmental factors on a phenotype, as shown in Figure B.4. This illustration has the same meaning as the statement P = G + E. The "paths" in this case express the extent to which genetic and environmental deviations cause phenotypic deviations. Thus, h_B is the path by which genetic deviations (G) from the population mean cause phenotypic deviations. In fact, the h_B path is the proportion of the phenotypic standard deviation (s_P) caused by the genetic standard deviation (s_G):

$$h_B = \frac{s_G}{s_P}$$

Remembering that the standard deviation is the square root of variance, you can see that the h_B path is, in fact, the square root of broad-sense heritability.

Multivariate Analyses

We have been focusing on the genetic-environmental analysis of only one phenotype, a univariate (one variable) approach. However, several phenotypes can be measured for each individual and subjected to multivariate quantitative genetic analysis. If two traits are measured for each individual in a population and a correlation is observed, this phenotypic correlation may be due to either genetic or environmental factors. Among the genetic causes of correlation, the same genes may influence both traits, an effect called *pleiotropy*. Genetic correlations can also result from temporary linkages due to recent admixtures of populations or nonrandom mating.

It is easy to visualize how environmental effects may give rise to correlations between traits, such as height and weight. A favorable diet, for example, may result in higher height and weight, whereas an unfavorable diet may be accompanied by lower values for both characters. At the psychological level, the phenotypic correlations among

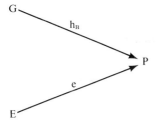

Figure B.4 Path model of the genetic and environmental components of the phenotypic value (P). See text for explanation.

measures of specific cognitive abilities may be due to environmental influences, such as the intellectual environment of the home or the quality of schooling, that affect various specific cognitive abilities in a similar way.

Why would we want to know the extent to which genetic and environmental factors contribute to the phenotypic correlation between two traits? When we study traits one at a time, many show genetic influence, but it is highly unlikely that each of these is influenced by a completely different set of genes. If the same genes affect different traits, we will observe a correlation among the traits. The same reasoning applies to environmental influences: They may affect several traits, thereby producing correlations among them. Thus, the importance of multivariate genetic-environmental analysis lies in its potential for revealing the genetic and environmental bases of phenotypic covariance. Development of multivariate quantitative genetic analysis is one of the most important advances in behavioral genetics during the past decade.

Just as quantitative genetics can be applied to the variance of a single behavior, it can also be applied to the correlation between two behaviors. In fact, any behavioral genetic method that can partition the variance of a single trait can also be applied to the partitioning of the covariance between two traits. Path analysis provides an easy way to visualize this analysis. Figure B.5 extends the path analysis of a single behavior (see Figure B.4) to the analysis of the correlation between two phenotypic traits P_X and P_Y. Just as the variance of a single trait (P_X) is due to an environmental path (e_X) and a genetic path (h_X), the phenotypic correlations between X and Y ($r_{P_XP_Y}$) may be caused by an environmental chain of paths ($e_Xe_Yr_E$) and a genetic chain of paths ($h_Xh_Yr_G$), where r_E and r_G are environmental and genetic correlations, respectively. These chains of paths add up to the phenotypic correlation ($r_{P_XP_Y}$):

$$r_{P_XP_Y} = h_Xh_Yr_G + e_Xe_Yr_E$$

Genetic and environmental chains are especially useful for investigating the causes of phenotypic correlations between traits. However, either genetic or environmental correlations (r_G and r_E) by themselves are also informative. The genetic correlation provides a measure of the extent to which two traits are influenced by the same genes. Likewise, the environmental correlation measures the extent to which two traits are affected by the same environmental influences. Whether one estimates genetic chains, genetic correlations, or both depends on the purpose of the investigator.

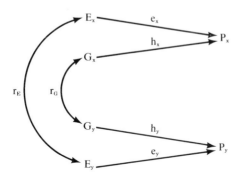

Figure B.5 Path diagram of the phenotypic correlation between two traits (P_X and P_Y) measured on an individual as a function of the genetic correlation (r_G) and the environmental correlation (r_E).

This path analysis merely states in more precise terms our previous assertion that the phenotypic correlation between two traits may be due to genetic or environmental influences. The phenotypic correlation by itself, however, does not provide a useful index of the importance of the genetic and environmental chains. Even when the phenotypic correlation between two traits is negligible, there may be substantial genetic and environmental chains of influence between the two traits if the genetic and environmental correlations are in opposite directions—that is, if one is positive and the other negative. For example, the same genes may affect specific cognitive abilities, leading to positive genetic correlation. However, environmentally, one might develop a few abilities to the exclusion of the others, leading to a negative environmental correlation. Moreover, even if two behaviors are both substantially heritable, the phenotypic correlation between them may be environmental in origin. For example, verbal ability and spatial ability are phenotypically correlated, and both show substantial heritability. However, it is possible that completely different sets of genes influence the two abilities. In other words, their genetic correlation could be 0. If this were the case, the environmental chain would be solely responsible for the phenotypic correlation.

Genetic and environmental chains can be estimated by methods analogous to those used to estimate heritability. When we consider the phenotypic covariance between different traits of relatives rather than for one trait, we need to introduce a new concept, *cross-covariance*. Rather than studying the covariance of character X in parents and character X in offspring, we consider the cross-covariance of character X in parents and character Y in offspring. Phenotypic cross-covariance between parents and offspring may be due to their genetic and environmental similarity. In fact, the components of cross-covariance between relatives are the same as those listed in Table B.4. This should not be surprising in view of the relationship between the univariate and multivariate analyses just described. Thus, the phenotypic cross-covariance for traits X and Y, for parents and offspring, involves half of the additive genetic covariance, as well as common environmental influences. Phenotypic cross-covariances of identical twins include all genetic sources of covariance in addition to shared environmental influences. The use of familial resemblance in univariate analyses of variance and in multivariate analyses of covariance is outlined in the following sections.

Resemblance of Relatives Revisited

Whether and how much genetic factors contribute to the variance of a trait (univariate analysis) or the covariance between traits (multivariate analysis) can be assessed from comparisons between the resemblances of different kinds of relatives.

Univariate Analysis

Regression and correlation are merely covariances divided by variances. If the covariance consists solely of the genetic component of variance, then the correlation between relatives estimates heritability. In this case, the correlation is found by dividing the genetic variance by the phenotypic variance. This is the definition of heritability.

Consider identical twins who have been separated from birth. As shown in Table B.4, identical twins share all genetic variance, plus common environmental influences. However, if they have been separated from birth, they do not have a common postnatal environment. Thus, their phenotypic covariance estimates V_G, and the correlation between them directly estimates heritability, V_G/V_P.

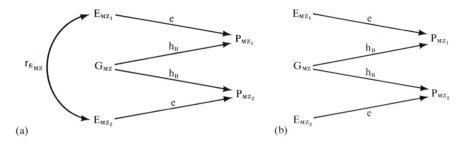

(a)   (b)

Figure B.6 Path diagram for identical twins reared (a) together and (b) apart in uncorrelated environments.

Each identical co-twin's phenotype is caused by genetic and environmental influences (shown as a path diagram in Figure B.6). However, identical co-twins have the same genotype. A useful feature of path analysis is the ability to trace the components of a correlation by following the paths. For identical twins reared together, one chain of paths from the phenotype of one identical twin to the phenotype of the co-twin is $(h_B)(h_B)$, or h_B^2. Another chain is $(e)(r_{E_{MZ}})(e)$ or $e^2(r_{E_{MZ}})$. (As illustrated in the following section on model-fitting, a single variable could be used to model shared environmental influences, E_C, with a corresponding environmental chain equal to c^2.) The correlation between two phenotypes, given an appropriate path model, is the sum of the chains of paths. Thus, the correlation between identical twins reared together is $h_B^2 + e^2(r_{E_{MZ}})$, which means that the correlation includes broad-sense heritability and environmental influences shared by MZ twins. This statement merely reiterates the point that identical twins reared together have genetic and environmental factors in common. However, as shown on the right half of Figure B.6, identical twins reared apart in uncorrelated environments (that is, $r_{E_{MZ}} = 0$) share only the genetic paths. Thus, their correlation directly estimates h_B^2 (broad-sense heritability).

It is in this sense that correlations can be used to imply causation, despite the revered rule to the contrary. If identical twins reared apart in uncorrelated environments correlate for some trait, how can such a correlation be explained other than by hereditary influence?

Another issue is that familial correlations represent components of variance; they are not squared to determine variance. Appendix A indicated that a correlation between X and Y is squared to determine the amount of variance in Y that can be predicted by X. In the case of a familial correlation, such as the correlation between identical twins reared apart, the correlation is not squared because at issue is the percentage of variance common to the twins, rather than the percentage of variance of twins' scores that can be predicted by the co-twins' scores (Jensen, 1971). The correlation for identical twins reared apart estimates heritability directly. For the same reason, the correlation for genetically unrelated children adopted together into the same adoptive family estimates the proportion of the variance due to shared environmental influences.

Multivariate Analysis

We have just seen in Figure B.6 that the univariate correlation for pairs of identical twins reared in uncorrelated environments estimates h_B^2. Figure B.7 extends this relationship of two characters, X and Y. We indicated earlier that the cross-covariance for trait X in one relative and trait Y in another has the same components of covariance as the univariate situation summarized in Table B.4. Similarly, cross-correlations for two characters for relatives have the same relationship to univariate familial correlations. Thus, as shown in Figure B.7, the cross-correlation for X and Y for separated identical twins is equivalent to the genetic chain discussed earlier:

$$r_{P_{x_1} P_{y_2}} = h_X h_Y r_G$$

In other words, if the phenotypic correlation between two traits is due entirely to their genetic correlation, then the cross-correlation for pairs of separated identical twins should be similar to the phenotypic correlation between X and Y observed within individuals. Of course, we do not need to find separated identical twins in order to conduct multivariate quantitative genetic analyses. As we have said, any behavioral genetic analysis of the variance of a single trait can be applied to the correlation among traits.

Model Fitting

Single correlations, such as the correlation for identical twins reared apart or the correlation for genetically unrelated children adopted together, may be quite sufficient for many purposes for estimating quantitative genetic parameters. Other behavioral genetic designs involve the comparison of two correlations, such as the twin method, which compares correlations for identical and fraternal twins, and the results of these studies can also be interpreted just by examining the correlations and calculating quantitative genetic parameters directly. However, even for such simple designs, fitting an explicit quantitative genetic model to observed data has many advantages over just calculating quantitative genetic parameters. Model fitting becomes particularly critical for the interpretation of data from various sources and designs—for example, when family, twin, and adoption data are included in the same analysis.

In quantitative genetics, model fitting basically involves solving a series of simultaneous equations in order to estimate genetic and environmental parameters that best fit observed familial correlations (Jinks & Fulker, 1970; Neale & Cardon, 1992). The major advantages of such model-fitting analyses include the following: Models make assumptions explicit. Model fitting tests the fit of a particular model with its set of assumptions. It is able to analyze data for several different familial relationships simultaneously. It provides appropriate estimates of quantitative genetic parameters and errors of estimate given the assumptions of the model. And it makes it possible to compare the fit

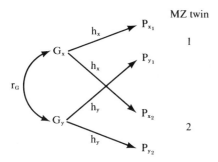

Figure B.7 Path diagram of the phenotypic correlation between two traits (P_x and P_y) measured on a pair of identical twins reared apart in uncorrelated environments.

of alternative models. Model fitting is as appropriate for the analysis of quantitative genetic analyses of animal designs (Mather & Jinks, 1982) as for human designs, although the following examples refer to human designs.

A complete description of model fitting is beyond the scope of this book; however, an excellent introduction to the topic that includes behavioral genetic examples is available (Loehlin, 1987). In order to provide an example of a simple model-fitting analysis, consider the model implied by the path diagrams in Figure B.6. The model contains two observations—correlations for identical twins reared apart and for identical twins reared together—and two *latent* variables—G and E. The goal is to solve the model for the unknown paths e and h. As explained in relation to Figure B.6, expectations can be written for the correlations for identical twins reared together (MZT) and for identical twins reared apart (MZA):

$$r_{MZT} = h_B^2 + r_{E_{MZ}} e_C^2$$

$$r_{MZA} = h_B^2$$

This model provides two equations with two unknowns. The assumptions of the model were mentioned earlier, such as the assumption that MZA do not share correlated environments, that is, that selective placement is unimportant for MZA. Given observed correlations for MZT and MZA, the equations can be solved by using simple algebra in order to estimate the genetic and environmental parameters. The MZA correlation estimates heritability, and the shared environment component is estimated as the extent to which the MZT correlation exceeds the MZA correlation.

Model-fitting analyses usually involve many more equations and parameters, in which case it becomes necessary to solve the equations by using computer programs. For example, equations could be added for fraternal twins reared together, nontwin siblings, and parents and offspring. Parameters could also be added to estimate assortative mating, to distinguish additive and nonadditive genetic variance, and to estimate parameters separately by gender or by age. It is important to have more equations than unknowns in the model, because such overdetermined designs make it possible to test the fit of the model to the data, in addition to estimating the parameters; if there are more unknowns than equations, the parameters cannot be estimated. The goodness-of-fit index is usually χ^2 (chi-square), which is a widely used statistic that indicates the statistical significance concerning the fit between expectations and observed data. The main analytic procedure used to solve complex series of equations is called *maximum-likelihood model fitting*. Such an approach employs computer programs that maximize the fit between the model and the data by finding the set of parameter estimates that yield the smallest possible discrepancies with the data. One of the most widely used maximum-likelihood computer programs is LISREL, which stands for LInear Structural RELations (Jöreskog & Sörbom, 1993).

As an example of model fitting involving several groups, consider an analysis involving twins, adoptive siblings, and parents and offspring. Familial correlations for the trait Sociable as measured by a self-report questionnaire called the Thurstone Temperament Schedule are presented in Table B.5. It is easy to see that some genetic influence is suggested by the pattern of correlations: For example, in two studies, identical twins (MZ, monozygotic) are more similar than fraternal twins (DZ, dizygotic), and adoptive

TABLE B.5

Correlations for the Trait Sociable in Two Twin Studies and an Adoption Study

Pairing	Correlation	Number of Pairs
1. MZ twins: Michigan	.47	45
2. DZ twins: Michigan	.00	34
3. MZ twins: Veterans	.45	102
4. DZ twins: Veterans	.08	119
5. Father-adopted child	.07	257
6. Mother-adopted child	−.03	271
7. Father-natural child	.22	56
8. Mother-natural child	.13	54
9. Adopted-natural child	−.05	48
10. Two adopted children	−.21	80

SOURCE: *Loehlin (1987).*

parents and their adopted children do not resemble each other as much as nonadoptive parents and offspring. Shared environmental influence does not appear to be important because adoptive relatives do not resemble each other. We could estimate each of these parameters from each informative comparison and average the estimates; for example, averaging the four correlations for adoptive relatives (lines 5, 6, 9, and 10) yields an average correlation (weighted by sample size) of −.01, a result indicating that shared environment accounts for essentially none of the variance for the Sociable scale. However, model fitting is much more informative, because as mentioned earlier, it tests an explicit model by using all of the data simultaneously and it can compare alternative models.

A path diagram of a model of causal paths that might underlie such patterns of correlations is shown in Figure B.8. The trait P is a score on the Sociable scale, measured on both individuals of each pair. The model proposes that correlations between relatives on the trait are due to three independent sources: additive effects of genes, A; nonadditive effects of genes due to dominance, D; and the environment common to pair members, E_C. The residual arrow to P allows for effects of nonshared environment unique to each individual and, in all but the MZ pairs, for genetic differences as well. The assumed genetic correlations (r_A and r_D) for the different types of relatives follow from Table B.4. The correlation for first-degree relatives for additive genetic variance is .50, and for MZ twins, 1.0. For nonadditive genetic variance due to dominance, correlations are 1.0 for MZ twins, .25 for genetically related siblings, including DZ twins, and .00 for parents and offspring. The genetic correlations are all 0 in the case of genetically unrelated relatives. These genetic correlations assume that assortative mating does not occur for the trait and that adopted children are randomly placed in adoptive homes with respect to the trait. Because the data set includes no genetically related relatives adopted apart (such as identical twins reared apart), all relatives can also resemble each other for reason of shared environment.

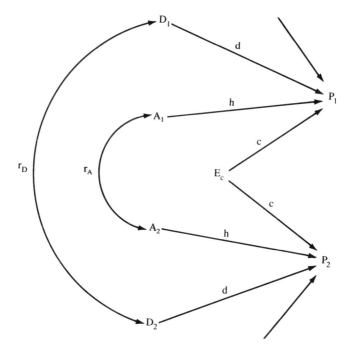

Figure B.8 Path model of genetic and environmental sources of correlation between two individuals. G, additive genetic effect; D, nonadditive genetic effect; E_C, shared environment; P, phenotypic score; 1,2, two individuals. (Adapted from Loehlin, 1987.)

From this model, Table B.6 shows equations for the correlations between test scores of members of the relatives whose data are listed in Table B.5. The model keeps open the possibility that shared environmental influence differs in magnitude for identical twins, siblings, and parents and their offspring by allowing different c paths for these relationships. The model, however, does not discriminate between environmental relationships of parents and adopted or natural children, between mothers' and fathers' relationships with their children, or between DZ twins and adoptive siblings; models could be constructed to permit these comparisons. There are ten observed correlations described in Table B.6 and five unknowns (h, d, and the three c parameters).

Five models with one to four unknowns were analyzed by using model-fitting procedures, with results shown in Table B.7. A "null model" that tests whether all ten correlations are equal can confidently be rejected because the model does not fit the data. A χ^2 of 36.02 with nine degrees of freedom (df; ten groups minus one parameter estimated) indicates a probability (p) value of less that .001. This p value means that the data significantly depart from this null model; fewer than 1 in 1000 times would such a result be observed by chance. None of the other models departs significantly from the data. Nonetheless, the fit of the various models can be compared directly when models are nested—that is, when one model includes a subset of the parameters in another model. In this case, the difference in χ^2 for the two models is itself distributed as χ^2 and may be

TABLE B.6

Equations for Correlations Between Pairs of Individuals in Different Relationships

Relationship	Table B.5 Pairings	Equation
MZ twins	1,3	$h^2 + d^2 + c_1^2$
DZ twins	2,4	$0.5h^2 + 0.25d^2 + c_2^2$
Parent, adopted child	5,6	c_3^2
Parent, natural child	7,8	$0.5h^2 + c_3^2$
Adoptive siblings	9,10	c_2^2

SOURCE: *Loehlin (1987).*

tested for significance. Model 2, which includes only additive genetic variance, is not improved significantly by adding any of the other parameters. That is, the fit is not improved significantly by adding a single shared environment parameter (model 3, which equates the three c parameters). The χ^2 difference between models 2 and 3 is 0.62, which with one degree of freedom yields a probability value greater than .40. Similarly, adding the dominance parameter (model 4) does not significantly improve the fit of model 2 (χ^2 difference = 1.49, df = 1, $p > .20$). Adding the three separate environmental parameters for parent and child, siblings, and MZ twins (model 5) yields a marginally better fit, although still not significantly better than model 2 (χ^2 difference = 6.49, df = 3, $p < .10$).

The rule of parsimony would suggest that the simplest model is best—model 2, which includes only additive genetic variance. However, because model 5 marginally improves the fit, Table B.8 lists parameter estimates for both models 2 and 5. Just because a parameter does not significantly improve the fit of the model does not mean that it must be dropped from the model. Keeping an eye on the basic data of the family correlations helps to add a dose of common sense and to avoid misinterpretation. As

TABLE B.7

Solutions of Equations in Table B.6 for Various Combinations of Parameters

Model	χ^2	df	p
1. all r's equal (null)	36.02	9	>0.001
2. h^2 only	9.09	9	0.44
3. $h^2 + c^2$	8.47	8	0.40
4. $h^2 + d^2$	7.60	8	0.48
5. $h^2 + c_1^2 + c_2^2 + c_3^2$	2.60	6	0.85

SOURCE: *Loehlin (1987).*

TABLE B.8

Parameter Estimates for Table B.7 Solutions

Model				
Model 2	$h^2 = .40$			
Model 5	$h^2 = .34$	$c^2_{p\text{-}c} = 0.02$	$c^2_{sib} = -0.13$	$c^2_{MZ} = 0.11$

Note: These parameter estimates are lower than those presented by Loehlin (1987), which were corrected for unreliability of measurement.
SOURCE: Loehlin (1987).

indicated below, it appears that, although shared environment has little effect on the similarity of parents and offspring, it has more of an effect on siblings, especially on identical twins. The single-parameter model yields a narrow heritability estimate of .40, which means that about 40 percent of the variance in the Sociable scale is due to additive genetic factors. Because the shared environment parameter is not included in this model, the rest of the variance is presumed to be due to nonshared environmental factors. The parameter estimates for model 5 are especially interesting in that they suggest that environmental effects on sociability are almost independent for parents and children, positively shared by identical twins, and negatively related for other siblings. This result suggests the following story: Parents' sociability has little environmental effect on their children's sociability; identical twins share some environmental effects, perhaps because of their considerable physical similarity; and other siblings, including fraternal twins, show contrast in terms of sociability. Additional research is needed to replicate this finding and to pin down the processes by which it occurs.

The results of model-fitting analyses can only be used to test the fit of alternative models. That is, as in tests of any scientific hypothesis, model fitting cannot prove that a particular model is correct. Another caveat should be mentioned. Although model fitting is sophisticated and elegant (the state of the art in behavioral genetic analysis), it has the disadvantage of being complex and sometimes seems to be a black box from which parameter estimates magically appear. We should not stand too much in awe of model fitting or allow it to obfuscate the basic simplicity of most behavioral genetic designs. For example, the twin design estimates genetic influence on the basis of the difference between identical and fraternal twin correlations. If the identical twin correlation does not exceed the fraternal twin correlation for a particular trait, there is no genetic influence, and model-fitting approaches must come to that conclusion or there is something wrong with the model. Model fitting can be best viewed as providing refined analyses of the basic data of behavioral genetics—resemblance for the relatives who vary in genetic and environmental relatedness.

GLOSSARY

additive genetic variance Individual differences caused by the independent effects of alleles or loci that "add up." In contrast to nonadditive genetic variance in which the effects of alleles or loci interact.

adoption studies A range of studies that use the separation of biological and social parentage brought about by adoption to assess the relative importance of genetic and environmental influences. Most commonly, the strategy involves a comparison of adoptees' resemblance to their biological parents who did not rear them and to their adoptive parents. May also involve the comparison of genetically related siblings and genetically unrelated (adoptive) siblings reared in the same family.

adoptive siblings Genetically unrelated children adopted by the same family and reared together.

allele An alternative form of a gene at a locus, for example, A_1 versus A_2.

allele sharing Presence of 0, 1, or 2 of the parents' alleles in two siblings (a sibling pair, or sib-pair).

allelic association An association between allelic frequencies and a phenotype. For example, the frequency of allele 4 of the apolipoprotein E gene is about 40% for individuals with Alzheimer's disease and 15% for control individuals who do not have the disorder.

amino acid One of the 20 building blocks of proteins, specified by a triplet code of DNA.

anticipation The severity of a disorder becomes greater or occurs at an earlier age in subsequent generations. In some disorders, this phenomenon is known to be due to the intergenerational expansion of DNA repeat sequences.

assortative mating Nonrandom mating that results in similarity between spouses. Assortative mating can be negative ("opposites attract") but is usually positive.

assortment Independent assortment is Mendel's second law of heredity. It states that the inheritance of one locus is not affected by the inheritance of another locus. Exceptions to the law occur when genes are inherited close together on the same chromosome. Such linkages make it possible to map genes to chromosomes.

autosome Any chromosome other than the X or Y sex chromosomes. Humans have 22 pairs of autosomal chromosomes and 1 pair of sex chromosomes.

band (chromosomal) A chromosomal segment defined by staining characteristics.

base pair (bp) One step in the spiral staircase of the double helix of DNA, consisting of adenine bonded to thymine or cytosine bonded to guanine.

carrier An individual who is heterozygous at a given locus for a normal and a mutant recessive allele and who appears normal phenotypically.

centimorgan (cM) Measure of genetic distance on a chromosome. Two loci are 1 cM apart if there is a 1 percent chance of recombination due to crossover in a single generation. In humans, 1 cM corresponds to approximately 1 million base pairs.

chromatid One member of one newly replicated chromosome in a chromosome pair, which may cross over with a chromatid from a homologous chromosome during meiosis.

chromosome A structure that is composed mainly of chromatin, which contains DNA, and resides in the nucleus of cells. Latin for "colored body" because chromosomes stain differently from the rest of the cell. See also *autosome*.

codon A sequence of three base pairs that codes for a particular amino acid or the end of a chain.

concordance Presence of a particular condition in two family members, such as twins.

correlation An index of resemblance that ranges from .00, indicating no resemblance, to 1.00, indicating perfect resemblance.

crossover See *recombination*.

developmental genetic analysis Analysis of change and continuity of genetic and environmental parameters during development. Applied to longitudinal data, assesses genetic and environmental influences on age-to-age change and continuity.

dichotomous trait See *qualitative disorder*.

dizygotic (DZ) Fraternal or nonidentical twins; literally, "two zygotes."

DNA (deoxyribonucleic acid) The double-stranded molecule that encodes genetic information. The two strands are held together by hydrogen bonds between two of the four bases, with adenine bonded to thymine and cytosine bonded to guanine.

DNA marker A polymorphism in DNA itself such as a restriction fragment length polymorphism (RFLP) and a simple-sequence repeat (SSR) polymorphism.

DNA sequence The order of base pairs on a single chain of the DNA double helix.

dominant An allele that produces a particular phenotype when present in the heterozygous state.

effect size The proportion of individual differences for the trait in the population accounted for by a particular factor. For example, heritability estimates the effect size of genetic differences among individuals.

electrophoresis A method used to separate DNA fragments by size. When an electrical charge is applied to DNA fragments in a gel, smaller fragments travel farther.

epistasis Nonadditive interaction between genes at different loci. The effect of one gene depends on that of another. Compare with *dominance*, which refers to nonadditive effects between alleles at the same locus.

equal environments assumption In twin studies, the assumption that environments are similar for identical and fraternal twins.

exon DNA sequence transcribed in messenger RNA and translated into protein. Compare with *intron*.

expanded triplet repeat A repeating sequence of three base pairs, such as the CGG repeat responsible for fragile X, that increases in number of repeats over several generations.

F_1 , F_2 The offspring in the first and second generations following mating between two inbred strains.

familial Among family members.

family study Assessing the resemblance between genetically related parents and off-spring and between siblings living together. Resemblance can be due to heredity or to shared family environment.

first-degree relative See *genetic relatedness.*

fragile X Fragile sites are breaks in chromosomes that occur when chromosomes are stained or cultured. Fragile X is a fragile site on the X chromosome that is the second most important cause of mental retardation in males after Down syndrome and is due to an expanded triplet repeat.

full siblings Individuals who have both biological (birth) parents in common.

gamete Mature reproductive cell (sperm or ovum) that contains a haploid (half) set of chromosomes.

gene A sequence of DNA bases that codes for a particular product. Includes DNA sequences that regulate transcription. See *allele* and *locus.*

gene map Visual representation of the relative distances between genes or genetic markers on chromosomes.

gene targeting Mutations that are created in a specific gene and can then be transferred to an embryo.

genetic anticipation See *anticipation.*

genetic counseling Conveys information about genetic risks and burdens and helps individuals to come to terms with the information and to make their own decisions concerning actions.

genetic relatedness The extent or degree to which relatives have genes in common. *First-degree relatives* of the proband (parents and siblings) are 50 percent similar genetically. *Second-degree relatives* of the proband (grandparents, aunts, and uncles) are 25 percent similar genetically. *Third-degree relatives* of the proband (first cousins) are 12.5 percent similar genetically.

genome All the DNA of an organism for one member of each chromosome pair. The human genome contains about 3 billion DNA base pairs.

genomic imprinting The process by which an allele at a given locus is expressed differently, depending on whether it is inherited from the mother or the father.

genotype The genetic constitution of an individual, or the combination of alleles at a particular locus.

half siblings Individuals who have just one biological (birth) parent in common.

Hardy-Weinberg equilibrium Allelic and genotypic frequencies remain the same generation after generation in the absence of forces such as natural selection that change these frequencies. If a two-allele locus is in Hardy-Weinberg equilibrium, the frequency of genotypes is $p^2 + 2pq + q^2$, where p and q are the frequencies of the two alleles.

heritability The proportion of phenotypic differences among individuals that can be attributed to genetic differences in a particular population. *Broad-sense heritability* involves all additive and nonadditive sources of genetic variance, whereas *narrow-sense heritability* is limited to additive genetic variance.

heterozygosity The presence of different alleles at a given locus on both chromosomes at a given locus.

homozygosity The presence of the same allele at a given locus on both members of a chromosome pair.

imprinting See *genomic imprinting*.

inbred strain study Comparing inbred strains, created by mating brothers and sisters for at least 20 generations. Differences between strains can be attributed to their genetic differences when the strains are reared in the same laboratory environment. Differences within strains estimate environmental influences, because all individuals within an inbred strain are virtually identical genetically.

inbreeding Mating between genetically related individuals.

index case See *proband*.

intron DNA sequence within a gene that is transcribed into messenger RNA but spliced out before translation into protein. Compare with *exon*.

kilobase (kb) 1000 base pairs of DNA.

knock out Inactivation of a gene by gene targeting.

liability-threshold model A model that assumes that dichotomous disorders are due to underlying genetic liabilities that are distributed normally. The disorder appears only when a threshold of liability is exceeded.

linkage Close proximity of loci on a chromosome. Linkage is an exception to Mendel's second law of independent assortment because closely linked loci are not inherited independently within families.

locus (plural, loci) The site of a specific gene on a chromosome. Latin for "place."

LOD score Log of the odds, a statistical term that indicates whether two loci are linked or unlinked. A LOD score of +3 or higher is commonly accepted as showing linkage and a score of −2 excludes linkage.

map unit See *centimorgan*.

mapping Linkage of DNA markers to a chromosome and to specific regions of chromosomes.

meiosis Cell division that occurs during gamete formation and results in halving the number of chromosomes so that each gamete contains only one member of each chromosome pair.

messenger RNA (mRNA) Processed RNA that leaves the nucleus of the cell and serves as a template for protein synthesis in the cell body.

mitosis Cell division that occurs in somatic cells in which a cell duplicates itself and its DNA.

model fitting In quantitative genetics, a method to test the goodness of fit between a model of genetic and environmental relatedness against observed data. Different models can be compared, and the best-fitting model is used to estimate genetic and environmental parameters.

molecular genetics The investigation of the effects of specific genes at the DNA level. In contrast to quantitative genetics, which investigates genetic and environmental components of variance.

monozygotic (MZ) Identical twins; literally, "one zygote."

morbidity risk estimate An incidence figure that is an estimate of the risk of being affected.

multiple-gene trait See *polygenic trait*.

multivariate genetic analysis Quantitative genetic analysis of the covariance between traits.

mutation A heritable change in DNA base pair sequences.

nonadditive genetic variance Individual differences due to the effects of alleles (dominance) or loci (epistasis) that interact with other alleles or loci. In contrast to additive genetic variance.

nondisjunction Uneven division of members of a chromosome pair during meiosis.

nonshared environment Environmental influences that contribute to differences between family members.

nucleus The part of the cell that contains chromosomes.

pedigree A family tree. Diagram depicting the genealogical history of a family, especially showing the inheritance of a particular condition in the family members.

phenotype The appearance of an individual that results from genotype and environment.

pleiotropy Multiple effects of a gene.

polygenic trait A trait influenced by many genes.

polymerase chain reaction (PCR) A method to amplify a particular DNA sequence.

polymorphism A locus with two or more alleles. Latin for "multiple forms."

population genetics The study of allelic and genotypic frequencies in populations and forces that change these frequencies, such as natural selection.

premutation Production of eggs or sperm with an unstable expanded number of repeats (up to 200 repeats for fragile X).

proband The index case from whom other family members are identified.

qualitative disorder An either-or trait, usually a diagnosis.

quantitative dimension Psychological and physical traits that are continuously distributed within a population, for example, general cognitive ability, height, and blood pressure.

quantitative genetics A theory of multiple-gene influences that, together with environmental variation, result in quantitative (continuous) distributions of phenotypes. Quantitative genetic methods, such as the twin and adoption methods for human analysis and inbred strain and selection methods for nonhuman analysis, estimate genetic and environmental contributions to phenotypic variance in a population.

quantitative trait loci (QTL) Genes of various effect sizes in multiple-gene systems that contribute to quantitative (continuous) variation in a phenotype.

recessive Alleles that produce a particular phenotype only when present in the homozygous state.

recombinant inbred strains Inbred strains derived from brother-sister matings from an initial cross of two inbred progenitor strains. Called *recombinant* because, in the F$_2$ and subsequent generations, chromosomes from the progenitor strains recombine and exchange parts. Used to map genes.

recombination During meiosis, chromosomes exchange parts by crossing over of chromatids.

restriction enzyme Recognizes specific short DNA sequences and cuts DNA at that site.

restriction fragment length polymorphism (RFLP) Variation in the length of DNA fragments generated after DNA is digested with a particular restriction enzyme. Caused by presence or absence of a particular sequence of DNA (restriction site) recognized by the restriction enzyme which cuts the DNA at that site.

second-degree relative See *genetic relatedness.*

segregation The process by which two alleles at a locus, one from each parent, separate during heredity. This is Mendel's law of segregation, or his first law of heredity.

selection study Breeding for a phenotype over several generations by selecting parents with high scores on the phenotype, mating them, and assessing their offspring to determine the response to selection. Bidirectional selection studies also select in the other direction, that is, for low scores.

selective placement In adoption studies, a method in which children are placed into adoptive families in which adoptive parents are similar to the children's biological parents.

sex chromosome See *autosome.*

sex-linked trait See *X-linked trait.*

shared environment Environmental factors responsible for resemblance between family members.

simple-sequence repeats (SSR) DNA markers that consist of two, three, or four DNA bases that repeat several times and are distributed throughout the genome for unknown reasons. The best studied is the two-base (dinucleotide) repeat, CA (cytosine followed by adenine).

sociobiology An extension of evolutionary theory that focuses on inclusive fitness and kin selection.

somatic cells All cells in the body except gametes.

synteny Loci on the same chromosome. Synteny homology refers to similar ordering of loci in chromosomal regions in different species.

third-degree relative See *genetic relatedness.*

transcription The synthesis of an RNA molecule from DNA in the cell nucleus.

transgenic Containing foreign DNA. For example, gene targeting can be used to replace a gene with a nonfunctional substitute in order to knock out the gene's functioning.

translation Assembly of amino acids into peptide chains on the basis of information encoded in messenger RNA. Occurs on ribosomes in the cell cytoplasm.

triplet code See *codon.*

triplet repeat See *expanded triplet repeat.*

trisomy Having three copies of a particular chromosome due to nondisjunction.

twin study Comparing the resemblance of identical and fraternal twins to estimate genetic and environmental components of variance.

variable expression A single genetic effect may result in variable manifestations in different individuals.

X-linked trait A phenotype controlled by a locus on the X chromosome.

zygote The cell, or fertilized egg. resulting from the union of a sperm and an egg.

REFERENCES

Agrawal, N., Sinha, S. N., & Jensen, A. R. (1984). Effects of inbreeding on Raven matrices. *Behavior Genetics, 14*, 579–585.

Ainsworth, M. D. S., Blehar, M. C., Waters, E., & Wall, S. (1978). *Patterns of attachment: A psychological study of the Strange Situation*. Hillsdale, NJ: Erlbaum.

Alarcón, M., DeFries, J. C., Light, J. G., & Pennington, B. F. (in press). A twin study of mathematics disability. *Journal of Learning Disabilities*.

Alexander, R. C., Coggiano, M., Daniel, D. G., et al. (1990). HLA antigens in schizophrenia. *Psychiatry Research, 31*, 221–233.

Allen, G. (1975). *Life science in the twentieth century*. New York: Wiley.

Allen, M. G. (1976). Twin studies of affective illness. *Archives of General Psychiatry, 33*, 1476–1478.

Allgulander, C., Nowak, J., & Rice, J. P. (1991). Psychopathology and treatment of 30,344 twins in Sweden. II. Heritability estimates of psychiatric diagnosis and treatment in 12,884 twin pairs. *Acta Psychiatrica Scandinavica, 83*, 12–15.

Anderson, L. T., & Ernst, M. (1994). Self-injury in Lesch-Nyhan disease. *Journal of Autism and Developmental Disorders, 24*, 67–81.

Andrews, G., Morris-Yates, A., Howie, P., & Martin, N. (1991). Genetic factors in stuttering confirmed. *Archives of General Psychiatry, 48*, 1034–1035.

Andrews, G., Stewart, G., Allen, R., & Henderson, A. S. (1990). The genetics of six neurotic disorders: A twin study. *Journal of Affective Disorders, 19*, 23–29.

Angleitner, A., Riemann, R., & Strelau, J. (1995). A study of twins using the self-report and peer-report NEO-FFI scales. Paper presented at the VIIth meeting of the International Society for the Study of Individual Differences, July 15–19, Warsaw, Poland.

Antonarakis, S. E., Blouin, J.-L., Pulver, A. E., Wolyniec, P., Lasseter, V. K., Nestadt, G., Kasch, L., Babb, R., Kazazian, H. H., Dombroski, B., et al. (1995). Correspondence. *Nature Genetics, 11*, 235–236.

Arvey, R. D., Bouchard, T. J., Segal, N. L., & Abraham, L. M. (1989). Job satisfaction: Environmental and genetic components. *Journal of Applied Psychology, 74*, 187–192.

Ashton, G. C. (1986). Blood polymorphisms and cognitive abilities. *Behavior Genetics, 16*, 517–529.

Bailey, A., Le Couteur, A., Gottesman, I. I., Bolton, P., Simonoff, E., Yuzda, E., & Rutter, M. (1995). Autism as a strongly genetic disorder: Evidence from a British twin study. *Psychological Medicine, 25*, 63–77.

Bailey, A., Phillips, W., & Rutter, M. (1996). Autism: Towards an integration of clinical, genetic, neuropsychological, and neurobiological perspectives. *Journal of Child Psychology and Psychiatry, 37*, 89–126.

Bailey, J. M. (1995). Sexual orientation revolution. *Nature Genetics, 1*, 354–355.

Bailey, J. M., & Pillard, R. C. (1991). A genetic study of male sexual orientation. *Archives of General Psychiatry, 48*, 1089–1096.

Bailey, J. M., Pillard, R. C., Neale, M. C., & Agyei, Y. (1993). Heritable factors influence sexual orientation in women. *Archives of General Psychiatry, 50*, 217–223.

Baker, L., Vernon, P. A., & Ho, H.-Z. (1991). The genetic correlation between intelligence and speed of information processing. *Behavior Genetics, 21*, 351–368.

Bakwin, H. (1971). Enuresis in twins. *American Journal of Diseases of Childhood, 21*, 222–225.

Bakwin, H. (1973). Reading disability in twins. *Developmental Medicine and Child Neurology, 15*, 184–187.

Baltes, P. B. (1993). The aging mind: Potential and limits. *Gerontologist, 33*, 580–594.

Barkley, R. A. (1990). *Attention-deficit hyperactivity disorder: A handbook for diagnosis and treatment*. New York: Guilford.

Barlow, D. P. (1995). Gametic imprinting in mammals. *Science, 270*, 1610–1613.

Baron, M., Freimer, N. F., Risch, N., Lerer, B., Alexander, J. R., Straub, R. E., Asokan, S., Das, K., Petersen, A., Amos, J., et al. (1993). Diminished support for linkage between manic depressive illness and X-chromosome markers in three Israeli pedigrees. *Nature Genetics, 3*, 49–55.

Baron, M., Gruen, R., Asnis, L., & Lord, S. (1985a). Familial transmission of schizo-typal and borderline personality disorders. *American Journal of Psychiatry, 142*, 927–934.

Baron, M., Gruen, R., Rainer, J. D., Kane, J., Asnis, L., & Lord, S. (1985b). A family study of schizophrenic and normal control probands: Implications for the spectrum concept of schizophrenia. *American Journal of Psychiatry, 142*, 447–455.

Bashi, J. (1977). Effects of inbreeding on cognitive performance. *Nature, 266*, 440–442.

Bellugi, U., Wang, P. P., & Jernigan, T. L. (1994). Williams syndrome: An unusual neuropsychological profile. In S. H. Broman & J. Grafman (Eds.), *Atypical cognitive deficits in developmental disorders* (pp. 22–83). Hillsdale, NJ: Erlbaum.

Bender, B. G., Linden, M. G., & Robinson, A. (1993). Neuropsychological impairment in 42 adolescents with sex chromosome abnormalities. *American Journal of Medical Genetics (Neuropsychiatric Genetics), 48*, 169–173.

Benjamin, J., Li, L., Patterson, C., Greenberg, B. D., Murphy, D. L., & Hamer, D. H. (1996). Population and familial association between the D4 dopamine receptor gene and measures of novelty seeking. *Nature Genetics, 12*, 81–84.

Bergeman, C. S. (in press). *Aging differently: Genetic and environmental influences on development in later life*. Newbury Park, CA: Sage.

Bergeman, C. S., Chipuer, H. M., Plomin, R., Pedersen, N. L., McClearn, G. E., Nesselroade, J. R., Costa, P. T., Jr., & McCrae, R. R. (1993). Genetic and environmental effects on openness to experience, agreeableness, and conscientiousness: An adoption/twin study. *Journal of Personality, 61*, 159–179.

Bergeman, C. S., Neiderhiser, J. M., Pedersen, N. L., & Plomin, R. (1996). *Genetic and environmental influences on social support in later life: A longitudinal analysis*. Manuscript submitted for publication.

Bergeman, C. S., & Plomin, R. (1988). Parental mediators of the genetic relationship between home environment and infant mental development. *British Journal of Developmental Psychology, 6*, 11–19.

Bergeman, C. S., & Plomin, R. (in press). Behavioral genetics. In J. Birren (Ed.), *Encyclopedia of aging* (Vol. 1). Orlando, FL: Academic Press.

Bergeman, C. S., Plomin, R., McClearn, G. E., Pedersen, N. L., & Friberg, L. (1988). Genotype-environment interaction in personality development: Identical twins reared apart. *Psychology and Aging, 3*, 399–406.

Bergeman, C. S., Plomin, R., Pedersen, N. L., & McClearn, G. E. (1991). Genetic mediation of the relationship between social support and psychological well-being. *Psychology and Aging, 6*, 640–646.

Bergeman, C. S., Plomin, R., Pedersen, N. L., McClearn, G. E., & Nesselroade, J. R. (1990). Genetic and environmental influences on social support: The Swedish Adoption/Twin Study of Aging (SATSA). *Journals of Gerontology: Psychological Sciences, 45*, P101–P106.

Berman, S. M., & Noble, E. P. (1995). Reduced visuospatial performance in children with the D2 dopamine receptor A1 allele. *Behavior Genetics, 25*, 45–58.

Bertelsen, A. (1985). Controversies and consistencies in psychiatric genetics. *Acta Psychiatrica Scandinavica, 71*, 61–75.

Bertelsen, A., Harvald, B., & Hauge, M. (1977). A Danish twin study of manic-depressive disorders. *British Journal of Psychiatry, 130*, 330–351.

Bessman, S. P., Williamson, M. L., & Koch, R. (1978). Diet, genetics, and mental retardation interaction between phenylketonuric heterozygous mother and fetus to produce nonspecific diminution of IQ: Evidence in support of the justification hypothesis. *Proceedings of the National Academy of Sciences, 78*, 1562–1566.

Biederman, J., Munir, K., & Knee, D. (1987). Conduct and oppositional disorder in clinically referred children with attention deficit disorder: A controlled family study. *Journal of the American Academy of Child and Adolescent Psychiatry, 26*, 724–727.

Biederman, J., Munir, K., Knee, D., Habelow, W., Armentano, M., Autor, S., Hoge, S. K., & Waternaux, C. (1986). A family study of patients with attention deficit disorder and normal controls. *Journal of Psychiatric Research, 20*, 263–274.

Bishop, D. V. M., North, T., & Donlan, C. (1995). Genetic basis of specific language impairment: Evidence from a twin study. *Developmental Medicine and Child Neurology, 37*, 56–71.

Bishop, J. E., & Waldholz, M. (1990). *Genome*. New York: Simon & Schuster.

Blazer, D. G., Kessler, R. C., McGonagle, K., & Swartz, M. S. (1994). The prevalence and distribution of major depression in a national community sample: The national comorbidity survey. *American Journal of Psychiatry, 151*, 979–986.

Bloom, F. E., & Kupfer, D. J. (Eds.). (1995). *Psychopharmacology: A fourth generation of progress*. New York: Raven Press.

Bock, G. R., & Goode, J. A. (Eds.). (1996). *Genetics of criminal and antisocial behaviour*. Chichester, UK: Wiley.

Bohman, M. (1996). Predisposition to criminality: Swedish adoption studies in retrospect. In G. R. Bock & J. A. Goode (Eds.), *Genetics of criminal and antisocial behaviour* (pp. 99–114). Chichester, UK: Wiley.

Bohman, M., Cloninger, C. R., Sigvardsson, S., & von Knorring, A. (1982). Predisposition to petty criminality in Swedish adoptees. I. Genetic and environmental heterogeneity. *Archives of General Psychiatry, 39*, 1233–1241.

Bohman, M., Cloninger, C. R., von Knorring, A., & Sigvardsson, S. (1984). An adoption study of somatoform disorders: III. Cross-fostering analysis and genetic relationship to alcoholism and criminality. *Archives of General Psychiatry, 41*, 872–878.

Bolton, P., Macdonald, H., Pickles, A., Rios, P., Goode, S., Crowson, M., Bailey, A., & Rutter, M. (1994). A case-control family history study of autism. *Journal of Child Psychology and Psychiatry, 35*, 877–900.

Böök, J. A. (1957). Genetical investigation in a north Swedish population: The offspring of first-cousin marriages. *Annals of Human Genetics, 21*, 191–221.

Boomsma, D. I., Orlebeke, J. F., Martin, N. G., Frants, R. R., & Clark, P. (1991). Alpha-1-antitrypsin and blood pressure. *Lancet, 337*, 1547.

Boomsma, D. I., & Somsen, R. J. M. (1991). Reaction times measured in a choice reaction time and a double task condition: A small twin study. *Personality and Individual Differences, 11*, 141–146.

Bouchard, T. J., Jr., Lykken, D. T., McGue, M., Segal, N. L., & Tellegen, A. (1990a). Sources of human psychological differences: The Minnesota Study of Twins Reared Apart. *Science, 250*, 223–228.

Bouchard, T. J., Jr., Lykken, D. T., Tellegen, A., & McGue, M. (in press). Genes, drives, environment, and experience: EPD theory—Revised. In C. P. Benbow & D. Lubinski (Eds.), *From psychometrics to giftedness: Essays in honor of Julian C. Stanley*. Baltimore: Johns Hopkins University Press.

Bouchard, T. J., Jr., & McGue, M. (1981). Familial studies of intelligence: A review. *Science, 212*, 1055–1059.

Bouchard, T. J., Jr., & Propping, P. (Eds.). (1993). *Twins as a tool of behavioral genetics*. New York: Wiley.

Bouchard, C., Tremblay, A., Despres, J., Nadeau, A., Lupien, P. J., Theriault, G., Dussault, J., Moorjani, S., Pinault, S., & Fournier, G. (1990b). The response to long-term overfeeding in identical twins. *New England Journal of Medicine, 322*, 1477–1482.

Bovet, D. (1977). Strain differences in learning in the mouse. In A. Oliverio (Ed.), *Genetics, environment and intelligence* (pp. 79–92). Amsterdam: North-Holland.

Bovet, D., Bovet-Nitti, F., & Oliverio, A. (1969). Genetic aspects of learning and memory in mice. *Science, 163*, 139–149.

Brandenburg, N. A., Friedman, R. M., & Silver, S. E. (1990). The epidemiology of childhood psychiatric disorders: Prevalence findings from recent studies. *Journal of the American Academy of Child and Adolescent Psychiatry, 29*, 76–83.

Bratko, D. (in press). Twin study of verbal and spatial abilities. *Personality and Individual Differences*.

Braungart, J. M. (1994). Genetic influence on "environmental" measures. In J. C. DeFries, R. Plomin, & D. W. Fulker (Eds.), *Nature and nurture during middle childhood* (pp. 233–248). Cambridge, MA: Blackwell.

Braungart, J. M., Fulker, D. W., & Plomin, R. (1992b). Genetic influence of the home environment during infancy: A sibling adoption study of the HOME. *Developmental Psychology, 28*, 1048–1055.

Braungart, J. M., Plomin, R., DeFries, J. C., & Fulker, D. W. (1992a). Genetic influence on tester-rated infant temperament as assessed by Bayley's Infant Behavior Record: Nonadoptive and adoptive siblings and twins. *Developmental Psychology, 28*, 40–47.

Bray, G. A. (1986). Effects of obesity on health and happiness. In K. D. Brownell & J. P. Foreyt (Eds.), *Handbook of eating disorders: Physiology, psychology, and treatment of obesity, anorexia, and bulimia* (pp. 1–44). New York: Basic Books.

Bregman, J. D., & Hodapp, R. M. (1991). Current developments in the understanding of mental retardation. *Journal of the American Academy of Child and Adolescent Psychiatry, 30*, 707–719.

Breitner, J. C., Gatz, M., Bergem, A. L., Christian, J. C., Mortimer, J. A., McClearn, G. E., Heston, L. L., Welsh, K. A., Anthony, J. C., Folstein, M. F., et al. (1993). Use of twin cohorts for research in Alzheimer's disease. *Neurology, 43*, 261–267.

Brennan, P. A., Mednick, S. A., & Jacobsen, B. (1996). Assessing the role of genetics in crime using adoption cohorts. In G. R. Bock & J. A. Goode (Eds.), *Genetics of criminal and antisocial behaviour* (pp. 115–128). Chichester, UK: Wiley.

Broadhurst, P. L. (1978). *Drugs and the inheritance of behavior*. New York: Plenum.

Brody, N. (1992). *Intelligence* (2nd ed.). New York: Academic Press.

Brooks, A., Fulker, D. W., & DeFries, J. C. (1990). Reading performance and general cognitive ability: A multivariate genetic analysis of twin data. *Personality and Individual Differences, 11*, 141–146.

Brunner, H. G. (1996). MAOA deficiency and abnormal behaviour: Perspectives on an association. In G. R. Bock & J. A. Goode (Eds.), *Genetics of criminal and antisocial behaviour* (pp. 155–164). Chichester, UK: Wiley.

Brunner, H. G., Nelen, M., Breakefield, X. O., Ropers, H. H., & von Ooost, B. A. (1993). Abnormal behavior associated with a point mutation in the structural gene for monoamine oxidase A. *Science, 26,* 578–580.

Bruun, K., Markkanen, T., & Partanen, J. (1966). *Inheritance of drinking behavior, a study of adult twins.* Helsinki, Finland: Finnish Foundation for Alcohol Research.

Burke, K. C., Burke, J. D., Roe, D. S., & Regier, D. A. (1991). Comparing age at onset of major depression and other psychiatric disorders by birth cohorts in five U.S. community populations. *Archives of General Psychiatry, 48,* 789–795.

Burks, B. (1928). The relative influence of nature and nurture upon mental development: A comparative study of foster parent–foster child resemblance and true parent–true child resemblance. *Yearbook of the National Society for the Study of Education, Part 1, 27,* 219–316.

Burt, C. (1966). The genetic determination of differences in intelligence: A study of monozygotic twins reared together and apart. *British Journal of Psychology, 57,* 137–153.

Buss, A. H., & Plomin, R. (1984). *Temperament: Early developing personality traits.* Hillsdale, NJ: Erlbaum.

Buss, D. M. (1991). Evolutionary personality psychology. *Annual Review of Psychology, 42,* 459–491.

Buss, D. M. (1994). *The evolution of desire: Strategies of human mating.* New York: Basic Books.

Cadoret, R. J., Cain, C. A., & Crowe, R. R. (1983). Evidence for gene-environment interaction in the development of adolescent antisocial behavior. *Behavior Genetics, 13,* 301–310.

Cadoret, R. J., O'Gorman, T. W., Heywood, E., & Troughton, E. (1985a). Genetic and environmental factors in major depression. *Journal of Affective Disorders, 9,* 155–164.

Cadoret, R. J., O'Gorman, T. W., Troughton, E., & Heywood, E. (1985b). Alcoholism and antisocial personality: Interrelationships, genetic and environmental factors. *Archives of General Psychiatry, 42,* 161–167.

Cadoret, R. J., & Stewart, M. (1991). An adoption study of attention deficit/ hyperactivity/aggression and their relationship to adult antisocial behavior. *Comprehensive Psychiatry, 32,* 73–82.

Cadoret, R. J., Yates, W. R., Troughton, E., Woodworth, G., & Stewart, M. A. (1995). Gene-environment interaction in the genesis of aggressivity and conduct disorders. *Archives of General Psychiatry, 52,* 916–924.

Caldwell, B. M., & Bradley, R. H. (1978). *Home Observation for Measurement of the Environment.* Little Rock: University of Arkansas.

Cambien, F., Poirier, O., Lecerf, L., Evans, A., Cambou, J., Arveiler, D., Luc, G., Bard, J. M., Bara, L., Ricard, S., Tiret, L., Amouyel, P., Alhenc-Gelas, F., & Soubrier, F. (1992). Deletion polymorphism in the gene coding for angiotensin-converting enzyme is a potent risk factor for myocardial infarction. *Nature, 359,* 641–644.

Cannon, T. D., Mednick, S. A., Parnas, J., Schulsinger, F., Praestholm, J., & Vestergaard, A. (1993). Developmental brain abnormalities in the offspring of schizophrenic mothers: I. Contributions of genetic and perinatal factors. *Archives of General Psychiatry, 50,* 551–564.

Canter, S. (1973). Personality traits in twins. In G. Claridge, S. Canter, & W. I. Hume (Eds.), *Personality differences and biological variations* (pp. 21–51). New York: Pergamon Press.

Cantwell, D. P. (1975). Genetic studies of hyperactive children: Psychiatric illness in biological and adopting parents. In R. R. Fieve, D. Rosenthal, & H. Brill (Eds.), *Genetic research in psychiatry* (pp. 273–280). Baltimore: Johns Hopkins University Press.

Capecchi, M. R. (1994). Targeted gene replacement. *Scientific American*, March, 52–59.

Capron, C., & Duyme, M. (1989). Assessment of effects of socio-economic status on IQ in a full cross-fostering study. *Nature, 340*, 552–553.

Capron, C., & Duyme, M. (in press). Effect of socio-economic status of biological and adoptive parents on WISC-R subtest scores of their French adopted children. *Intelligence*.

Cardon, L. R. (1994a). Specific cognitive abilities. In J. C. DeFries, R. Plomin, & D. W. Fulker (Eds.), *Nature and nurture during middle childhood* (pp. 57–76), Cambridge, MA: Blackwell.

Cardon, L. R. (1994b). Height, weight, and obesity. In J. C. DeFries, R. Plomin, & D. W. Fulker (Eds.), *Nature and nurture during middle childhood* (pp. 165–172). Cambridge, MA: Blackwell.

Cardon, L. R. (1995). Genetic influences on body mass index in early childhood. In J. R. Turner, L. R. Cardon, & J. K. Hewitt (Eds.), *Behavior genetic approaches in behavioral medicine* (pp. 133–143). New York: Plenum.

Cardon, L. R., & Fulker, D. W. (1993). Genetics of specific cognitive abilities. In R. Plomin & G. E. McClearn (Eds.), *Nature, nurture, and psychology* (pp. 99–120). Washington, DC: American Psychological Association.

Cardon, L. R., Smith, S. D., Fulker, D. W., Kimberling, W. J., Pennington, B. F., & DeFries, J. C. (1994). Quantitative trait locus for reading disability on chromosome 6. *Science, 266*, 276–279.

Carey, G. (1986). Sibling imitation and contrast effects. *Behavior Genetics, 16*, 319–341.

Carey, G. (1992). Twin imitation for antisocial behavior: Implications for genetic and family environment research. *Journal of Abnormal Psychology, 101*, 18–25.

Carey, G., & DiLalla, D. L. (1994). Personality and psychopathology: Genetic perspectives. *Journal of Abnormal Psychology, 103*, 32–43.

Carey, G., & Gottesman, I. I. (1981). Twin and family studies of anxiety, phobic and obsessive disorders. In D. F. Klein & J. G. Rabkin (Eds.), *Anxiety: New research and changing concepts* (pp. 117–136). New York: Raven Press.

Carmelli, D., Swan, G. E., & Cardon, L. R. (1995). Genetic mediation in the relationship of education to cognitive function in older people. *Psychology and Aging, 10*, 48–53.

Carr, J. (1995). *Down's syndrome children growing up: A longitudinal perspective.* Cambridge: Cambridge University Press.

Carroll, J. B. (1993). *Human cognitive abilities.* New York: Cambridge University Press.

Carter, A. S., Pauls, D. L., & Leckman, J. F. (1995). The development of obsessionality: Continuities and discontinuities. In D. Cicchetti & D. J. Cohen (Eds.), *Developmental psychopathology: Vol. 2. Risk, disorder, and adaptation* (pp. 609–632). New York: Wiley.

Caspi, A., & Moffitt, T. E. (1995). The continuity of maladaptive behavior: From description to understanding of antisocial behavior. In D. Cicchetti & D. J. Cohen (Eds.), *Developmental psychopathology: Vol. 2. Risk, disorder, and adaptation* (pp. 472–511). New York: Wiley.

Casto, S. D., DeFries, J. C., & Fulker, D. W. (1995). Multivariate genetic analysis of Wechsler Intelligence Scale for Children—Revised (WISC-R) factors. *Behavior Genetics, 25*, 25–32.

Cattell, R. B. (1982). *The inheritance of personality and ability*. New York: Academic Press.

Chawla, S. (1993). Demographic aging and development. *Generations, 17*, 20–23.

Cherny, S. S., Fulker, D. W., Emde, R. N., Robinson, J., Corley, R. P., Reznick, J. S., Plomin, R., & DeFries, J. C. (1994). Continuity and change in infant shyness from 14 to 20 months. *Behavior Genetics, 24*, 365–379.

Chipuer, H. M., & Plomin, R. (1992). Using siblings to identify shared and nonshared HOME items. *British Journal of Developmental Psychology, 10*, 165–178.

Chipuer, H. M., Plomin, R., Pedersen, N. L., McClearn, G. E., & Nesselroade, J. R. (1992). Genetic influence on family environment: The role of personality. *Developmental Psychology, 29*, 110–118.

Chipuer, H. M., Rovine, M. J., & Plomin, R. (1990). LISREL modeling: Genetic and environmental influences on IQ revisited. *Intelligence, 14*, 11–29.

Christiansen, K. O. (1977). A preliminary study of criminality among twins. In S. A. Mednick & K. O. Christiansen (Eds.), *Biosocial bases of criminal behavior* (pp. 89–108). New York: Gardner.

Chua, S. C., Jr., Chung, W. K., Wu-Peng, X. S., Zhang, Y., Liu, S.-M., Tartaglia, L., & Leibel, R. L. (1996). Phenotypes of mouse *diabetes* and Rat *fatty* due to mutations in the OB (leptin) receptor. *Science, 271*, 994–996.

Cicchetti, D., & Beeghly, M. (Eds.). (1990). *Children with Down syndrome: A developmental perspective*. Cambridge: Cambridge University Press.

Claridge, G., & Hewitt, J. K. (1987). A biometrical study of schizotypy in a normal population. *Personality and Individual Differences, 8*, 303–312.

Clarke, B. (1975). The causes of biological diversity. *Scientific American, 233*, 50–60.

Clementz, B. A., McDowell, J. E., & Zisook, S. (1994). Saccadic system functioning among schizophrenic patients and their first-degree biological relatives. *Journal of Abnormal Psychology, 103*, 277–287.

Clifford, C. A., Murray, R. M., & Fulker, D. W. (1984). Genetic and environmental influences on obsessional traits and symptoms. *Psychological Medicine, 14*, 791–800.

Cloninger, C. R. (1987a). A systematic method for clinical description and classification of personality variants. A proposal. *Archives of General Psychiatry, 44*, 573–588.

Cloninger, C. R. (1987b). Neurogenetic adaptive mechanisms in alcoholism. *Science, 236*, 410–416.

Cloninger, C. R., Bohman, M., & Sigvardsson, S. (1981). Inheritance of alcohol abuse: Cross-fostering analysis of adopted men. *Archives of General Psychiatry, 38*, 861–868.

Cloninger, C. R., Sigvardsson, S., Bohman, M., & von Knorring, A. L. (1982). Predisposition to petty criminality in Swedish adoptees: II. Cross fostering analysis of gene-environment interaction. *Archives of General Psychiatry, 39*, 1242–1247.

Cloninger, C. R., Sigvardsson, S., von Knorring, A., & Bohman, M. (1984). An adoption study of somatoform disorders: II. Identification of two discrete somatoform disorders. *Archives of General Psychiatry, 41*, 863–871.

Cohen, B. H. (1964). Family patterns of mortality and life span. *Quarterly Review of Biology, 39*, 130–181.

Cohen, P., Cohen, J., Kasen, S., Velez, C. N., Hartmark, C., Johnson, J., Rjas, M., Brook, J., & Streuning, E. L. (1993). An epidemiological study of disorders in late childhood and adolescence. I. Age- and gender-specific prevalence. *Journal of Child Psychology and Psychiatry, 34*, 851–867.

Colletto, G. M., Cardon, L. R., & Fulker, D. W. (1993). A genetic and environmental time series analysis of blood pressure in male twins. *Genetic Epidemiology, 10*, 533–538.

Collins, A. C. (1981). A review of research using short-sleep and long-sleep mice. In G. E. McClearn, R. A. Deitrich, & V. G. Erwin (Eds.), Development of animal models as pharmacogenetic tools. *USDHHS-NIAAA Research Monograph No. 6* (pp. 161–170). Washington, DC: U.S. Government Printing Office.

Coon, H., Fulker, D. W., DeFries, J. C., & Plomin, R. (1990). Home environment and cognitive ability of 7-year-old children in the Colorado Adoption Project: Genetic and environmental etiologies. *Developmental Psychology, 26*, 459–468.

Cooper, R. M., & Zubek, J. P. (1958). Effects of enriched and restricted early environments on the learning ability of bright and dull rats. *Canadian Journal of Psychology, 12*, 159–164.

Corder, E. H., Saunders, A. M., Risch, N. J., Strittmatter, W. J., Shmechel, D. E., Gaskell, P. C., Jr., Rimmler, J. B., Locke, P. A., Conneally, P. M., Schmader, K. E., et al. (1994). Protective effect of apolipoprotein E type 2 allele for late onset Alzheimer disease. *Nature Genetics, 7*, 180–184.

Corder, E. H., Saunders. A. M., Strittmatter, W. J., Shmechel, D. E., Gaskell, P. C., Small, G. W., Roses, A. D., Haines, J. L., & Pericak-Vance, M. A. (1993). Gene dose of apolipoprotein E type 4 allele and the risk of Alzheimer's disease in late onset families. *Science, 261*, 921–923.

Costa, P. T., & McCrae, R. R. (1994). Stability and change in personality from adolescence through adulthood. In C. F. Halverson, Jr., G. A. Kohnstamm, & R. P. Martin (Eds.), *The developing structure of temperament and personality from infancy to adulthood* (pp. 139–150). Hillsdale, NJ: Erlbaum.

Cotton, N. S. (1979). The familial incidence of alcoholism. *Journal of Studies of Alcohol, 40*, 89–116.

Crabbe, J. C., Belknap, J. K., & Buck, K. J. (1994). Genetic animal models of alcohol and drug abuse. *Science, 264*, 1715–1723.

Crabbe, J. C., & Harris, R. A. (Eds.). (1991). *The genetic basis of alcohol and drug actions.* New York: Plenum.

Crabbe, J. C., Kosubud, A., Young, E. R., Tam, B. R., & McSwigan, J. D. (1985). Bidirectional selection for susceptibility to ethanol withdrawal seizures in *Mus musculus. Behavior Genetics, 15*, 521–536.

Crow, T. J. (1985). The two syndrome concept: Origins and current states. *Schizophrenia Bulletin, 11*, 471–486.

Crow, T. J. (1994). Con: The demise of the Kraepelinian binary system as a prelude to genetic advance. In E. S. Gershon & C. R. Cloninger (Eds.), *Genetic approaches to mental disorders* (pp. 163–192). Washington, DC: American Psychiatric Press.

Crowe, R. R. (1972). The adopted offspring of women criminal offenders: A study of their arrest records. *Archives of General Psychiatry, 27*, 600–603.

Crowe, R. R. (1974). An adoption study of antisocial personality. *Archives of General Psychiatry, 31*, 785–791.

Crowe, R. R. (1994). The Iowa linkage study of panic disorder. In E. S. Gershon & C. R. Cloninger (Eds.), *Genetic approaches to mental disorders* (pp. 291–309). Washington, DC: American Psychiatric Press.

Crowe, R. R., Noyes, R., Jr., Pauls, D. L., & Slyman, D. (1983). A family study of panic disorder. *Archives of General Psychiatry, 40,* 1065–1069.

Cumings, J. L., & Benson, D. F. (1992). *Dementia: A clinical approach* (2nd ed.). Boston: Butterworth.

Daniels, J. K., Owen, M. J., McGuffin, P., Thompson, L., Detterman, D. K., Chorney, M. J., Chorney, K., Smith, D., Skuder, P., Vignetti, S., McClearn, G. E., & Plomin, R. (1994). IQ and variation in the number of fragile X CGG repeats: No association in a normal sample. *Intelligence, 19,* 45–50.

Darwin, C. (1859). *On the origin of species by means of natural selection, or the preservation of favoured races in the struggle for life.* London: John Murray. (New York: Modern Library, 1967.)

Darwin, C. (1871). *The descent of man and selection in relation to sex.* London: John Murray. (New York: Modern Library, 1967.)

Darwin, C. (1896). *Journal of researches into the natural history and geology of the countries visited during the voyage of H. M. S. Beagle round the world under the command of Capt. Fitz Roy, T. N.* New York: Appleton.

Davidson, J. R. T., Hughes, D., Blazer, D. G., & George, L. (1991). Posttraumatic stress disorder in the community: An epidemiological study. *Psychological Medicine, 21,* 713–721.

Davidson, J. R. T., Smith, R. D., & Kudler, H. S. (1989). Familial psychiatric illness in chronic posttraumatic stress disorder. *Comprehensive Psychiatry, 30,* 339–345.

Dawkins, R. (1976). *The selfish gene.* New York: Oxford University Press.

Decker, S. N., & Vandenberg, S. G. (1985). Colorado Twin Study of Reading Disability. In D. B. Gray & J. F. Kavanagh (Eds.), *Biobehavioral measures of dyslexia* (pp. 123–135). Parkton, MD: York Press.

DeFries, J. C., & Alarcón, M. (1996). Genetics of specific reading disability. *Mental Retardation and Developmental Disabilities Research Reviews, 2,* 39–47.

DeFries, J. C., & Fulker, D. W. (1985). Multiple regression analysis of twin data. *Behavior Genetics, 5,* 467–473.

DeFries, J. C., & Fulker, D. W. (1988). Multiple regression analysis of twin data: Etiology of deviant scores versus individual differences. *Acta Geneticae Medicae et Gemellolgiae, 37,* 205–216.

DeFries, J. C., Fulker, D. W., & LaBuda, M. C. (1987). Evidence for a genetic aetiology in reading disability of twins. *Nature, 329,* 537–539.

DeFries, J. C., Gervais, M. C., & Thomas, E. A. (1978). Response to 30 generations of selection for open-field activity in laboratory mice. *Behavior Genetics, 8,* 3–13.

DeFries, J. C., & Gillis, J. J. (1993). Genetics of reading disability. In R. Plomin & G. E. McClearn (Eds.), *Nature, nurture, and psychology* (pp. 121–137). Washington, DC: American Psychological Association.

DeFries, J. C., Johnson, R. C., Kuse, A. R., McClearn, G. E., Polovina, J., Vandenberg, S. G., & Wilson, J. R. (1979). Familial resemblance for specific cognitive abilities. *Behavior Genetics, 9,* 23–43.

DeFries, J. C., Plomin, R., & Fulker, D. W. (1994). *Nature and nurture during middle childhood.* Cambridge, MA: Blackwell.

DeFries, J. C., Vandenberg, S. G., & McClearn, G. E. (1976). Genetics of specific cognitive abilities. *Annual Review of Genetics, 10,* 179–207.

DeFries, J. C., Vogler, G. P., & LaBuda, M. C. (1986). Colorado Family Reading Study: An overview. In J. L. Fuller & E. C. Simmel (Eds.), *Perspectives in behavior genetics* (pp. 29–56). Hillsdale, NJ: Erlbaum.

DeLisi, L. E., Mirsky, A. F., Buchsbaum, M. S., van Kammen, D. P., Berman, K. F., Caton, C., Kafka, M. S., Ninan, P. T., Phelps, B. H., Karoum, F., et al. (1984). The Genain quadruplets 25 years later: A diagnostic and biochemical followup. *Psychiatric Research, 13,* 59–76.

Detterman, D. K. (1986). Human intelligence is a complex system of separate processes. In R. J. Sternberg & D. K. Detterman (Eds.), *What is intelligence? Contemporary viewpoints on its nature and measurement* (pp. 57–61). Norwood, NJ: Ablex.

Detterman, D. K. (1990). CAT: Computerized abilities test for research and teaching. *MicroPsych Network, 4,* 51–62.

de Waal, F. (1996). *Good natured: The origins of right and wrong in humans and other animals.* Cambridge, MA: Harvard University Press.

DiLalla, L. F., & Gottesman, I. I. (1989). Heterogeneity of causes for delinquency and criminality: Lifespan perspectives. *Development and Psychopathology, 1,* 339–349.

Dobzhansky, T. (1964). *Heredity and the nature of man.* New York: Harcourt, Brace & World.

Donaldson, M. D. C., Chu, C. E., Cooke, A., Wilson, A., Greene, S. A., & Stephenson, J. B. P. (1994). The Prader-Willi syndrome. *Archives of Disease in Childhood, 70,* 58–63.

Dunn, J., & Plomin, R. (1986). Determinants of maternal behavior toward three-year-old siblings. *British Journal of Developmental Psychology, 4,* 127–137.

Dunn, J., & Plomin, R. (1990). *Separate lives: Why siblings are so different.* New York: Basic Books.

Dutch-Belgian Fragile X Consortium. (1994). Fmr1 knockout mice: A model to study fragile X mental retardation. *Cell, 78,* 23–33.

Dworkin, R. H. (1979). Genetic and environmental influences on person-situation interactions. *Journal of Research in Personality, 13,* 279–293.

Dworkin, R. H., & Lenzenweger, M. F. (1984). Symptoms and the genetics of schizophrenia: Implications for diagnosis. *American Journal of Psychiatry, 14,* 1541–1546.

Dykens, E. M., Hodapp, R. M., & Leckman, J. F. (1994). *Behavior and development in fragile X syndrome.* London: Sage.

Eaves, L. J. (1976). A model for sibling effects in man. *Heredity, 36,* 205–214.

Eaves, L. J., & Eysenck, H. J. (1976). Genetical and environmental components of inconsistency and unrepeatability in twins' responses to a neuroticism questionnaire. *Behavior Genetics, 6,* 145–160.

Eaves, L. J., Eysenck, H. J., & Martin, N. G. (1989). *Genes, culture and personality: An empirical approach.* London: Academic Press.

Eaves, L. J., Kendler, K. S., & Schulz, S. C. (1986). The familial sporadic classification: Its power for the resolution of genetic and environmental etiological factors. *Journal of Psychiatric Research, 20,* 115–130.

Eaves, L. J., Silberg, J. L., Hewitt, J. K., Meyer, J., Rutter, M., Simonoff, E., Neale, M., & Pickles, A. (1993). Genes, personality and psychopathology: A latent class analysis of liability to symptoms of attention-deficit hyperactivity disorder in twins. In R. Plomin & G. E. McClearn (Eds.), *Nature, nurture, and psychology* (pp. 285–303). Washington, DC: American Psychological Association.

Eaves, L. J., Silberg, J. L., Meyer, J. M., Maes, H. H., Simonoff, E., Pickles, A., Rutter, M., Neale, M. C., Reynolds, C. A., Erikson, M. T., et al. (1996). Genetics and developmental psychopathology: 2. The main effects of genes and environment on behavioral problems in the Virginia Twin Study of Adolescent Behavioral Development. Manuscript submitted for publication.

Ebstein, R. P., Novick, O., Umansky, R., Priel, B., Osher, Y., Blaine, D., Bennett, E. R., Nemanov, L., Katz, M., & Belmaker, R. H. (1995). Dopamine D4 receptor (D4DR) exon III polymorphism associated with the human personality trait of novelty seeking. *Nature Genetics, 12,* 78–80.

Egeland, J. A., Gerhard, D. S., Pauls, D. L., Sussex, J. N., & Kidd, K. K. (1987). Bipolar affective disorders linked to DNA markers on chromosome 11. *Nature, 325,* 783–787.

Ehrman, L., & Parsons, P. A. (1981). *The genetics of behavior.* Sunderland, MA: Sinauer Associates.

Ehrman, L., & Seiger, M. B. (1987). Diversity in Hawaiian drosophilids: A tribute to Dr. Hampton L. Carson upon his retirement. *Behavior Genetics, 17,* 537–565.

Eisensmith, R. C., & Woo, S. L. C. (1992). Molecular basis of phenylketonuria and related hyperphenylalaninemias: Mutations and polymorphisms in the human phenylalanine hydroxylase gene. *Human Mutation, 1,* 13–23.

El Abd, S., Turk, J., & Hill, P. (1995). Annotation: Psychological characteristics of Turner syndrome. *Journal of Child Psychology and Psychiatry, 36,* 1109–1125.

Emery, A. E. H. (1993). *Duchenne muscular dystrophy* (2nd ed.). Oxford: Oxford University Press.

Erlenmeyer-Kimling, L. (1972). Gene-environment interactions and the variability of behavior. In L. Ehrman, G. S. Omenn, & E. Caspari (Eds.), *Genetics, environment, and behavior* (pp. 181–208). San Diego: Academic Press.

Erlenmeyer-Kimling, L., & Jarvik, L. F. (1963). Genetics and intelligence: A review. *Science, 142,* 1477–1479.

Erlenmeyer-Kimling, L., Squires-Wheeler, E., Adamo, U. H., Bassett, A. S., Cornblatt, B. A., Kestenbaum, C. J., Rock, D., Roberts, S. A., & Gottesman, I. I. (1995). The New York high-risk project: Psychoses and cluster A personality disorders in off-spring of schizophrenic parents at 23 years of follow-up. *Archives of General Psychiatry, 52,* 857–865.

Eysenck, H. J. (1952). *The scientific study of personality.* London: Routledge & Kegan Paul.

Eysenck, H. J. (1983). A biometrical-genetical analysis of impulsive and sensation-seeking behavior. In M. Zuckerman (Ed.), *Biological bases of sensation seeking, impulsivity, and anxiety* (pp. 1–36). Hillsdale, NJ: Erlbaum.

Fagard, R., Bielen, E., & Amery, A. (1991). Heritability of aerobic power and anaerobic energy generation during exercise. *Journal of Applied Physiology, 70,* 357–362.

Falconer, D. S. (1960). *Introduction to quantitative genetics.* New York: Ronald Press. (London: Longman, 1981.)

Falconer, D. S. (1965). The inheritance of liability to certain diseases estimated from the incidence among relatives. *Annals of Human Genetics, 29,* 51–76.

Falconer, D. S., & Mackay, T. F. C. (1996). *Introduction to quantitative genetics* (4th ed.). Harlow, UK: Longman.

Fantino, E., & Logan, C. A. (1979). *The experimental analysis of behavior.* San Francisco: Freeman.

Faraone, S. V., Biederman, J., Keenan, K., & Tsuang, M. T. (1991). Separation of DSM-III attention deficit disorder and conduct disorder: Evidence from a family-genetic study of American child psychiatric patients. *Psychological Medicine, 21,* 109–121.

Farmer, A. E., McGuffin, P., & Gottesman, I. I. (1987). Twin concordance for DSM-III schizophrenia: Scrutinizing the validity of the definition. *Archives of General Psychiatry, 44,* 634–641.

Felsenfeld, S. (1994). Developmental speech and language disorders. In J. C. DeFries, R. Plomin, & D. W. Fulker (Eds.), *Nature and nurture during middle childhood* (pp. 102–119). Oxford: Blackwell.

Ferner, R. E. (1994). Intellect in neurofibromatosis 1. In S. M. Huson & R. A. C. Hughes (Eds.), *The neurofibromatoses: A pathogenetic and clinical overview* (pp. 233–252). London: Chapman & Hall.

Feskens, E. J. M., Havekes, L. M., Kalmijn, S., de Knijff, P., Launer, L. J., & Kromhout, D. (1994). Apolipoprotein e4 allele and cognitive decline in elderly men. *British Medical Journal, 309,* 1202–1206.

Fichter, M. M., & Noegel, R. (1990). Concordance for bulimia nervosa in twins. *International Journal of Eating Disorders, 9,* 255–263.

Fischer, P. J., & Breakey, W. R. (1991). The epidemiology of alcohol, drug, and mental disorders among homeless persons. *American Psychologist, 46,* 1115–1128.

Fisher, R. A. (1918). The correlation between relatives on the supposition of Mendelian inheritance. *Transactions of the Royal Society of Edinburgh, 52,* 399–433.

Fisher, R. A. (1930). *The genetical theory of natural selection.* Oxford: Clarendon Press.

Fletcher, R. (1990). *The Cyril Burt scandal: Case for the defense.* New York: Macmillan.

Flint, J., Corley, R., DeFries, J. C., Fulker, D. W., Gray, J. A., Miller, S., & Collins, A. C. (1995a). A simple genetic basis for a complex psychological trait in laboratory mice. *Science, 269,* 1432–1435.

Flint, J., Wilkie, A. O. M., Buckle, V. J., Winter, R. M., Holland, A. J., & McDermid, H. E. (1995b). The detection of subtelomeric chromosomal rearrangements in idiopathic mental retardation. *Nature Genetics, 9,* 132–137.

Folstein, S., & Rutter, M. (1977). Genetic influences and infantile autism. *Nature, 265,* 726–728.

Freeman, F. N., Holzinger, K. J., & Mitchell, B. (1928). The influence of environment on the intelligence, school achievement, and conduct of foster children. *Yearbook of the National Society for the Study of Education, 27,* 103–217.

Fulker, D. W. (1979). Nature and nurture: Heredity. In H. J. Eysenck (Ed.), *The structure and measurement of intelligence* (pp. 102–132). New York: Springer-Verlag.

Fulker, D. W., Cherny, S. S., & Cardon, L. R. (1993). Continuity and change in cognitive development. In R. Plomin & G. E. McClearn (Eds.), *Nature, nurture, and psychology* (pp. 77–97). Washington, DC: American Psychological Association.

Fulker, D. W., DeFries, J. C., & Plomin, R. (1988). Genetic influence on general mental ability increases between infancy and middle childhood. *Nature, 336,* 767–769.

Fulker, D. W., Eysenck, S. B. G., & Zuckerman, M. (1980). A genetic and environmental analysis of sensation seeking. *Journal of Research in Personality, 14,* 261–281.

Fuller, J. L. (1983). Sociobiology and behavior genetics. In J. L. Fuller & E. C. Simmel (Eds.), *Behavior genetics: Principles and applications* (pp. 435–477). Hillsdale, NJ: Erlbaum.

Fuller, J. L., & Thompson, W. R. (1960). *Behavior genetics.* New York: Wiley.

Fuller, J. L., & Thompson, W. R. (1978). *Foundations of behavior genetics.* St. Louis: Mosby.

Fyer, A. J., Mannuzza, S., Gallops, M. S., Martin, L. Y., Aaronson, C., Gorman, J. M., Liebowitz, M. R., & Klein, D. F. (1990). Familial transmission of simple phobias and fears. *Archives of General Psychiatry, 47,* 252–256.

Galton, F. (1865). Hereditary talent and character. *Macmillan's Magazine, 12,* 157–166, 318–327.

Galton, F. (1869). *Hereditary genius: An inquiry into its laws and consequences.* London: Macmillan. (Cleveland, OH: World, 1962.)

Galton, F. (1876). The history of twins as a criterion of the relative powers of nature and nurture. *Royal Anthropological Institute of Great Britain and Ireland Journal, 6,* 391–406.

Galton, F. (1883). *Inquiries into human faculty and its development.* London: Macmillan.

Galton, F. (1889). *Natural inheritance.* London: Macmillan.

Gatz, M., Pedersen, N. L., Plomin, R., Nesselroade, J. R., & McClearn, G. E. (1992). The importance of shared genes and shared environments for symptoms of depression in older adults. *Journal of Abnormal Psychology, 10,* 701–708.

Ge, X., Conger, R. D., Cadoret, R.J., Neiderhiser, J. M., Yates, W., Troughton, E., & Stewart, M. A. (1996). The developmental interface between nature and nurture: A mutual influence model of child antisocial behavior and parenting. *Developmental Psychology, 32,* 574–589.

Gelernter, J., Goldman, D., & Risch, N. (1993). The A1 allele at the D_2 dopamine receptor gene and alcoholism: A reappraisal. *Journal of the American Medical Association, 269,* 1673–1677.

Gershon, E. S., & Cloninger, C. R. (Eds.). (1994). *Genetic approaches to mental disorders.* Washington, DC: American Psychiatric Press.

Gholson, B., & Barker, B. (1985). Kuhn, Lakatos, and Laudan: Applications in the history of physics and psychology. *American Psychologist, 40,* 755–769.

Ghosh, S., Palmer, S. M., Rodrigues, N. R., Cordell, H. J., Hearne, C. M., Cornall, R. J., Prins, J.-B., McShane, P., Lathrop, G. M., et al. (1993). Polygenic control of autoimmune diabetes in nonobese diabetic mice. *Nature Genetics, 4,* 404–409.

Gibson, J. B., Harrison, G. A., Clarke, V. A., & Hiorns, R. W. (1973). IQ and ABO blood groups. *Nature, 246,* 498–500.

Gillis, J. J, Gilger, J. W., Pennington, B. F., & DeFries, J. C. (1992). Attention deficit disorder in reading-disabled twins: Evidence for a genetic etiology. *Journal of Abnormal Child Psychology, 20,* 303–315.

Goldberg, L. R. (1990). An alternative description of personality: The big five factor structure. *Journal of Personality and Social Psychology, 59,* 1216–1229.

Goldsmith, H. H. (1983). Genetic influences on personality from infancy to adulthood. *Child Development, 54,* 331–355.

Goldsmith, H. H. (1993). Nature-nurture and the development of personality: Introduction. In R. Plomin & G. E. McClearn (Eds.), *Nature, nurture, and psychology* (pp. 155–160). Washington, DC: American Psychological Association.

Goldsmith, H. H., Buss, A. H., Plomin, R., Rothbart, M. K., Thomas, A., Chess, S., Hinde, R. A., & McCall, R. B. (1987). Roundtable: What is temperament? Four approaches. *Child Development, 58,* 505–529.

Goldsmith, H. H., & Campos, J. J. (1986). Fundamental issues in the study of early temperament: The Denver twin temperament study. In M. E. Lamb, A. L., Brown, & B. Rogoff (Eds.), *Advances in developmental psychology* (pp. 231–283). Hillsdale, NJ: Erlbaum.

Goldsmith, H. H., & Gottesman, I. I. (1981). Origins of variation in behavioral style: A longitudinal study of temperament in young twins. *Child Development, 52,* 91–103.

Goodman, R., & Stevenson, J. (1989). A twin study of hyperactivity. II. The aetiological role of genes, family relationships, and perinatal adversity. *Journal of Child Psychology and Psychiatry, 30,* 691–709.

Goodwin, F. K., & Jamison, K. R. (1990). *Manic-depressive illness.* New York: Oxford University Press.

Gottesman, I. I. (1991). *Schizophrenia genesis: The origins of madness.* New York: Freeman.

Gottesman, I. I., & Bertelsen, A. (1989). Confirming unexpressed genotypes for schizophrenia. *Archives of General Psychiatry, 46,* 867–872.

Gottesman, I. I., & Shields, J. (1972). A polygenic theory of schizophrenia. *International Journal of Mental Health, 1,* 107–115.

Greenspan, R. J. (1995). Understanding the genetic construction of behavior. *Scientific American, 272,* 72–78.

Grice, D. E., Leckman, J. F., Pauls, D. L., Kurlan, R., Kidd, K. K., Pakstis, A. J., Change, F. M., Buxbaum, J. D., Cohen, D. J., & Gelernter, J. (1996). Linkage disequilibrium between an allele at the dopamine D4 receptor locus and Tourette syndrome, by the transmission-disequilibrium test. *American Journal of Human Genetics, 59,* 644–652.

Grigorenko, E. L., Wood, F. B., Meyer, M. S., Hart, L. A., Speed, W. C., Schuster, A., & Pauls, D. L. (1996). Susceptibility loci for distinct components of dyslexia on chromosomes 6 and 15. *American Journal of Human Genetics, 59,* A219 abstract).

Grilo, C. M., & Pogue-Geile, M. F. (1991). The nature of environmental influences on weight and obesity: A behavior genetic analysis. *Psychological Bulletin, 10,* 520–537.

Grove, W. M., Eckert, E. D., Heston, L., Bouchard, T. J., Jr., Segal, N., & Lykken, D. T. (1990). Heritability of substance abuse and antisocial behavior: A study of monozygotic twins reared apart. *Biological Psychiatry, 27,* 1293–1304.

Gusella, J. F., Tanzi, R. E., Anderson, M. A., Hobbs, W., Gibbons, K., Raschtchian, R., Gilliam, T. C., & Wallace, M. R. (1984). DNA markers for nervous system diseases. *Science, 225,* 1320–1326.

Gusella, J. F., Wexler, N. S., Conneally, P. M., Naylor, S. L., Anderson, M. A., Tanzi, R. E., Watkins, P. C., & Ottina, K. (1983). A polymorphic DNA marker genetically linked to Huntington's disease. *Nature, 306,* 234–238.

Guze, S. B. (1993). Genetics of Briquet's syndrome and somatization disorder: A review of family, adoption, and twin studies. *Annals of Clinical Psychiatry, 5,* 225–230.

Guze, S. B., Cloninger, C. R., Martin, R. L., & Clayton, P. J. (1986). A follow-up and family study of Briquet's syndrome. *British Journal of Psychiatry, 149,* 17–23.

Hagerman, R. (1995). Lessons from fragile X syndrome. In G. O'Brien & W. Yule (Eds.), *Behavioural phenotypes* (pp. 59–74). London: MacKeith Press.

Hagerman, R. J., Hull, C. E., Safanda, J. F., Carpenter, I., Staley, L. W., O'Connor, R. A., Seydel, C., Mazzocco, M. M. M., Snow, K., Thibodeau, S. N., et al. (1994). High functioning fragile X males: Demonstration of an unmethylated fully expanded FMR-1 mutation associated with protein expression. *American Journal of Medical Genetics, 51,* 298–308.

Halaas, J. L., Gajiwala, K. S., Maffei, M., Cohen, S. L., Chait, B. T., Rabinowitz, D., Lallone, R. L., Burley, S. K., & Friedman, J. M. (1995). Weight-reducing effects of the plasma protein encoded by the *obese* gene. *Science, 269,* 543–546.

Hall, L. L. (Ed.). (1996). *Genetics and mental illness: Evolving issues for research and society.* New York: Plenum.

Hallgren, B. (1957). Enuresis, a clinical and genetic study. *Acta Psychiatrica et Neurologia Scandinavica Supplement No. 114.*

Hallgren, B. (1960). Nocturnal enuresis in twins. *Acta Psychiatric Scandinavica, 35,* 73–90.

Hamer, D. H., Hu, S., Magnuson, V. L., Hu, N., & Pattatucci, A. M. L. (1993). A linkage between DNA markers on the X chromosome and male sexual orientation. *Science, 261,* 321–327.

Hamilton, D. W. (1968). The genetical theory of social behaviour (I and II). *Journal of Theoretical Biology*, 7, 1–52.

Hardy, J. A., & Allsop, D. (1991). Amyloid deposition as the central event in the aetiology of AD. *TIPS*, 12, 383–388.

Hardy, J. A., & Higgins, G. A. (1992). Alzheimer's disease: The amyloid cascade hypothesis. *Science*, 256, 184–187.

Hardy, J. A., & Hutton, M. (1995). Two new genes for Alzheimer's disease. *Trends in Neurosciences*, 18, 436.

Harper, P. S. (1992). Insurance and genetic testing. *Lancet*, 341, 224–227.

Harrington, R., Rutter, M., & Fombonne, E. (in press). Developmental pathways in depression: Multiple meanings, antecedents and endpoints. *Development and Psychopathology*.

Harris, J. R., Pedersen, N. L., Stacey, C., McClearn, G. E., & Nesselroade, J. R. (1992). Age differences in the etiology of the relationship between life satisfaction and self-rated health. *Journal of Aging and Health*, 4, 349–368.

Harter, S. (1983). Developmental perspectives on the self-system. In E. M. Hetherington (Ed.), *Handbook of child psychology: Socialization, personality, and social development* (Vol. 4, pp. 275–385). New York: Wiley.

Hartl, D. L., & Clark, A. G. (1989). *Principles of population genetics* (2nd ed.). Sunderland, MA: Sinauer Associates.

Hearnshaw, L. S. (1979). *Cyril Burt, psychologist*. Ithaca, NY: Cornell University Press.

Heath, A. C., Jardine, R., & Martin, N. G. (1989). Interactive effects of genotype and social environment on alcohol consumption in female twins. *Journal of Studies on Alcohol*, 50, 38–48.

Heath, A. C., & Madden, P. F. (1995). Genetic influences on smoking behavior. In J. R. Turner, L. R. Cardon, & J. K. Hewitt (Eds.), *Behavior genetic approaches in behavioral medicine* (pp. 45–66). New York: Plenum.

Heath, A. C., & Martin, N. (1993). Genetic models for the natural history of smoking: Evidence for a genetic influence on smoking persistence. *Addictive Behaviors*, 18, 19–34.

Hebebrand, J. (1992). A critical appraisal of X-linked bipolar illness: Evidence for the assumed mode of inheritance is lacking. *British Journal of Psychiatry*, 160, 7–11.

Henderson, A. S., Easteal, S., Jorm, A. F., Mackinnon, A. J., Korten, A. E., Christensen, H., Croft, L., & Jacomb, P. A. (1995). Apolipoprotein E allele ε4, dementia, and cognitive decline in a population sample. *Lancet*, 346, 1387–1390.

Henderson, N. D. (1967). Prior treatment effects on open field behavior of mice—A genetic analysis. *Animal Behaviour*, 15, 365–376.

Henderson, N. D. (1972). Relative effects of early rearing environment on discrimination learning in housemice. *Journal of Comparative and Physiological Psychology*, 72, 505–511.

Herrnstein, R. J., & Murray, C. (1994). *The bell curve: Intelligence and class structure in American life*. New York: Free Press.

Hershberger, S. L., Lichtenstein, P., & Knox, S. S. (1994). Genetic and environmental influences on perceptions of organizational climate. *Journal of Applied Psychology*, 79, 24–33.

Hershberger, S. L., Plomin, R., & Pedersen, N. L. (in press). Traits and metatraits: Their reliability, stability, and shared genetic influence. *Journal of Personality and Social Psychology*.

Heston, L. L. (1966). Psychiatric disorders in foster home reared children of schizophrenic mothers. *British Journal of Psychiatry*, 112, 819–825.

Hetherington, E. M., & Clingempeel, W. G. (1992). Coping with marital transitions: A family systems perspective. *Monographs of the Society for Research in Child Development*, Nos. 2–3, Serial No. 227.

Hetherington, E. M., Reiss, D., & Plomin, R. (Eds.). (1994). *Separate social worlds of siblings: Impact of the nonshared environment on development*. Hillsdale, NJ: Erlbaum.

Hewitt, J. K., & Turner, J. R. (1995). Behavior genetic studies of cardiovascular responses to stress. In J. R. Turner, L. R. Cardon, & J. K. Hewitt (Eds.), *Behavior genetic approaches in behavioral medicine* (pp. 87–103). New York: Plenum.

Hilbert, P., Lindpaintner, K., Beckmann, J. S., Serikawa, T., Soubrier, F., Dubay, C., Cartwright, P., DeGouyon, B., Julier, C., Takahasi, S., et al. (1991). Chromosomal mapping of two genetic loci associated with blood-pressure regulation in hereditary hypertensive rats. *Nature*, *353*, 521–529.

Ho, H.-Z., Baker, L., & Decker, S. N. (1988). Covariation between intelligence and speed-of-cognitive processing: Genetic and environmental influences. *Behavior Genetics*, *18*, 247–261.

Hobbs, H. H., Russell, D. W., Brown, M. S., & Goldstein, J. L. (1990). The LDL receptor locus in familial hypercholesterolemia: Mutational analysis of a membrane protein. *Annual Review of Genetics*, *24*, 133–170.

Hodgkinson, S., Mullan, M., & Murray, R. M. (1991). The genetics of vulnerability to alcoholism. In P. McGuffin & R. Murray (Eds.), *The new genetics of mental illness* (pp. 182–197). London: Mental Health Foundation.

Hollister, J. M., Mednick, S. A., Brennan, P., & Cannon, T. D. (1994). Impaired autonomic nervous system habituation in those at genetic risk for schizophrenia. *Archives of General Psychiatry*, *51*, 552–558.

Hotta, Y., & Benzer, S. (1970). Genetic dissection of the *Drosophila* nervous system by means of mosaics. *Proceedings of the National Academy of Sciences*, *67*, 1156–1163.

Howie, P. (1981). Concordance for stuttering in monozygotic and dizygotic twin pairs. *Journal of Speech and Hearing*, *5*, 343–348.

Hu, S., Patatucci, A. M. L., Patterson, C., Li, L., Fulker, D. W., Cherny, S. S., Kruglyak, L., & Hamer, D. H. (1995). Linkage between sexual orientation and chromosome Xq28 in males but not in females. *Nature Genetics*, *11*, 248–256.

Hudson, T. J., Stein, L. D., Gerety, S. S., Ma, J., Castle, A. B., Silva, J., Slonim, D. K., Baptisa, R., Kruglyak, L., Xu, S. H., et al. (1995). An STS-based map of the human genome. *Science*, *270*, 1945–1954.

Husén, T. (1959). *Psychological twin research*. Stockholm: Almqvist & Wiksell.

Iacono, W. G., & Grove, W. M. (1993). Schizophrenia revised: Toward an integrative genetic model. *Psychological Science*, *4*, 273–276.

Jacomb, P. A., Jorm, A. F., Croft, L., Gao, X., & Easteal, S. (in press). Further evidence associating HLA but not CTGB33 with variation in IQ scores. *Intelligence*.

Jang, K. L. (1993). *A behavioural genetic analysis of personality, personality disorder, the environment, and the search for sources of nonshared environmental influences*. Unpublished doctoral dissertation, University of Western Ontario, London, Ontario.

Jary, M. L., & Stewart, M. A. (1985). Psychiatric disorder in the parents of adopted children with aggressive conduct disorder. *Neuropsychobiology*, *13*, 7–11.

Jensen, A. R. (1971). A note on why genetic correlations are not squared. *Psychological Bulletin*, *75*, 223–224.

Jensen, A. R. (1974). The problem of genotype-environment correlation in the estimation of heritability from monozygotic and dizygotic twins. *Acta Geneticae Medicae et Gemellogiae*, *25*, 86–99.

Jensen, A. R. (1978). Genetic and behavioral effects of nonrandom mating. In R. T. Osborne, C. E. Noble, & N. Weyl (Eds.), *Human variation: The biopsychology of age, race, and sex* (pp. 51–105). New York: Academic Press.

Jensen, A. R. (1987). The *g* beyond factor analysis. In R. R. Ronning, J. A. Clover, J. C. Conoley, & J. C. Witt (Eds.), *The influence of cognitive psychology on testing* (pp. 87–142). Hillsdale, NJ: Erlbaum.

Jensen, A. R. (1993). Psychometric *g* and achievement. In B. R. Gifford (Ed.), *Policy perspective on educational testing* (pp. 117–227). Boston: Kluwer.

Jensen, A. R. (1994). Phlogiston, animal magnetism, and intelligence. In D. K. Detterman (Ed.), *Current topics in human intelligence: Vol. 4. Theories of intelligence* (pp. 257–284). Norwood, NJ: Ablex.

Jinks, J. L., & Fulker, D. W. (1970). Comparison of the biometrical genetical, MAVA, and classical approaches to the analysis of human behavior. *Psychological Bulletin, 75*, 311–349.

Johnson, C. A., Ahern, F. M., & Johnson, R. C. (1976). Level of functioning of siblings and parents of probands of varying degrees of retardation. *Behavior Genetics, 6*, 473–477.

Jones, P. B., & Murray, R. M. (1991). Aberrant neurodevelopment as the expression of schizophrenia genotype. In P. McGuffin & R. Murray (Eds.), *The new genetics of mental illness* (pp. 112–129). Oxford: Butterworth-Heinemann.

Jöreskog, K. G., & Sörbom, D. (1983). *LISREL 8 users reference guide*. Chicago: Scientific Software International.

Joynson, R. B. (1989). *The Burt affair*. London: Routledge.

Kallmann, F. J. (1952). Twin and sibship study of overt male homosexuality. *American Journal of Human Genetics, 4*, 136–146.

Kallmann, F. J. (1955). Genetic aspects of mental disorders in later life. In O. J. Kaplan (Ed.), *Mental disorders in later life* (pp. 26–46). Stanford, CA: Stanford University Press.

Kamin, L. J. (1974). *The science and politics of IQ*. Potomac, MD: Erlbaum.

Keating, M. T., & Sanguinetti, M. C. (1996). Molecular genetic insights into cardiovascular disease. *Science, 272*, 681–685.

Keller, L. M., Bouchard, T. J., Arvey, R. D., Segal, N. L., & Dawis, R. V. (1992). Work values: Genetic and environmental influences. *Journal of Applied Psychology, 77*, 79–88.

Kelsoe, J. R., Ginns, E. I., Egeland, J. A., Goldstein, A. M., Bale, S. J., Pauls, D. L., Long, R. T., Conte, G., Gerhard, D. S., Housman, D. E., & Paul, S. M. (1989). Re-evaluation of the linkage relationship between chromosome 11q loci and the gene for bipolar affective disorder in the Old Order Amish. *Nature, 325*, 238–243.

Kendler, K. S. (1988). Familial aggregation of schizophrenia and schizophrenia spectrum disorder. *Archives of General Psychiatry, 45*, 377–383.

Kendler, K. S., & Eaves, L. J. (1986). Models for the joint effects of genotype and environment on liability to psychiatric illness. *American Journal of Psychiatry, 143*, 279–289.

Kendler, K. S., Gruenberg, A. M., & Kinney, D. K. (1994). Independent diagnoses of adoptees and relatives, as defined by DSM-III, in the provincial and national samples of the Danish adoption study of schizophrenia. *Archives of General Psychiatry, 51*, 456–468.

Kendler, K. S., & Hewitt, J. (1992). The structure of self-report schizotypy in twins. *Journal of Personality Disorders, 6*, 1–17.

Kendler, K. S., Kessler, R. C., Walters, E. E., MacLean, C., Neale, M. C., Heath, C., & Eaves, L. J. (1995). Stressful life events, genetic liability, and onset of an episode of major depression in women. *American Journal of Psychiatry, 152*, 833–842.

Kendler, K. S., McLean, C., Neale, M., Kessler, R., Heath, A., & Eaves, L. J. (1991). The genetic epidemiology of bulimia nervosa. *American Journal of Psychiatry, 148*, 1627–1637.

Kendler, K. S., Neale, M. C., Kessler, R. C., Heath, A. C., & Eaves, L. J. (1992a). A population-based twin study of major depression in women: The impact of varying definitions of illness. *Archives of General Psychiatry, 49*, 257–266.

Kendler, K. S., Neale, M. C., Kessler, R. C., Heath, A. C., & Eaves, L. J. (1992b). Major depression and generalized anxiety disorder: Same genes (partly) different environments? *Archives of General Psychiatry, 49*, 716–722.

Kendler, K. S., Neale, M. C., Kessler, R. C., Heath, A. C., & Eaves, L. J. (1992c). The genetic epidemiology of phobias in women: The interrelationship of agoraphobia, social phobia, situational phobia, and simple phobia. *Archives of General Psychiatry, 49*, 273–281.

Kendler, K. S., Neale, M. C., Kessler, R. C., Heath, A. C., & Eaves, L. J. (1993a). A test of the equal-environment assumption in twin studies of psychiatric illness. *Behavior Genetics, 23*, 21–27.

Kendler, K. S., Neale, M. C., Kessler, R. C., Heath, A. C., & Eaves, L. J. (1993c). A twin study of recent life events and difficulties. *Archives of General Psychiatry, 50*, 789–796.

Kendler, K. S., Neale, M. C., MacLean, C. J., Heath, A. C., Eaves, L. J., & Kessler, R. C. (1993b). Smoking and major depression: A causal analysis. *Archives of General Psychiatry, 50*, 36–43.

Kendler, K. S., Neale, M. C., Prescott, C. A., Kessler, R. C., Heath, A. C., Corley, L. A., & Eaves, L. J. (1996). Childhood parental loss and alcoholism in women: A causal analysis using a twin-family design. *Psychological Medicine, 26*, 79–95.

Kenrick, D. T., & Funder, D. C. (1988). Profiting from controversy: Lessons from the person-situation debate. *American Psychologist, 43*, 23–34.

Kessler, R. C., Kendler, K. S., Heath, A., Neale, M. C., & Eaves, L. J. (1992). Social support, depressed mood, and adjustment to stress: A genetic epidemiologic investigation. *Journal of Personality and Social Psychology, 62*, 257–272.

Kessler, R., McGonagle, K. A., Zhao, S., Nelson, C. B., Hughes, M., Eshleman, S., Wittchen, H. U., & Kendler, K. S. (1994). Lifetime and 12-month prevalence of DSM-III-R psychiatric disorders in the United States: Results from the National Comorbidity Study. *Archives of General Psychiatry, 51*, 8–19.

Kety, S. S. (1987). The significance of genetic factors in the etiology of schizophrenia: Results from the national study of adoptees in Denmark. *Journal of Psychiatric Research, 21*, 423–430.

Kety, S. S., Wender, P. H., Jacobsen, B., Ingraham, L. J., Jansson, L., Faber, B., & Kinney, D. K. (1994). Mental illness in the biological and adoptive relatives of schizophrenic adoptees: Replication of the Copenhagen study in the rest of Denmark. *Archives of General Psychiatry, 51*, 442–455.

Kidd, K. (1983). Recent progress on the genetics of stuttering. In C. Ludlow & J. Cooper (Eds.), *Genetic aspects of speech and language disorders* (pp. 197–213). New York: Academic Press.

Klein, R., & Mannuzza, S. (1991). Long-term outcome of hyperactive children: A review. *Journal of the American Academy of Child and Adolescent Psychiatry, 30*, 383–387.

Kohnstamm, G. A., Bates, J. E., & Rothbart, M. K. (1989). *Temperament in childhood.* New York: Wiley.

Kringlen, E., & Cramer, G. (1989). Offspring of monozygotic twins discordant for schizophrenia. *Archives of General Psychiatry, 46,* 873–877.

Kuehn, M. R., Bradley, A., Robertson, E. J., & Evans, M. J. (1987). A potential animal model for Lesch-Nyhan syndrome through introduction of HPRT mutations into mice. *Nature, 326,* 295–298.

Kung, C., Chang, S. Y., Satow, Y., Van Houten, J., & Hansma, H. (1975). Genetic dissection of behavior in *Paramecium. Science, 188,* 898–904.

Lack, D. (1953). *Darwin's finches.* Cambridge: Cambridge University Press.

Lawrence, P. A. (1992). *The making of a fly: The genes of animal design.* Oxford: Blackwell.

Leahy, A. M. (1935). Nature-nurture and intelligence. *Genetic Psychology Monographs, 17,* 236–308.

Le Couteur, A., Bailey, A., Goode, S., Pickles, A., Robertson, S., Gottesman, I. I., & Rutter, M. (1996). A broader phenotype of autism: The clinical spectrum in twins. *Journal of Child Psychology and Psychiatry, 37,* 785–801.

Lerner, I. M. (1968). *Heredity, evolution, and society.* San Francisco: Freeman.

LeVay, S., & Hamer, D. H. (1994). Evidence for a biological influence in male homosexuality. *Scientific American, 270,* 44–49.

Levy, D. L., Holzman, P. S., Matthysee, S., & Mendell, N. R. (1993). Eye tracking dysfunction and schizophrenia: A critical perspective. *Schizophrenia Bulletin, 19,* 461–536.

Lewis, B., & Thompson, L. A. (1992). A study of developmental speech and language disorders in twins. *Journal of Speech and Hearing Research, 35,* 1086–1094.

Lichtenstein, P., Harris, J. R., Pedersen, N. L., & McClearn, G. E. (1992a). Socioeconomic status and physical health, how are they related? An empirical study based on twins reared apart and twins reared together. *Social Science and Medicine, 36,* 441–450.

Lichtenstein, P., Pedersen, N. L., & McClearn, G. E. (1992b). The origins of individual differences in occupational status and educational level: A study of twins reared apart and together. *Acta Sociologica, 35,* 13–31.

Lidsky, A. S., Robson, K., Chandra, T., Barker, P., Ruddle, F., & Woo, S. L. C. (1984). The PKU locus in man is on chromosome 12. *American Journal of Human Genetics, 36,* 527–533.

Lifton, R. P. (1996). Molecular genetics of human blood pressure variation. *Science, 272,* 676–680.

Light, J. G., & DeFries, J. C. (1995). Comorbidity of reading and mathematics disabilities: Genetic and environmental etiologies. *Journal of Learning Disabilities, 28,* 96–106.

Light, J. G., Pennington, B. F., Gilger, J. W., & DeFries, J. C. (1995). Reading disability and hyperactivity disorder: Evidence for a common genetic etiology. *Developmental Neuropsychology, 11,* 323–335.

Lilienfeld, S. O. (1992). The association between antisocial personality and somatization disorders: A review and integration of theoretical models. *Clinical Psychology Review, 12,* 641–662.

Lipovechaja, N. G., Kantonistowa, N. S., & Chamaganova, T. G. (1978). The role of heredity and environment in the determination of intellectual function. *Medicinskie, Probleing Formirovaniga Livenosti, 1,* 48–59.

Loehlin, J. C. (1987). *Latent variable models: An introduction to factor, path, and structural analysis.* Hillsdale, NJ: Erlbaum.

Loehlin, J. C. (1989). Partitioning environmental and genetic contributions to behavioral development. *American Psychologist, 44,* 1285–1292.

Loehlin, J. C. (1992). *Genes and environment in personality development.* Newbury Park, CA: Sage.

Loehlin, J. C., Horn, J. M., & Willerman, L. (1989). Modeling IQ change: Evidence from the Texas Adoption Project. *Child Development, 60,* 993–1004.

Loehlin, J. C., Horn, J. M., & Willerman, L. (1990). Heredity, environment, and personality change: Evidence from the Texas Adoption Study. *Journal of Personality, 58,* 221–243.

Loehlin, J. C., & Nichols, R. C. (1976). *Heredity, environment, and personality.* Austin: University of Texas Press.

Loehlin, J. C., Willerman, L., & Horn, J. M. (1982). Personality resemblance between unwed mothers and their adopted-away offspring. *Journal of Personality and Social Psychology, 42,* 1089–1099.

Loehlin, J. C., Willerman, L., & Horn, J. M. (1987). Personality resemblance in adoptive families: A 10-year follow-up. *Journal of Personality and Social Psychology, 53,* 961–969.

Loranger, A. W., Oldham, J. M., & Tulis, E. H. (1982). Familial transmission of DSM-III borderline personality disorders. *Archives of General Psychiatry, 39,* 795–799.

Luo, D., Petrill, S. A., & Thompson, L.A. (1994). An exploration of genetic g: Hierarchical factor analysis of cognitive data from the Western Reserve Twin Project. *Intelligence, 18,* 335–348.

Lykken, D. T. (1982). Research with twins: The concept of emergenesis. *Psychophysiology, 19,* 361–373.

Lyons, M. J. (1996). A twin study of self-reported criminal behaviour. In G. R. Bock & J. A. Goode (Eds.), *Genetics of criminal and antisocial behaviour* (pp. 1–75). Chichester, UK: Wiley.

Lyons, M. J., Goldberg, J., Eisen, S. A., True, W., Tsuang, M. T., Meyer, J. M., & Henderson, W. G. (1993). Do genes influence exposure to trauma: A twin study of combat. *American Journal of Medical Genetics (Neuropsychiatric Genetics), 48,* 22–27.

Lyons, M. J., True, W. R., Eisen, S. A., Goldberg, J., Meyer, J. M., Faraone, S. V., Eaves, L. J., & Tsuang, M. T. (1995). Differential heritability of adult and juvenile antisocial traits. *Archives of General Psychiatry, 52,* 906–915.

Lytton, H. (1977). Do parents create or respond to differences in twins? *Developmental Psychology, 13,* 456–459.

Lytton, H. (1980). *Parent-child interaction: The socialization process observed in twin and singleton families.* New York: Plenum.

Lytton, H. (1991). Different parental practices—different sources of influence. *Behavioral and Brain Sciences, 14,* 399–400.

MacGillivray, I., Campbell, D. M., & Thompson, B. (Eds.). (1988). *Twinning and twins.* Chichester, UK: Wiley.

Mack, K. J., & Mack, P. A. (1992). Introduction of transcription factors in somatosensory cortex after tactile stimulation. *Molecular Brain Research, 12,* 141–149.

Mackintosh, N. J. (Ed.). (1995). *Cyril Burt: Fraud or framed?* Oxford: Oxford University Press.

Mandoki, M. W., Sumner, G. S., Hoffman, R. P., & Riconda, D. L. (1991). A review of Klinefelter's syndrome in children and adolescents. *Journal of the American Academy of Child and Adolescent Psychiatry, 30,* 167–172.

Manke, B., McGuire, S., Reiss, D., Hetherington, E. M., & Plomin, R. (1995). Genetic contributions to children's extrafamilial social interactions: Teachers, friends, and peers. *Social Development, 4,* 238–256.

Manuck, S. B. (1994). Cardiovascular reactivity in cardiovascular disease: "Once more unto the breach." *International Journal of Behavioral Medicine, 1,* 4–31.

Marks, I. M. (1986). Genetics of fear and anxiety disorders. *British Journal of Psychiatry, 149,* 406–418.

Marks, I. M., & Nesse, R. M. (1994). Fear and fitness: An evolutionary analysis of anxiety disorders. *Etiology and Sociobiology, 15,* 247–261.

Martin, N. G., & Eaves, L. J. (1977). The genetical analysis of covariance structure. *Heredity, 38,* 79–95.

Martin, N. G., Jardine, R., & Eaves, L. J. (1984). Is there only one set of genes for different abilities? A reanalysis of the National Merit Scholarship Qualifying Tests (NMSQT) data. *Behavior Genetics, 14,* 355–370.

Mascie-Taylor, C. G. N., Gibson, J. B., Hiorns, R. W., & Harrison, G. A. (1985). Associations between some polymorphic markers and variation in IQ and its components in Otmoor villagers. *Behavior Genetics, 15,* 371–383.

Matheny, A. P., Jr. (1980). Bayley's Infant Behavior Record: Behavioral components and twin analysis. *Child Development, 51,* 1157–1167.

Matheny, A. P., Jr. (1989). Children's behavioral inhibition over age and across situations: Genetic similarity for a trait during change. *Journal of Personality, 57,* 215–235.

Matheny, A. P., Jr. (1990). Developmental behavior genetics: Contributions from the Louisville Twin Study. In M. E. Hahn, J. K. Hewitt, N. D. Henderson, & R. H. Benno (Eds.), *Developmental behavior genetics: Neural, biometrical, and evolutionary approaches* (pp. 25–39). New York: Oxford University Press.

Matheny, A. P., Jr., & Dolan, A. B. (1975). Persons, situations, and time: A genetic view of behavioral change in children. *Journal of Personality and Social Psychology, 14,* 224–234.

Mather, K., & Jinks, J. K. (1982). *Biometrical genetics: The study of continuous variation* (3rd ed.). New York: Chapman & Hall.

McCartney, K., Harris, M. J., & Bernieri, F. (1990). Growing up and growing apart: A developmental meta-analysis of twin studies. *Psychological Bulletin, 107,* 226–237.

McClearn, G. E. (1963). The inheritance of behavior. In L. J. Postman (Ed.), *Psychology in the making.* New York: Knopf.

McClearn, G. E., (1976). Experimental behavioural genetics. In D. Barltrop (Ed.), *Aspects of genetics in paediatrics* (pp. 31–39). London: Fellowship of Postdoctorate Medicine.

McClearn, G. E., & DeFries, J. C. (1973). *Introduction to behavioral genetics.* San Francisco: Freeman.

McClearn, G. E., & Rodgers, D. A. (1959). Differences in alcohol preference among inbred strains of mice. *Quarterly Journal of Studies on Alcohol, 52,* 62–67.

McFarlane, A. C. (1989). The aetiology of post-traumatic morbidity: Predisposing, precipitating and perpetuating factors. *British Journal of Psychiatry, 154,* 221–228.

McGue, M. (1993). From proteins to cognitions: The behavioral genetics of alcoholism. In R. Plomin & G. E. McClearn (Eds.), *Nature, nurture, and psychology* (pp. 245–268). Washington, DC: American Psychological Association.

McGue, M. (1994). Genes, environment, and the etiology of alcoholism. In R. Zucker, G. Boyd, & J. Howard (Eds.), *Development of alcohol-related problems: Exploring the biopsychosocial matrix of risk.* National Institute on Alcoholism and Alcohol Abuse Research Monograph No. 26 (pp. 1–40). Rockville, MD: National Institute on Alcoholism and Alcohol Abuse.

McGue, M., Bacon, S., & Lykken, D. T. (1993b). Personality stability and change in early adulthood: A behavioral genetic analysis. *Developmental Psychology, 29,* 96–109.

McGue, M., & Bouchard, T. J., Jr. (1989). Genetic and environmental determinants of information processing and special mental abilities: A twin analysis. In R. J. Sternberg (Ed.), *Advances in the psychology of human intelligence* (Vol. 5, pp. 7–45). Hillsdale, NJ: Erlbaum.

McGue, M., Bouchard, T. J., Jr., Iacono, W. G., & Lykken, D. T. (1993a). Behavioral genetics of cognitive ability: A life-span perspective. In R. Plomin & G. E. McClearn (Eds.), *Nature, nurture, and psychology* (pp. 59–76). Washington, DC: American Psychological Association.

McGue, M., & Gottesman, I. I. (1989). Genetic linkage in schizophrenia: Perspectives from genetic epidemiology. *Schizophrenia Bulletin, 15,* 453–464.

McGue, M., Hirsch, B., & Lykken, D. T. (1993c). Age and the self-perception of ability: A twin study analysis. *Psychology and Aging, 8,* 72–80.

McGue, M., & Lykken, D. T. (1992). Genetic influence on risk of divorce. *Psychological Science, 3,* 368–373.

McGue, M., Pickens, R. W., & Svikis, D. S. (1992). Sex and age effects on the inheritance of alcohol problems: A twin study. *Journal of Abnormal Psychology, 101,* 3–17.

McGue, M., Vaupel, J. W., Holm, N., & Harvald, B. (1993d). Longevity is moderately heritable in a sample of Danish twins born 1870–1880. *Journals of Gerontology, 48,* B237–B244.

McGuffin, P., Farmer, A. E., & Gottesman, I. I. (1987). Is there really a split in schizophrenia? The genetic evidence. *British Journal of Psychiatry, 50,* 581–592.

McGuffin, P., & Gottesman, I. I. (1985). Genetic influences on normal and abnormal development. In M. Rutter & L. Hersov (Eds.), *Child and adolescent psychiatry: Modern approaches* (2nd ed., pp. 17–33). Oxford: Blackwell Scientific.

McGuffin, P., & Katz, R. (1986). Nature, nurture, and affective disorder. In J. W. F. Deakin (Ed.), *The biology of depression* (pp. 26–51). London: Gaskell Press.

McGuffin, P., Katz, R., & Rutherford, J. (1991). Nature, nurture, and depression: A twin study. *Psychological Medicine, 21,* 329–335.

McGuffin, P., Katz, R., Rutherford, J., Watkins, S., Farmer, A. E., & Gottesman, I. I. (1993). Twin studies as vital indicators of phenotypes in molecular genetic research. In T. J. Bouchard & P. Propping (Eds.), *Twins as a tool of behavioral genetics* (pp. 243–256). Chichester, UK: Wiley.

McGuffin, P., Katz, R., Watkins, S., & Rutherford, J. (1996). A hospital-based twin register of the heritability of DSM-IV unipolar depression. *Archives of General Psychiatry, 53,* 129–136.

McGuffin, P., Owen, M. J., O'Donovan, M. C., Thapar, A., & Gottesman, I. I. (1994). *Seminars in psychiatric genetics.* London: Gaskell Press.

McGuffin, P., Sargeant, M., Hetti, G., Tidmarsh, S., Whatley, S., & Marchbanks, R. M. (1990). Exclusion of a schizophrenia susceptibility gene from the chromosome 5q11-q13 region. New data and a reanalysis of previous reports. *American Journal of Human Genetics, 47,* 524–535.

McGuffin, P., & Sturt, E. (1986). Genetic markers in schizophrenia. *Human Heredity, 16,* 461–465.

McGuire, S., Neiderhiser, J. M., Reiss, D., Hetherington, E. M., & Plomin, R. (1994). Genetic and environmental influences on perceptions of self-worth and competence in adolescence: A study of twins, full siblings, and step siblings. *Child Development, 65,* 785–799.

McGuire, T. R. (1984). Learning in three species of Diptera: The blow fly *Phormia regina,* the fruit fly *Drosophila melanogaster,* and the house fly *Musca domestica. Behavior Genetics, 14,* 479–526.

McIntire, S. L., Jorgensen, E., & Horvitz, H. R. (1993). Genes required for GABA function in *Caenorhabditis elegans. Nature, 364,* 334–341.

McMahon, R. C. (1980). Genetic etiology in the hyperactive child syndrome: A critical review. *American Journal of Orthopsychiatry, 50,* 145–150.

Medical Research Council Working Party on Phenylketonuria. (1993). Phenylketonuria due to phenylalanine hydroxylase deficiency: An unfolding story. *British Medical Journal, 306,* 115–119.

Medlund, P., Cederlof, R., Floderus-Myrhed, B., Friberg, L., & Sorensen, S. (1977). A new Swedish twin registry. *Acta Medica Scandinavica Supplementum, 60,* 1–11.

Mednick, S. A., Gabrielli, W. F., & Hutchings, B. (1984). Genetic factors in criminal behavior: Evidence from an adoption cohort. *Science, 224,* 891–893.

Mello, C. V., Vicario, D. S., & Clayton, D. F. (1992). Song presentation induces gene expression in the songbird forebrain. *Proceedings of the National Academy of Sciences USA, 89,* 6818–6821.

Mendel, G. J. (1866). Versuche ueber Pflanzenhybriden. *Verhandlungen des Naturforschunden Vereines in Bruenn, 4,* 3–47.

Mendlewicz, J., & Rainer, J. D. (1977). Adoption study supporting genetic transmission in manic-depressive illness. *Nature, 268,* 326–329.

Merikangas, K. R. (1990). The genetic epidemiology of alcoholism. *Psychological Medicine, 20,* 11–22.

Merriman, C. (1924). The intellectual resemblance of twins. *Psychological Monographs, 33,* 1–58.

Meyer, J. M. (1995). Genetic studies of obesity across the life span. In J. R. Turner, L. R. Cardon, & J. K. Hewitt (Eds.), *Behavior genetic approaches to behavioral medicine* (pp. 145–166). New York: Plenum.

Mineka, S., Davidson, M., Cook, M., & Keir, R. (1984). Observational conditioning of snake fear in rhesus monkeys. *Journal of Abnormal Psychology, 93,* 355–372.

Moffitt, T. E. (1993). Adolescence-limited and life-course-persistent antisocial behavior: A developmental taxonomy. *Psychological Review, 100,* 674–701.

Moises, H. W., Yang, L., Kristbjarnarson, H., Wiese, C., Byerley, W., Macciardi, F., Arolt, V., Blackwood, D., Liu, X., Sjögren, B., et al. (1995). An international two-stage genome-wide search for schizophrenia susceptibility genes. *Nature Genetics, 11,* 321–324.

Morgan, T. H., Sturtevant, A. H., Muller, H. J., & Bridges, C. B. (1915). *The mechanism of Mendelian heredity.* New York: Holt.

Morris, M. J., & Harper, P. S. (1991). Prediction and prevention in Huntington's disease. In P. McGuffin & R. Murray (Eds.), *The new genetics and mental illness* (pp. 281–298). Oxford: Butterworth-Heinemann.

Morris-Yates, A., Andrews, G., Howie, P., & Henderson, S. (1990). Twins: A test of the equal environments assumption. *Acta Psychiatrica Scandinavica, 8,* 322–326.

Mosher, L. R., Pollin, W., & Stabenau, J. R. (1971). Identical twins discordant for schizophrenia: Neurological findings. *Archives of General Psychiatry, 24,* 422–430.

Müller-Hill, B. (1988). *Murderous science*. Oxford: Oxford University Press.

Murray, A., Youings, S., Dennis, N., Latsky, L., Linehan, P., McKechnie, N., Macpherson, J., Pound, M., & Jacobs, P. (1996). Population screening at the *FRAXA* and *FRAXE* loci: Molecular analyses of boys with learning disabilities and their mothers. *Human Molecular Genetics, 5*, 727–735.

Murray, R. M., Lewis, S. W., & Reveley, A. M. (1985). Towards an aetiological classification of schizophrenia. *Lancet, 1*, 1023–1026.

National Foundation for Brain Research. (1992). *The care of disorders of the brain*. Washington, DC: National Foundation for Brain Research.

Neale, M. C., & Cardon, L. R. (1992). *Methodology for genetic studies of twins and families*. Dordrecht, Netherlands: Kluwer.

Neale, M. C., & Stevenson, J. (1989). Rater bias in the EASI temperament scales: A twin study. *Journal of Personality and Social Psychology, 56*, 446–455.

Neiderhiser, J. M. (1994). Family environment in early childhood, outcomes in middle childhood, and genetic mediation. In J. C. DeFries, R. Plomin, & D. W. Fulker (Eds.), *Nature and nurture in middle childhood* (pp. 249–261). Cambridge, MA: Blackwell.

Neiderhiser, J. M., & McGuire, S. (1994). Competence during middle childhood. In J. C. DeFries, R. Plomin, & D. W. Fulker (Eds.), *Nature and nurture during middle childhood* (pp. 141–151). Cambridge, MA: Blackwell.

Neiderhiser, J. M., O'Connor, T. G., Chipuer, H. M., Reiss, D., Hetherington, M., & Plomin, R. (1996). Adding stepfamilies to genetic designs. Manuscript submitted for publication.

Neiswanger, K., Hill, S. Y., & Kaplan, B. B. (1995). Association and linkage studies of TaqI A alleles at the dopamine D2 receptor gene in samples of female and male alcoholics. *American Journal of Medical Genetics (Neuropsychiatric Genetics), 60*, 267–271.

Nelkin, D., & Tancredi, L. (1989). *Dangerous diagnostics: The social power of biological information*. New York: Basic Books.

Nelson, R. J., Demas, G. E., Huang, P. L., Fishman, M. C., Dawson, V. L., Dawson, T. M., & Snyder, S. H. (1995). Behavioural abnormalities in male mice lacking neuronal nitric oxide synthase. *Nature, 378*, 383–386.

Nichols, P. L. (1984). Familial mental retardation. *Behavior Genetics, 14*, 161–170.

Nichols, R. C. (1978). Twin studies of ability, personality, and interests. *Homo, 29*, 158–173.

Nigg, J. T., & Goldsmith, H. H. (1994). Genetics of personality disorders: Perspectives from personality and psychopathology research. *Psychological Bulletin, 115*, 346–380.

Noyes, R., Jr., Clarkson, C., Crowe, R. R., Yates, W. R., & McClesney, C. M. (1987). A family study of generalized anxiety disorder. *American Journal of Psychiatry, 144*, 1019–1024.

Noyes, R., Jr., Crowe, R. R., Harris, E. L., Hamra, B. J., McChesney, C. M., & Chaudhry, D. R. (1986). Relationship between panic disorder and agoraphobia: A family study. *Archives of General Psychiatry, 43*, 227–232.

O'Connor, T. G., Hetherington, E. M., Reiss, D., & Plomin, R. (1995). A twin-sibling study of observed parent-adolescent interactions. *Child Development, 6*, 812–824.

Owen, M. J., Liddell, M. B., & McGuffin, P. (1994). Alzheimer's disease: An association with apolipoprotein e4 may help unlock the puzzle. *British Medical Journal, 308*, 672–673.

Parnas, J., Cannon, T. D., Jacobsen, B., Schulsinger, H., Schulsinger, F., & Mednick, S. A. (1993). Lifetime DSM-III-R diagnostic outcomes in the offspring of schizophrenic mothers: Results from the Copenhagen high-risk study. *Archives of General Psychiatry, 50,* 707–714.

Pauls, D. L., Leckman, J. F., & Cohen, D. J. (1993). Familial relationship between Gilles de la Tourette's syndrome, attention deficit disorder, learning disabilities, speech disorders, and stuttering. *Journal of the American Academy of Child and Adolescent Psychiatry, 32,* 1044–1050.

Pauls, D. L., Raymond, C., Stevenson, J., & Leckman, J. F. (1990). A family study of Gilles de la Tourette syndrome. *American Journal of Human Genetics, 48,* 154–163.

Pauls, D. L., Towbin, K. E., Leckman, J. F., Zahner, G. E. P., & Cohen, D. J. (1986). Gilles de la Tourette's syndrome and obsessive compulsive disorder. *Archives of General Psychiatry, 43,* 1180–1182.

Pedersen, N. L. (1996). Gerontological behavioral genetics. In J. E. Birren & K. W. Schaie (Eds.), *Handbook of the psychology of aging* (4th ed., pp. 59–77). San Diego: Academic Press.

Pedersen, N. L., Gatz, M., Plomin, R., Nesselroade, J. R., & McClearn, G. E. (1989a). Individual differences in locus of control during the second half of the life span for identical and fraternal twins reared apart and reared together. *Journal of Gerontology, 44,* 100–105.

Pedersen, N. L., Lichtenstein, P., Plomin, R., DeFaire, U., McClearn, G. E., & Matthews, K. A. (1989b). Genetic and environmental influences for Type A-like measures and related traits: A study of twins reared apart and twins reared together. *Psychosomatic Medicine, 51,* 428–440.

Pedersen, N. L., Plomin, R., & McClearn, G. E. (1994). Is there G beyond g? (Is there genetic influence on specific cognitive abilities independent of genetic influence on general cognitive ability?) *Intelligence, 18,* 133–143.

Pedersen, N. L., Plomin, R., Nesselroade, J. R., & McClearn, G. E. (1992). A quantitative genetic analysis of cognitive abilities during the second half of the life span. *Psychological Science, 3,* 346–353.

Peltonen, L. (1995). All out for chromosome six. *Nature, 378,* 665–666.

Pervin, L. A. (Ed.). (1990). *Handbook of personality: Theory and research.* New York: Guilford.

Peterson, B. S., Leckman, J. F., & Cohen, D. J. (1995). Tourette's syndrome: A genetically predisposed and an environmentally specified developmental psychopathology. In D. Cicchetti & D. J. Cohen (Eds.), *Developmental psychopathology: Vol. 2. Risk, disorder, and adaptation* (pp. 213–242). New York: Wiley.

Peto, R., Lopez, A. D., Boreham, J., Thun, M., & Heath, C. (1992). Mortality from tobacco in developed countries: Indirect estimation from national vital statistics. *Lancet, 339,* 1268–1278.

Petrill, S. A., Luo, D., Thompson, L. A., & Detterman, D. K. (in press-a). The independent prediction of general intelligence by elementary cognitive tasks: Genetic and environmental influences. *Behavior Genetics.*

Petrill, S., & Plomin, R. (in press-d). DNA markers associated with general and specific cognitive abilities. *Behavior Genetics* (abstract).

Petrill, S., Plomin, R., McClearn, G. E., Smith, D. L.,Vignetti, S., Chorney, M. J., Chorney, K., Thompson, L. A., Detterman, D. K., Benbow, C., Lubinski, D., Daniels, J., Owen, M. J., & McGuffin, P. (in press-b). No association between general cognitive ability and the A1 allele of the D2 dopamine receptor gene. *Behavior Genetics.*

Petrill, S., Saudino, K. J., Cherny, S. S., Emde, R. N., Hewitt, J. K., Kagan, J., & Plomin, R. (in press-c). Exploring the genetic etiology of low general cognitive ability from 14 to 36 months. *Developmental Psychology.*

Petrill, S. A., Thompson, L. A., & Detterman, D. K. (1995). The genetic and environmental variance underlying elementary cognitive tasks. *Behavior Genetics, 25,* 199–209.

Phillips, K., & Matheny, A. P., Jr. (1995). Quantitative genetic analysis of injury liability in infants and toddlers. *American Journal of Medical Genetics (Neuropsychiatric Genetics), 60,* 64–71.

Phillips, T. J., & Crabbe, J. C. (1991). Behavioral studies of genetic differences in alcohol action. In J. C. Crabbe & R. A. Harris (Eds.), *The genetic basis of alcohol and drug actions* (pp. 25–104). New York: Plenum.

Phillips, T. J., Huson, M., Gwiazdon, C., Burkhart-Kasch, S., & Shen, E. H. (1995). Effects of acute and repeated ethanol exposures on the locomotor activity of BXD recombinant inbred mice. *Alcoholism: Clinical and Experimental Research, 19,* 269–278.

Pickering, T. G. (1991). *Ambulatory monitoring and blood pressure variability.* London: Science Press.

Pickles, A., Bolton, P., MacDonald, H., Bailey, A., Le Couteur, A., Sim, L., & Rutter, M. (1995). Latent class analysis of recurrence risk for complex phenotypes with selection and measurement error: A twin and family history study of autism. *American Journal of Human Genetics, 57,* 717–726.

Pike, A., Hetherington, E. M., Reiss, D., & Plomin, R. (1996a). Adolescents' nonshared experience of parental negativity: In the eye of the beholder? Manuscript submitted for publication.

Pike, A., McGuire, S., Hetherington, E. M., Reiss, D., & Plomin, R. (1996b). Family environment and adolescent depressive symptoms and antisocial behavior: A multivariate genetic analysis. *Developmental Psychology, 32,* 590–603.

Pike, A., Reiss, D., Hetherington, E. M., & Plomin, R. (1996c). Using MZ differences in the search for nonshared environmental effects. *Journal of Child Psychology and Psychiatry, 37,* 695–704.

Pinker, S. (1994). *The language instinct: The new science of language and mind.* London: Penguin.

Plomin, R. (1986). *Development, genetics, and psychology.* Hillsdale, NJ: Erlbaum.

Plomin, R. (1987). Developmental behavioral genetics and infancy. In J. Osofsky (Ed.), *Handbook of infant development* (2nd ed., pp. 363–417). New York: Wiley-Interscience.

Plomin, R. (1988). The nature and nurture of cognitive abilities. In R. J. Sternberg (Ed.), *Advances in the psychology of human intelligence* (pp. 1–33). Hillsdale, NJ: Erlbaum.

Plomin, R. (1991). Genetic risk and psychosocial disorders: Links between the normal and abnormal. In M. Rutter & P. Casaer (Eds.), *Biological risk factors for psychosocial disorders* (pp. 101–138). Cambridge: Cambridge University Press.

Plomin, R. (1993). Nature and nurture: Perspective and prospective. In R. Plomin & G. E. McClearn (Eds.), *Nature, nurture, and psychology* (pp. 459–485). Washington, DC: American Psychological Association.

Plomin, R. (1994a). *Genetics and experience: The developmental interplay between nature and nurture.* Newbury Park, CA: Sage.

Plomin, R. (1994b). The Emanuel Miller Memorial Lecture 1993: Genetic research and identification of environmental influences. *Journal of Child Psychology and Psychiatry, 35,* 817–834.

Plomin, R. (1995a). Molecular genetics and psychology. *Current Directions in Psychological Science, 4,* 114–117.

Plomin, R. (1995b). Genetics, environmental risks, and protective factors. In J. R. Turner, L. R. Cardon, & J. K. Hewitt (Eds.), *Behavior genetic approaches to behavioral medicine* (pp. 217–235). New York: Plenum.

Plomin, R. (1995c). Genetics and children's experiences in the family. *Journal of Child Psychology and Psychiatry, 36,* 33–68.

Plomin, R., Chipuer, H. M., & Loehlin, J. C. (1990a). Behavioral genetics and personality. In L. A. Pervin (Ed.), *Handbook of personality: Theory and research* (pp. 225–243). New York: Guilford.

Plomin, R., Chipuer, H. M., & Neiderhiser, J. M. (1994b). Behavioral genetic evidence for the importance of nonshared environment. In E. M. Hetherington, D. Reiss, & R. Plomin (Eds.), *Separate social worlds of siblings: Impact of nonshared environment on development* (pp. 1–31). Hillsdale, NJ: Erlbaum.

Plomin, R., Coon, H., Carey, G., DeFries, J. C., & Fulker, D. W. (1991). Parent-offspring and sibling adoption analyses of parental ratings of temperament in infancy and childhood. *Journal of Personality, 59,* 705–732.

Plomin, R., Corley, R., DeFries, J. C., & Fulker, D. W. (1990c). Individual differences in television viewing in early childhood: Nature as well as nurture. *Psychological Science, 1,* 371–377.

Plomin, R., & DeFries, J. C. (1979). Multivariate behavioral genetic analysis of twin data on scholastic abilities. *Behavior Genetics, 9,* 505–517.

Plomin, R., DeFries, J. C., & Fulker, D. W. (1988a). *Nature and nurture during infancy and early childhood.* New York: Cambridge University Press.

Plomin, R., DeFries, J. C., & Loehlin, J. C. (1977). Genotype-environment interaction and correlation in the analysis of human behavior. *Psychological Bulletin, 84,* 309–322.

Plomin, R., DeFries, J. C., & McClearn, G. E. (1980, 1990). *Behavioral genetics: A primer.* New York: Freeman.

Plomin, R., Emde, R. N., Braungart, J. M., Campos, J., Corley, R., Fulker, D. W., Kagan, J., Reznick, J. S., Robinson, J., Zahn-Waxler, C., & DeFries, J. C. (1993). Genetic change and continuity from fourteen to twenty months: The MacArthur Longitudinal Twin Study. *Child Development, 64,* 1354–1376.

Plomin, R., & Foch, T. T. (1980). A twin study of objectively assessed personality in childhood. *Journal of Personality and Social Psychology, 39,* 680–688.

Plomin, R., Foch, T. T., & Rowe, D. C. (1981). Bobo clown aggression in childhood: Environment, not genes. *Journal of Research in Personality, 15,* 331–342.

Plomin, R., & Hershberger, S. (1991). Genotype-environment interaction. In T. D. Wachs & R. Plomin (Eds.), *Conceptualization and measurement of organism-environment interaction* (pp. 29–43). Washington, DC: American Psychological Association.

Plomin, R., Lichtenstein, P., Pedersen, N. L., McClearn, G. E., & Nesselroade, J. R. (1990b). Genetic influence on life events during the last half of the life span. *Psychology and Aging, 5,* 25–30.

Plomin, R., Loehlin, J. C., & DeFries, J. C. (1985). Genetic and environmental components of "environmental" influences. *Developmental Psychology, 21,* 391–402.

Plomin, R., & McClearn, G. E. (1990). Human behavioral genetics of aging. In J. E. Birren & K. W. Schaie (Eds.), *Handbook of the psychology of aging* (pp. 66–77). New York: Academic Press.

Plomin, R., & McClearn, G. E. (Eds.). (1993a). *Nature, nurture, and psychology.* Washington, DC: American Psychological Association.

Plomin, R., & McClearn, G. E. (1993b). Quantitative trait loci (QTL) analysis and alcohol-related behaviors. *Behavior Genetics, 23,* 197–211.

Plomin, R., McClearn, G. E., Pedersen, N. L., Nesselroade, J. R., & Bergeman, C. S. (1988b). Genetic influence on childhood family environment perceived retrospectively from the last half of the life span. *Developmental Psychology, 24,* 738–745.

Plomin, R., McClearn, G. E., Smith, D. L., Skuder, P., Vignetti, S., Chorney, M. J., Chorney, K., Kasarda, S., Thompson, L. A., Detterman, D. K., Petrill, S. A., Daniels, J., Owen, M. J., & McGuffin, P. (1995). Allelic associations between 100 DNA markers and high versus low IQ. *Intelligence, 21,* 31–48.

Plomin, R., & Nesselroade, J. R. (1990). Behavioral genetics and personality change. *Journal of Personality, 58,* 191–220.

Plomin, R., Pedersen, N. L., Lichtenstein, P., & McClearn, G. E. (1994a). Variability and stability in cognitive abilities are largely genetic later in life. *Behavior Genetics, 24,* 207–215.

Plomin, R., Reiss, D., Hetherington, E. M., & Howe, G. (1994c). Nature and nurture: Genetic influence on measures of the family environment. *Developmental Psychology, 30,* 32–43.

Plomin, R., & Rende, R. D. (1991). Human behavioral genetics. *Annual Review of Psychology, 42,* 161–190.

Plomin, R., & Saudino, K. J. (1994). Quantitative genetics and molecular genetics. In J. E. Bates & T. D. Wachs (Eds.), *Temperament: Individual differences at the interface of biology and behavior* (pp. 143–171). Washington, DC: American Psychological Association.

Pogue-Geile, M. F., & Rose, R. J. (1985). Developmental genetic studies of adult personality. *Developmental Psychology, 21,* 547–557.

Pollen, D. A. (1993). *Hannah's heirs: The quest for the genetic origins of Alzheimer's disease.* New York: Oxford University Press.

Price, R. A., Kidd, K. K., Cohn, D. J., Pauls, D. L., & Leckman, J. F. (1985). A twin study of Tourette syndrome. *Archives of General Psychiatry, 42,* 815–820.

Price, R. A., Kidd, K. K., & Weissman, M. M. (1987). Early onset (under age 30 years) and panic disorder as markers for etiologic homogeneity in major depression. *Archives of General Psychiatry, 44,* 434–440.

Propping, P. (1987). Single gene effects in psychiatric disorders. In F. Vogel & K. Sperling (Eds.), *Human genetics: Proceedings of the 7th International Congress, Berlin* (pp. 452–457). New York: Springer.

Raine, A. (1993). *The psychopathology of crime: Criminal behavior as a clinical disorder.* San Diego: Academic Press.

Rasmussen, S. A., & Tsuang, M. T. (1984). The epidemiology of obsessive compulsive disorder. *Journal of Clinical Psychiatry, 45,* 450–457.

Ratcliffe, S. G. (1994). The psychological and psychiatric consequences of sex chromosome abnormalities in children, based on population studies. In F. Poustka (Ed.), *Basic approaches to genetic and molecular-biological developmental psychiatry* (pp. 92–122). Quintessenz Library of Psychiatry.

Reed, E. W., & Reed, S. C. (1965). *Mental retardation: A family study.* Philadelphia: Saunders.

Reich, J., & Yates, W. (1988). Family history of psychiatric disorders in social phobia. *Comprehensive Psychiatry, 2,* 72–75.

Reich, T., & Cloninger, R. (1990). Time-dependent model of the familial transmission of alcoholism. In *Banbury Report 33: Genetics and biology of alcoholism* (pp. 55–73). Cold Spring Harbor, NY: Cold Spring Harbor Laboratory Press.

Reiss, D., Hetherington, E. M., Plomin, R., Howe, G. W., Simmens, S. J., Henderson, S. H., O'Connor, T. J., Bussell, D. A., Anderson, E. R., & Law, T. (1995). Genetic questions for environmental studies: Differential parenting and psychopathology in adolescence. *Archives of General Psychiatry, 52,* 925–936.

Reiss, D., Plomin, R., Hetherington, E. M., Howe, G., Rovine, M., Tryon, A., & Stanley, M. (1994). The separate worlds of teenage siblings: An introduction to the study of the nonshared environment and adolescent development. In E. M. Hetherington, D. Reiss, & R. Plomin (Eds.), *Separate social worlds of siblings: Impact of nonshared environment on development* (pp. 63–109). Hillsdale, NJ: Erlbaum.

Rende, R. D., Slomkowski, C. L., Stocker, C., Fuller, D. W., & Plomin, R. (1992). Genetic and environmental influences on maternal and sibling interaction in middle childhood: A sibling adoption study. *Developmental Psychology, 28,* 484–490.

Ricciuti, A. E. (1993). Child-mother attachment: A twin study. Poster presented at the *Sixtieth Anniversary Meeting of the Society for Research in Child Development.* March 25–28, New Orleans.

Riese, M. L. (1990). Neonatal temperament in monozygotic and dizygotic twin pairs. *Child Development, 61,* 1230–1237.

Risch, N., Ghosh, S., & Todd, J. A. (1993). Statistical evaluation of multiple-locus linkage data in experimental species and its relevance to human studies: Application to nonobese diabetic (NOD) mouse and human insulin-dependent diabetes mellitus (IDDM). *American Journal of Human Genetics, 53,* 702–714.

Roberts, C. A., & Johansson, C. B. (1974). The inheritance of cognitive interest styles among twins. *Journal of Vocational Behavior, 4,* 237–243.

Robins, L. N., & Price, R. K. (1991). Adult disorders predicted by childhood conduct problems: Results from the NIMH epidemiologic catchment area project. *Psychiatry, 54,* 116–132.

Robins, L. N., & Regier, D. A. (1991). *Psychiatric disorders in America.* New York: Free Press.

Robinson, J. L., Kagan, J., Reznick, J. S., & Corley, R. (1992). The heritability of inhibited and uninhibited behavior: A twin study. *Developmental Psychology, 28,* 1030–1037.

Rose, R. J. (1992). Genes, stress, and cardiovascular reactivity. In J. R. Turner, A. Sherwood, & K. C. Light (Eds.), *Individual differences in cardiovascular response to stress* (pp. 87–102). New York: Plenum.

Rose, R. J., & Ditto, W. B. (1983). A developmental-genetic analysis of common fears from early adolescence to early adulthood. *Child Development, 54,* 361–368.

Rosenthal, D., Wender, P. H., Kety, S. S., Schulsinger, F., Welner, J., & Ostergaard, L. (1968). Schizophrenics' offspring reared in adoptive homes. *Journal of Psychiatric Research, 6,* 377–391.

Rosenthal, D., Wender, P. H., Kety, S. S., Welner, J., & Schulsinger, F. (1971). The adopted-away offspring of schizophrenics. *American Journal of Psychiatry, 128,* 307–311.

Roush, W. (1995). Conflict marks crime conference. *Science, 269,* 1808–1809.

Rowe, D. C. (1981). Environmental and genetic influences on dimensions of perceived parenting: A twin study. *Developmental Psychology, 17,* 203–208.

Rowe, D. C. (1983a). Biometric models of self-reported delinquent behavior: A twin study. *Behavior Genetics, 13,* 473–489.

Rowe, D. C. (1983b). A biometrical analysis of perceptions of family environment: A study of twin and singleton sibling kinships. *Child Development, 54,* 416–423.

Rowe, D. C. (1986). Genetic and environmental components of antisocial pairs: A study of 275 twin pairs. *Criminology, 24,* 513–532.

Rowe, D. C. (1987). Resolving the person-situation debate: Invitation to an interdisciplinary dialogue. *American Psychologist, 42,* 218–227.

Rowe, D. C. (1994). *The limits of family influence: Genes, experience, and behavior.* New York: Guilford.

Rowe, D. C., & Linver, M. R. (1995). Smoking and addictive behaviors: Epidemiological, individual, and family factors. In J. R. Turner, L. R. Cardon, & J. K. Hewitt (Eds.), *Behavior genetic approaches in behavioral medicine* (pp. 67–84). New York: Plenum.

Rush, A. J., & Weissenburger, J. E. (1994). Melancholic symptom features and DSM-IV. *American Journal of Psychiatry, 151,* 489–498.

Rutherford, J., McGuffin, P., Katz, R. J., & Murray, R. M. (1993). Genetic influences on eating attitudes in a normal female twin population. *Psychological Medicine, 23,* 425–436.

Rutter, M. (1996a). Introduction: Concepts of antisocial behavior, of cause, and of genetic influences. In G. R. Bock & J. A. Goode (Eds.), *Genetics of criminal and antisocial behaviour* (pp. 1–15). Chichester, UK: Wiley.

Rutter, M. (1996b). Concluding remarks. In G. R. Bock & J. A. Goode (Eds.), *Genetics of criminal and antisocial behaviour* (pp. 265–271). Chichester, UK: Wiley.

Rutter, M. (1996c). Autism research: Prospects and priorities. *Journal of Autism and Developmental Disorders, 26,* 257–275.

Rutter, M., Bailey, A., Bolton, P., & Le Couteur, A. (1993). Autism: Syndrome definition and possible genetic mechanisms. In R. Plomin & G. E. McClearn (Eds.), *Nature, nurture, and psychology* (pp. 269–284). Washington, DC: American Psychological Association.

Rutter, M., Bailey, A., Simonoff, E., & Pickles, A. (in press-a). Genetic influences and autism. In F. Volkmar & D. Cohen (Eds.), *Handbook of autism.* New York: Wiley.

Rutter, M., Dunn, J., Plomin, R., Simonoff, E., Pickles, A., Maughan, B., Ormel, J., Meyer, J., & Eaves, L. J. (in press-c). Integrating nature and nurture: Implications of person-environment correlations and interactions for developmental psychopathology. *Development and Psychopathology.*

Rutter, M., & Giller, H. (1983). *Juvenile delinquency: Trends and perspectives.* Harmondsworth, UK: Penguin.

Rutter, M., Macdonald, H., Le Couteur, A., Harrington, R., Bolton, P., & Bailey, A. (1990). Genetic factors in child psychiatric disorders: II. Empirical findings. *Journal of Child Psychology and Psychiatry, 31,* 39–83.

Rutter, M., Maughan, B., Myer, J., Pickles, A., Silberg, J., Simonoff, E., & Taylor, E. (in press-b). Heterogeneity of antisocial behavior: Causes, continuities, and consequences. In W. Osgood (Ed.), *Motivation and delinquency.* Lincoln: University of Nebraska Press.

Rutter, M., & Pickles, A. (1991). Person-environment interactions: Concepts, mechanisms, and implications for data analysis. In T. D. Wachs & R. Plomin (Eds.), *Conceptualization and measurement of organism-environment interaction* (pp. 105–147). Washington, DC: American Psychological Association.

Rutter, M., & Redshaw, J. (1991). Annotation: Growing up as a twin: Twin-singleton differences in psychological development. *Journal of Child Psychology and Psychiatry, 32,* 885–895.

Rutter, M., Simonoff, E., & Plomin, R. (1996). Genetic influences on mild mental retardation: Concepts, findings, and research implications. *Journal of Biosocial Science, 28,* 509–526.

Saudino, K. J., & Eaton, W. O. (1991). Infant temperament and genetics: An objective twin study of motor activity level. *Child Development, 62,* 1167–1174.

Saudino, K. J., McGuire, S., Reiss, D., Hetherington, E. M., & Plomin, R. (in press-a). Parent ratings of EAS temperaments in twins, full siblings, half siblings, and step siblings. *Journal of Personality and Social Psychology.*

Saudino, K. J., Pedersen, N. L., Lichtenstein, P., McClearn, G. E., & Plomin, R. (in press-b). Can personality explain genetic influences on life events? *Journal of Personality and Social Psychology.*

Saudino, K. J., & Plomin, R. (in press). Cognitive and temperamental mediators of genetic influences on the home environment during infancy. *Merrill-Palmer Quarterly.*

Saudino, K. J., Plomin, R., & DeFries, J. C. (1996). Tester-rated temperament at 14, 20, and 24 months: Environmental change and genetic continuity. *British Journal of Developmental Psychology.*

Saudino, K. J., Plomin, R., Pedersen, N. L., & McClearn, G. E. (1994). The etiology of high and low cognitive ability during the second half of the life span. *Intelligence, 19,* 359–371.

Saudou, F., Amara, D. A., Dierich, A., LeMeur, M., Ramboz, S., Segu, L., Buhot, M. C., & Hen, R. (1994). Enhanced aggressive behavior in mice lacking 5-HT$_{1B}$ receptor. *Science, 265,* 1875–1878.

Scarr, S. (1992). Developmental theories for the 1990s: Development and individual differences. *Child Development, 63,* 1–19.

Scarr, S., & Carter-Saltzman, L. (1979). Twin method: Defense of a critical assumption. *Behavior Genetics, 9,* 527–542.

Scarr, S., & McCartney, K. (1983). How people make their own environments: A theory of genotype → environment effects. *Child Development, 54,* 424–435.

Scarr, S., Webber, P. I., & Wittig, M. A. (1981). Personality resemblance among adolescents and their parents in biologically related and adoptive families. *Journal of Personality and Social Psychology, 40,* 885–898.

Scarr, S., & Weinberg, R. A. (1978a). The influence of "family background" on intellectual attainment. *American Sociological Review, 43,* 674–692.

Scarr, S., & Weinberg, R. A. (1978b, April). Attitudes, interests, and IQ. *Human Nature,* pp. 29–36.

Schaffer, H. R. (1996). *Social development: A textbook.* Oxford: Blackwell.

Schmitz, S. (1994). Personality and temperament. In J. C. DeFries, R. Plomin, & D. W. Fulker (Eds.), *Nature and nurture during middle childhood* (pp. 120–140). Cambridge, MA: Blackwell.

Schoenfeldt, L. F. (1968). The hereditary components of the Project TALENT two-day test battery. *Measurement and Evaluation in Guidance, 1,* 130–140.

Schulsinger, F. (1972). Psychopathy: Heredity and environment. *International Journal of Mental Health, 1,* 190–206.

Schwab, S. G., Albus, M., Hallmayer, J., Honig, S., Borrmann, M., Lichtermann, D., Ebstein, R. P., Ackenheil, M., Lerer, B., Risch, N., Maier, W., & Wildenauer, D. B. (1995). Evaluation of a susceptibility gene for schizophrenia on chromosome 6p by multipoint affected sib-pair linkage analysis. *Nature Genetics, 22,* 325–327.

Scott, J. P., & Fuller, J. L. (1965). *Genetics and the social behavior of the dog.* Chicago: University of Chicago Press.

Seale, T. W. (1991). Genetic differences in response to cocaine and stimulant drugs. In J. C. Crabbe & R. A. Harris (Eds.), *The genetic basis of alcohol and drug actions* (pp. 279–321). New York: Plenum.

Shapiro, B. L. (1994). The environmental basis of the Down syndrome phenotype. *Developmental Medicine and Child Neurology, 36*, 84–90.

Sherman, P. W. (1977). Nepotism and the evolution of alarm calls. *Science, 197*, 1246–1253.

Sherrington, R., Brynjolfsson, J., Petursson, H., Potter, M., Dudleston, K., Barraclough, B., Wasmuth, J., Dobbs, M., & Gurling, H. (1988). Localisation of susceptibility locus for schizophrenia on chromosome 5. *Nature, 336*, 164–167.

Sherrington, R., Rogaev, E. I., Liang, Y., Rogaeva, E. A., Levesque, G., Ikeda, M., Chi, H., Lin, C., Li, G., Holman, K., et al. (1995). Cloning of a gene bearing missense mutations in early-onset familial Alzheimer's disease. *Nature, 375*, 754–760.

Shields, J. (1962). *Monozygotic twins brought up apart and brought up together*. London: Oxford University Press.

Siever, L. J., Silverman, K. M., Horvath, T. B., Klar, H., Coccaro, E., Keefe, R. S. E., Pinkham, L., Rinaldi, P., Mohs, R. C., & Davis, K. L. (1990). Increased morbid risk for schizophrenia-related disorders in relatives of schizotypal personality disordered patients. *Archives of General Psychiatry, 47*, 634–640.

Silberg, J., Meyer, J., Pickles, A., Simonoff, E., Eaves, L., Hewitt, J. K., Maes, H., & Rutter, M. (1996). Heterogeneity among juvenile antisocial behaviours: Findings from the Virginia Twin Study of Adolescent Behavioural Development. In G. R. Bock & J. A. Goode (Eds.), *Genetics of criminal and antisocial behaviour* (pp. 76–92). Chichester, UK: Wiley.

Silberg, J. L., Rutter, M. L., Meyer, J., Maes, H., Hewitt, J., Simonoff, E., Pickles, A., Loeber, R., & Eaves, L. (1996). Genetic and environmental influences on the covariation between hyperactivity and conduct disturbance in juvenile twins. *Journal of Child Psychology and Psychiatry, 37*, 803–816.

Silva, A. J., Paylor, R., Wehner, J. M., & Tonegawa, S. (1992). Impaired spatial learning in α-calcium-calmodulin kinase mutant mice. *Science, 257*, 206–211.

Sing, C. F., & Boerwinkle, E. A. (1987). Genetic architecture of inter-individual variability in apolipoprotein, lipoprotein and lipid phenotypes. In G. Bock & G. M. Collins (Eds.), *Molecular approaches to human polygenic disease* (pp. 99–122). Chichester, UK: Wiley.

Siomi, H., Choi, M., Siomi, M. C., Nussbaum, R. L., & Dreyfuss, G. (1994). Essential role for KH domains in RNA binding: Impaired RNA binding by a mutation in the KH domain of MR1 that causes fragile X syndrome. *Cell, 77*, 33–39.

Skodak, M., & Skeels, H. M. (1949). A final follow-up on one hundred adopted children. *Journal of Genetic Psychology, 75*, 84–125.

Skoog, I., Nilsson, L., Palmertz, B., Andreasson, L. A., & Svanborg, A. (1993). A population-based study of dementia in 85-year-olds. *New England Journal of Medicine, 328*, 153–158.

Slater, E., & Cowie, V. (1971). *The genetics of mental disorder*. London: Oxford University Press.

Slater, E., & Shields, J. (1969). Genetical aspects of anxiety. In M. H. Lader (Ed.), *Studies of anxiety* (pp. 62–71). Headley, UK: Ashford.

Smalley, S. L., Asarnow, R. F., & Spence, M. A. (1988). Autism and genetics: A decade of research. *Archives of General Psychiatry, 45*, 953–961.

Smith, C. (1974). Concordance in twins: Methods and interpretation. *American Journal of Human Genetics, 26*, 454–466.

Smith, E. M., North, C. S., McColl, R. E., & Shea, J. M. (1990). Acute postdisaster psychiatric disorders: Identification of persons at risk. *American Journal of Psychiatry, 147*, 202–206.

Smith, I., Beasley, M. G., Wolff, O. H., & Ades, A. E. (1991). Effect on intelligence of relaxing the low phenylalanine diet in phenylketonuria. *Archives of Disease in Childhood, 66*, 311–316.

Snieder, H., van Doornen, L. J. P., & Boomsma, D. I. (1995). Developmental genetic trends in blood pressure levels and blood pressure reactivity to stress. In J. R. Turner, L. R. Cardon, & J. K. Hewitt (Eds.), *Behavior genetic approaches in behavioral medicine* (pp. 105–130). New York: Plenum.

Snyderman, M., & Rothman, S. (1988). *The IQ controversy, the media and publication.* New Brunswick, NJ: Transaction.

Sokol, D. K., Moore, C. A., Rose, R. J., Williams, C. J., Reed, T., & Christian, J. C. (1995). Intrapair differences in personality and cognitive ability among young monozygotic twins distinguished by chorion type. *Behavior Genetics, 25*, 457–466.

Somes, G. W., Harshfield, G. A., Alpert, B. S., Goble, M. M., & Schieken, R. M. (in press). Genetic influences on ambulatory blood pressure patterns: The Medical College of Virginia twin study. *American Journal of Hypertension.*

Spearman, C. (1904). "General intelligence," objectively determined and measured. *American Journal of Psychology, 15*, 201–293.

Spelt, J. R., & Meyer, J. M. (1995). Genetics and eating disorders. In J. R. Turner, L. R. Cardon, & J. K. Hewitt (Eds.), *Behavior genetic approaches in behavioral medicine* (pp. 167–185). New York: Plenum.

Spitz, H. H. (1988). Wechsler subtest patterns of mentally retarded groups: Relationship to *g* and to estimates of heritability. *Intelligence, 12*, 279–297.

Sprott, R. L., & Staats, J. (1975). Behavioral studies using genetically defined mice— A bibliography. *Behavior Genetics, 5*, 27–82.

Spuhler, J. N. (1968). Assortative mating with respect to physical characteristics. *Eugenics Quarterly, 15*, 128–140.

Steffenburg, S., Gillberg, C., Hellgren, L., Anderson, L., Gillberg, I., Jakobsson, G., & Bohman, M. (1989). A twin study of autism in Denmark, Finland, Iceland, Norway, and Sweden. *Journal of Child Psychology and Psychiatry, 30*, 405–416.

Stent, G. S. (1963). *Molecular biology of bacterial viruses.* New York: Freeman.

Stevenson, J., Graham, P., Fredman, G., & McLoughlin, V. (1987). A twin study of genetic influences on reading and spelling ability and disability. *Journal of Child Psychology and Psychiatry, 28*, 229–247.

St. George-Hyslop, P., Haines, J., Rogaev, E., Mortilla, M., Vaula, G., Pericak-Vance, M., Foncin, J.-F., Montesi, M., Bruni, A., Sorbi, S., et al. (1992). Genetic evidence for a novel familial Alzheimer's disease locus on chromosome 14. *Nature Genetics, 2*, 330–334.

Stone, W. S., & Gottesman, I. I. (1993). A perspective on the search for the causes of alcoholism: Slow down the rush to genetic judgements. *Neurology, Psychiatry and Brain Research, 1*, 123–132.

Straub, R. E., MacLean, C. J., O'Neill, F. A., Burke, J., Murphy, B., Duke, F., Shinkwin, R., Webb, B. T., Zhang, J., Walsh, D., & Kendler, K. S. (1995). A potential vulnerability locus for schizophrenia on chromosome 6p24-22: Evidence for genetic heterogeneity. *Nature Genetics, 11*, 287–293.

Stunkard, A. J., Foch, T. T., & Hrubec, Z. (1986). A twin study of human obesity. *Journal of the American Medical Association, 256*, 51–54.

Sturtevant, A. H. (1915). Experiments on sex recognition and the problem of sexual selection in *Drosophila. Journal of Animal Behavior, 5,* 351–366.

Tambs, K., Sundet, J. M., & Magnus, P. (1986). Genetic and environmental contribution to the covariation between the Wechsler Adult Intelligence Scale (WAIS) subtests: A study of twins. *Behavior Genetics, 16,* 475–491.

Tambs, K., Sundet, J. M., Magnus, P., & Berg, K. (1989). Genetic and environmental contributions to the covariance between occupational status, educational attainment, and IQ: A study of twins. *Behavior Genetics, 19,* 209–222.

Taylor, E. (1995). Dysfunctions of attention. In D. Cicchetti & D. J. Cohen (Eds.), *Developmental psychopathology. Vol. 2. Risk, disorder, and adaptation* (pp. 243–273). New York: Wiley.

Tellegen, A., Lykken, D. T., Bouchard, T. J., Wilcox, K., Segal, N., & Rich, A. (1988). Personality similarity in twins reared together and apart. *Journal of Personality and Social Psychology, 54,* 1031–1039.

Tesser, A. (1993). On the importance of heritability in psychological research: The case of attitudes. *Psychological Review, 100,* 129–142.

Thapar, A., Hervas, A., & McGuffin, P. (1995). Childhood hyperactivity scores are highly heritable and show sibling competition effects: Twin study evidence. *Behavior Genetics, 25,* 537–544.

Theis, S. V. S. (1924). *How foster children turn out.* Publication No. 165. New York: State Charities Aid Association.

Thompson, L. A., Detterman, D. K., & Plomin, R. (1991). Associations between cognitive abilities and scholastic achievement: Genetic overlap but environmental differences. *Psychological Science, 2,* 158–165.

Thorndike, E. L. (1905). Measurement of twins. *Archives of Philosophy, Psychology, and Scientific Methods, 1,* 1–64.

Thorndike, R. (1985). The central role of general ability in prediction. *Multivariate Behavioral Research, 20,* 241–254.

Tienari, P., Wynne, L. C., Moring, J., Lahti, I., Naarala, M., Sorri, A., Wahlberg, K. E., Saarento, O., Seitamaa, M., Kaleva, M., et al. (1994). The Finnish adoptive family study of schizophrenia: Implications for family research. *British Journal of Psychiatry Supplement No. 23.*

Torgersen, S. (1979). The nature and origin of common phobic fears. *British Journal of Psychiatry, 134,* 343–352.

Torgersen, S. (1983). Genetic factors in anxiety disorders. *Archives of General Psychiatry, 40,* 1085–1089.

Torgersen, S. (1986). Genetic factors in moderately severe and mild affective disorders. *Archives of General Psychiatry, 43,* 222–226.

Torgersen, S. (1990). A twin-study perspective of the comorbidity of anxiety and depression. In J. D. Maser & C. R. Cloninger (Eds.), *Comorbidity of mood and anxiety disorders* (pp. 367–378). Washington, DC: American Psychiatric Press.

Torgersen, S., & Psychol, C. (1980). The oral, obsessive, and hysterical personality syndromes: A study of hereditary and environmental factors by means of the twin method. *Archives of General Psychiatry, 37,* 1272–1277.

Torgersen, S., & Psychol, C. (1984). Genetic and nosological aspects of schizotypal and borderline personality disorders: A twin study. *Archives of General Psychiatry, 41,* 546–554.

Torrey, E. F. (1990). Offspring of twins with schizophrenia. *Archives of General Psychiatry, 47,* 976–977.

Torrey, E. F., Bowler, A. E., Taylor, E. H., & Gottesman, I. I. (1994). *Schizophrenia and manic-depressive disorder*. New York: Basic Books.

Treasure, J. L., & Holland, A. J. (1991). Genes and the aetiology of eating disorders. In P. McGuffin & R. Murray (Eds.), *The new genetics of mental illness* (pp. 198–211). Oxford: Butterworth-Heinemann.

True, W. R., Rice, J., Eisen, S. A., Heath, A. C., Goldberg, J., Lyons, M. J., & Nowak, J. (1993). A twin study of genetic and environmental contributions to liability for posttraumatic stress symptoms. *Archives of General Psychiatry, 50,* 257–264.

Tsuang, M., & Faraone, S. D. (1990). *The genetics of mood disorders*. Baltimore: Johns Hopkins University Press.

Tsuang, M. T., Lyons, M. J., Eisen, S. A., True, W. T., Goldberg, J., & Henderson, W. (1992). A twin study of drug exposure and initiation of use. *Behavior Genetics, 22,* 756 (abstract).

Turner, J. R. (1994). *Cardiovascular reactivity and stress: Patterns of physiological response*. New York: Plenum.

Turner, J. R., Cardon, L. R., & Hewitt, J. K. (Eds.). (1995). *Behavior genetic approaches in behavioral medicine*. New York: Plenum.

Tyler, A., Ball, D., & Crawford, D. (1992). Presymptomatic testing for Huntington's disease in the U.K. *British Medical Journal, 304,* 1593–1596.

Uhl, G., Blum, K., Nobel, E. P., & Smith, S. (1993). Substance abuse vulnerability and D_2 dopamine receptor gene and severe alcoholism. *Trends in Neuroscience, 16,* 83–88.

U.S. Bureau of the Census. (1995). *Sixty-five plus in America*. Washington, DC: U.S. Government Printing Office.

Vandenberg, S. G. (1971). What do we know today about the inheritance of intelligence and how do we know it? In R. Cancro (Ed.), *Genetic and environmental influences* (pp. 182–218). New York: Grune & Stratton.

Vandenberg, S. G. (1972). Assortative mating, or who marries whom? *Behavior Genetics, 2,* 127–157.

Vernon, P. A. (1989). The heritability of measures of speed of information-processing. *Personality and Individual Differences, 10,* 573–576.

Vernon, P. A. (Ed.). (1993). *Biological approaches to the study of human intelligence*. Norwood, NJ: Ablex.

Vieland, V. J., Knowles, J. A., Fyer, A. J., Stefanovich, M., Freimer, N. F., Lish, J., Adams, P., Woodley, K., Rassnick, H., Heiman, G. A., Whie, P., Das, K., Klein, D. F., Ott, J., Weissman, M. M., & Gilliam, T. C. (1994). Linkage study of panic disorder: A preliminary report. In E. S. Gershon & C. R. Cloninger (Eds.), *Genetic approaches to mental disorders* (pp. 345–354). Washington, DC: American Psychiatric Press.

Von Knorring, A. L., Cloninger, C. R., Bohman, M., & Sigvardsson, S. (1983). An adoption study of depressive disorders and substance abuse. *Archives of General Psychiatry, 40,* 943–950.

Vrendenberg, K., Flett, G. L., & Krames, L. (1993). Analog versus clinical depression: A clinical reappraisal. *Psychological Bulletin, 113,* 327–344.

Wadsworth, S. J. (1994). School achievement. In J. C. DeFries, R. Plomin, & D. W. Fulker (Eds.), *Nature and nurture during middle childhood* (pp. 86–101). Oxford: Blackwell.

Wahlsten, D. (1990). Insensitivity of the analysis of variance to heredity-environment interaction. *Behavioral and Brain Sciences, 13,* 109–161.

Wahlström, J. (1990). Gene map of mental retardation. *Journal of Mental Deficiency Research, 34*, 11–27.

Waller, N. G., & Shaver, P. R. (1994). The importance of nongenetic influence on romantic love styles: A twin-family study. *Psychological Science, 5*, 268–274.

Ward, M. J., Vaughn, B. E., & Robb, M. D. (1988). Social-emotional adaptation and infant-mother attachment in siblings: Role of the mother in cross-sibling consistency. *Child Development, 59*, 643–651.

Warren, S. T., & Nelson, D. L. (1994). Advances in molecular analysis of fragile X syndrome. *Journal of the American Medical Association, 271*, 536–542.

Watson, J. B. (1930). *Behaviorism*. New York: Norton.

Watson, J. D., & Crick, F. H. C. (1953). Genetical implications of the structure of deoxyribonucleic acid. *Nature, 171*, 964–967.

Watt, N. F., Anthony, E. J., Wynne, L. C., & Rolf, J. E. (1984). *Children at risk for schizophrenia: A longitudinal perspective*. Cambridge: Cambridge University Press.

Wehner, J. M., Bowers, B. J., & Paylor, R. (1996). The use of null mutant mice to study complex learning and memory processes. *Behavior Genetics, 26*, 301–312.

Weiner, J. (1994). *The beak of the finch*. New York: Vintage Books.

Weiss, P. (1982). *Psychogenetik: Humangenetik in psychologie and psychiatrie*. Jena: Fischer.

Weiss, D. S., Marmar, C. R., Schlenger, W. E., Fairbank, J. A., Jordan, B. K., Hough, R. L., & Kulka, R. A. (1992). The prevalence of lifetime and partial posttraumatic stress disorder in Vietnam theater veterans. *Journal of Traumatic Stress, 5*, 365–376.

Weissman, M. M., Fendrich, M., Warner, V., & Wickramaratne, P. (1992). Incidence of psychiatric disorder in offspring at high and low risk for depression. *Journal of the American Academy of Child and Adolescent Psychiatry, 31*, 640–648.

Weissman, M. M., Warner, V., Wickramaratne, P., & Prusoff, B. A. (1988). Early-onset major depression in parents and their children. *Journal of Affective Disorders, 15*, 269–277.

Wender, P. H., Kety, S. S., Rosenthal, D., Schulsinger, F., Ortmann, J., & Lunde, I. (1986). Psychiatric disorders in the biological and adoptive families of adopted individuals with affective disorders. *Archives of General Psychiatry, 43*, 923–939.

Wender, P. H., Rosenthal, D., Kety, S. S., Schulsinger, F., & Welner, J. (1974). Crossfostering: A research strategy for clarifying the role of genetic and experimental factors in the etiology of schizophrenia. *Archives of General Psychiatry, 30*, 121–128.

Wilson, E. O. (1975). *Sociobiology: The new synthesis*. Cambridge, MA: Belknap Press.

Wilson, R. S. (1983). The Louisville Twin Study: Developmental synchronies in behavior. *Child Development, 54*, 298–316.

Wilson, R. S., & Matheny, A. P., Jr. (1986). Behavior genetics research in infant temperament: The Louisville Twin Study. In R. Plomin & J. Dunn (Eds.), *The study of temperament: Changes, continuities, and challenges* (pp. 81–97). Hillsdale, NJ: Erlbaum.

Wright, A. (1990, July 9–16). Achilles' helix. *The New Republic*, pp. 21–31.

Wright, S. (1921). Systems of mating. *Genetics, 6*, 111–178.

Wu, C.-L., & Melton, D. W. (1993). Production of a model for Lesch-Nyhan syndrome in hypoxanthine phosphoribosyltransferase-deficient mice. *Nature Genetics, 366*, 742–745.

Yin, J. C. P., Vecchio, M. D., Zhou, H., & Tully, T. (1995). CRAB as a memory modulator: Induced expression of a dCREB2 activator isoform enhances long-term memory in *Drosophila. Cell, 8*, 107–115.

Young, J. P. R., Fenton, G. W., & Lader, M. H. (1971). The inheritance of neurotic traits: A twin study of the Middlesex Hospital Questionnaire. *British Journal of Psychiatry, 119*, 393–398.

Zahn-Waxler, C., Robinson, J., & Emde, R. N. (1992). The development of empathy in twins. *Developmental Psychology, 28*, 1038–1047.

Zhang, Y., Proenca, R., Maffei, M., Barone, M., Leopold, L., & Friedman, J. M. (1994). Positional cloning of the mouse obese gene and its human homologue. *Nature, 372*, 425–432.

Zori, R. T., Hendrickson, J., Woolven, S., Whidden, E. M., Gray, B., & Williams, C. A. (1992). Angelman syndrome: Clinical profile. *Journal of Child Neurology, 7*, 270–280.

Zuckerman, M. (1994). *Behavioral expressions and biosocial bases of sensation seeking.* Cambridge: Cambridge University Press.

NAME INDEX

Murray, C., 137
Murray, R. M., 176, 185, 209, 228

National Foundation for Brain
    Research, 170
Neale, M. C., 200, 305
Neidershiser, J. M., 77, 206, 249, 259
Neiswanger, K., 228
Nelkin, D., 105
Nelson, D. L., 113
Nelson, R. J., 93, 215
Nesse, R. M., 236
Nesselroade, J. R., 202
Nichols, P. L., 109f
Nichols, R. C., 73, 156, 158, 165,
    195f
Nigg, J. T., 208f, 211
Noble, E. P., 167
Noegel, R., 187
North, T., 125
Nowak, J., 184
Noyes, R., 184, 193

O'Connor, T. G., 265
Oldham, J. M., 209
Oliverio, A., 135, 138
Orel, V., 10
Owen, M. J., 127

Parnas, J., 170
Parsons, P. A., 238
Partanen, J., 156
Pauls, D. L., 184, 191
Paylor, R., 93
Pearson, K., 28
Pedersen, N. L., 76, 78, 140f, 142, 147,
    157, 163 202, 230, 262
Peltonen, L., 177
Pervin, L. A., 195
Peterson, B. S., 191
Peto, R., 228
Petrill, S. A., 110, 160, 163, 167
Phillips, K., 268
Phillips, T. J., 95, 99, 225
Phillips, W., 188
Pickens, R. W., 224
Pickering, T. G., 219
Pickles, A., 188, 269
Pike, A., 252, 262
Pillard, R. C., 206
Pinker, S., 236

Plomin, R., 4, 76, 78, 103, 109,
    123, 141, 146f, 150, 153,
    156, 158, 163f, 166f, 195,
    197, 200ff, 215, 217, 219,
    227, 230f, 249ff, 254ff,
    258ff, 266ff, 274, 277,
    295f
Pogue-Geile, M. F., 202, 219ff
Pollen, D. A., 127
Pollin, W., 172
Postman, L. J., 134
Price, R. A., 180, 191
Price, R. K., 190
Propping, P., 73f, 103, 152
Psychol, C., 208f

Raine, A., 212
Rainer, J. D., 182
Rasmussen, S. A., 209
Ratcliffe, S. G., 119, 215
Redshaw, J., 74
Reed, E. W., 109
Reed, S. C., 109, 172
Regier, D. A., 211
Reich, J., 184
Reich, T., 224
Reiss, D., 251
Rende, R. D., 217, 264
Reveley, A. M., 176
Ricciuti, A. E., 205
Rice, J. P., 184
Riemann, R., 199
Riese, M. L., 201f
Risch, N., 94, 228
Robb, M. D., 205
Roberts, C. A., 207
Robins, L. N., 190, 211
Robinson, A., 119
Robinson, J. L., 200, 205
Rodgers, D. A., 225
Rose, R. J., 184, 202, 219
Rosenthal, D., 173
Rothbart, M. K., 199
Rothman, S., 137, 153
Roush, W., 213
Rovine, M. J., 141
Rowe, D. C., 190, 195, 200, 204, 229,
    266f, 269
Rush, A. J., 180
Rutherford, J., 181, 187, 222, 268
Rutter, M., 74, 109, 180, 186ff, 211,
    214, 245, 269

SUBJECT INDEX